Universitext

Springer
Berlin
Heidelberg
New York
Barcelona
Hong Kong
London
Milan
Paris
Singapore
Tokyo

Raymond Séroul

Programming for Mathematicians

Translated from the French
by Donal O'Shea

With 40 Figures

Springer

Raymond Séroul
Université Louis Pasteur
U.F.R. de Mathématiques et d'Informatique
7, rue René Descartes
67084 Strasbourg, France
e-mail: seroul@math.u-strasbg.fr

Translator:

Donal O'Shea
Department of Mathematics
Mount Holyoke College
Clapp Laboratory
South Hadley, MA 01075-1461, USA
e-mail: doshea@mtholyoke.edu

Title of the French original edition:
math-info. Informatique pour mathématiciens. © InterEditions, Paris, 1995.

Library of Congress Cataloging-in-Publication Data applied for

Die Deutsche Bibliothek – CIP-Einheitsaufnahme
Séroul, Raymond: Programming for mathematicians / Raymond Séroul. Transl. from the French by Donal O'Shea. – Berlin; Heidelberg; New York; Barcelona; Hong Kong; London; Milan; Paris; Singapore; Tokyo: Springer, 2000
(Universitext) ISBN 3-540-66422-X

Mathematics Subject Classification (1991): 04-04, 68-01

ISBN 3-540-66422-X Springer-Verlag Berlin Heidelberg New York

Printed in Germany

Cover design: *design & production* GmbH, Heidelberg
Photocomposed from the translator's LaTeX files after editing and reformatting by the author using a Springer LaTeX macro-package.

Printed on acid-free paper SPIN: 10574792 41/3143Ko – 5 4 3 2 1 0

Preface

> *We are as dwarves sitting on the shoulders of giants. We see things that are deeper or further, not by the penetration of our own vision or by our own height, but because they support us and lend us their height.*
>
> *Bernard de Chartres (XIIth century)*

I became interested in computer science in the 1980's after twenty years of practicing mathematics when the first microcomputers appeared. At this time, there was much hype about how easy programming was. Supposedly, one could learn it in eight days!

I examined the available literature and became quite disillusioned. The books through which I leafed were nothing but dismal treatments of syntax and those containing programs did not explain how these programs were created. There were several more ambitious books, very formal, which showed how one could deduce an algorithm beginning with an invariant of some loop. I did not find these books very convincing: their approach seemed too formal, too heavy, and too cut off from reality. But perhaps I was too old? However, discussions with my colleagues convinced me that professional programmers do not work in this way.

There was a rift: I would have to leave a world which I knew well to find my way, most of the time without a guide, in a universe with laws and values that were foreign to me and that I needed to understand.[1]

After an apprenticeship of several years, I began to teach programming to students in their third year of mathematics (which we call the Licence). I was very surprised to find that many were frankly hostile to computer science. After talking to them, I realized that their hostility was a reaction to the way that they had been taught: how can one learn the subtleties of a language or conduct a refined analysis of the performance of an algorithm when one cannot even write a program[2] several lines in length? My goal became to entice the students by showing them that they could very quickly write interesting programs knowing only very few techniques from computer science.

---

[1] Any individual who has been exiled or who has radically changed profession will know what I am talking about.

[2] They reminded me of some of the unhappy practices of the "new mathematics" which pretended to teach ring structures to students who had not yet mastered their multiplication tables.

The modern mathematician is condemned to program. But how does one learn to program? This is as difficult as learning how to do proofs! And, in both cases, there is no *method*.

Little by little, my ideas became clearer and my background as a mathematician began to help. This book is the fruit of that apprenticeship and that synthesis. It is the book that I would have liked to have had when I began. It is meant for those who want an introduction to programming without renouncing their mathematical background and who want to harmoniously integrate this new discipline into their way of thought.

However, a mathematician can rapidly convert the ability to create proofs to the art of programming if the following two conditions are fullfilled:

- he or she has a clear idea of what a computer is, what it can do, and what it cannot do.
- he or she learns to think *dynamically*.

Mathematical results can be roughly divided into two classes:

- *static results:* Cramer's formula, "there exist x such that $P(x)$", the definition of the greatest common divisor and the Bézout theorem $1 = au + bv$, etc.
- *dynamic results:* Gaussian pivoting, an explicit construction of an element x in a set E satisfying property P, Euclid's and Blankinship's algorithms that compute the GCD and the numbers u, v in Bézout's theorem, etc.

A traditional mathematician thinks statically; *systematic dynamical* thinking is a new discipline whose development has coincided with that of computer science. Converting from static to dynamical thinking is not easy (and some will never make the transition).

Experience shows that writing a good program is even more difficult than writing a proof because the objects that one considers are like tiny bars of wet soap: they slide between one's fingers. Programming is often synonymous with "fooling around" which makes it distasteful and often paralyzing for mathematicians acustomed to other standards.

We use a very stripped down programming language (assignments, tests and loops) and three standard programming techniques:

- recycling and rewriting code
- descending analysis and successive refinement
- the use of sequences to allow one to transform a dynamical problem into a static problem more familiar to mathematicians.

I place equal emphasis on the aesthetic side. A mathematician always experiences profound satisfaction on studying a beautiful theory or proof. This is why he or she practices mathematics. I want to convince mathematicians that they will experience the same feelings upon analyzing and setting up an algorithm, even a very simple one.

The use of computers creates new or, in some instances, resuscitates old problems in mathematics (how much space should one reserve in memory for the divisors of a given integer n?) and computer science (investigating algorithms, proving programs). This book bears witness to this interaction: to the *pleasure* there is in programming mathematics and to the *pleasure* of reasoning when one programs.

Since for me programming does not mean denying my first love, the rigorous side takes priority over a more detailed study of a programming language. Whenever possible, I take the opportunity to present mathematics that my students do not know. For this reason, one finds many mathematical results and proofs in this book.

Most of all, this book is aimed at undergraduates with some mathematical background. It will also be of interest to those with some mathematical training who wish (or need) to begin programming.

For historical reasons, Pascal is the language used. *But this book is absolutely not a treatise on the Pascal language.* Conversion to a more modern language is instantaneous: one need only have assignments, tests and loops.

Acknowledgements

A book is like an iceberg. The author is the only part that emerges. Out of sight are those who inspired and who helped in other ways. Among this latter group are the following whom I would especially like to thank.

- The IREM at Strasbourg which allowed me access to the columns of the review *l'Ouvert*, thereby allowing me to refine my ideas.
- All the colleagues and friends who have followed the courses in programming that I have taught, those with whom I have discussed matters about which I care deeply, and those who have given me advice about this book.
- The students majoring in mathematics at Strasbourg, on whom I tested and refined my ideas for over ten years. Without their feedback this book would never have seen the light of day.
- Finally, I would like to thank the translator, Donal O'Shea, of Mount Holyoke College for his excellent work.

Raymond Séroul
Université Louis Pasteur,
UFR de Mathématiques et d'Informatique
7 rue René Descartes, 67084 Strasbourg CEDEX (France)
seroul@math.u-strasbg.fr

Contents

1. Programming Proverbs

Programmers enjoy creating *proverbs* which reflect their professional experience.

1.1. Above all, no tricks!

This proverb, perhaps the most important of all, always shocks mathematicians who have never programmed. It also shocks beginning programmers who always have a tendency to be too clever by half.

You should not conclude from this that computer scientists are imbeciles. You will understand the wisdom of this advice when you begin to write complicated programs. What counts most is that a program be clear, that it function immediately (or almost immediately) and be easy to maintain.

A computer scientist spends much time *maintaining* programs. This means modifying programs which were written, in general, by another person who is no longer around. Professional programs are very long: often several hundred pages. Reading a program is very difficult (as is reading a proof reduced to simple calculations with all explanations suppressed). Moreover, time counts: a programmer who modifies a program simply cannot afford to spend hours asking what the code under his or her eyes means.

The more brutish (that is, the more limpid and straightforward) the code, the more certain it is, and the simpler it is to modify.

Another big objection to tricks is that most of the time they are useless. In general, a program spends 80 % of the time in less than 20 % of the code. Consequently, a trick has a very strong chance of appearing in a part of the code where the machine is only waiting (as, for example, when one is entering data via the keyboard). It is stupid and suicidal to jeopardize a program by using a trick to gain a few milliseconds in a program which runs a thousand or hundred thousand times slower because it is waiting for you.

Be brutal: always choose the most straightforward code, even if it requires several supplementary lines. Modesty always pays in the long run.

This said, recourse to a trick is sometimes indispensible, for example in a loop solicited at all hours in a program that is too slow. If this is the case, *document it*! Call it to your reader's attention, explain in detail what you did and why (the machine is too slow, there is not enough memory, etc.). Do not forget that this reader could be you in a couple of months with a different machine and a different point of view.

Which would you prefer: a program easy to adjust and that you will have the pleasure of optimizing, or an unreadable program, full of faults, which will not work without hours and hours of debugging?

1.2. Do not chewing gum while climbing stairs

This proverb, inspired by an American presidential campaign, expresses an idea full of good sense: it is better to do only one thing at a time. This is why one breaks a program into procedures and independent modules.

1.3. Name that which you still don't know

This proverb applies each time that you encounter a subproblem in the interior of a problem. Refuse to solve the subproblem (the preceding proverb), leave it to one side and advance. How? By giving a name to what you do not yet know (a piece of code, a function). This allows you to end the work (*i.e.* the problem). For the subproblem, see the following proverb.

1.4. Tomorrow, things will be better; the day after, better still

No, this is not a call to indolence! On the contrary, it is an extraordinarily effective technique. Apply this proverb each time that you encounter a blockage caused by the appearance of a new problem inside the problem that you are trying to solve: apply the preceding proverb, putting off the solution of the new problem until tomorrow by *giving it a name* as a procedure or function, the code for which you will write later.

Always separate what is urgent from what is not; learn to distinguish between essentials and accessories; do not drown prematurely in details. The details you can handle *tomorrow*, which generally means several minutes or hours; it is not a question of postponing all the work twenty-four hours, as the truly lazy would like to believe!

This technique allows you to advance a little at a time. You certainly have practiced this in mathematics: "I will first prove my theorem by provisionally allowing lemmas 1, 2 and 3."

This proverb also guards against a common beginner's fault: the desire to act immediately by prematurely writing very technical code. The right attitude is to resist this: it is necessary to cultivate a certain nonchalance by giving some orders today; the rest will keep until tomorrow. "Make haste slowly" says another proverb.

1.5. Never execute an order before it is given

Beginners are always in a hurry: they write and they write. The result is code that is too rich, too technical, and perforce incomprehensible. There will always be errors in such a jumble!

This fault is easy to diagnose. If understanding the code requires you to highlight and explain a piece of the code, you can be certain that you are lacking a procedure (the order, if you will) at this point.

Replace this part of the code by a procedure call. You will "execute" this order later, when you write the code for the procedure. You need to pace yourself ...

1.6. Document today to avoid tears tomorrow

Imagine a proof reduced to calculations and some logical symbols: it is unreadable, hence *useless*. When you program, take time to explain very precisely what you are doing.

- First of all, you must understand what it is you want to do: *what is well-conceived can be clearly stated*. If you cannot explain your code to a friend, you can be certain that your ideas need sharpening and that your program is probably incorrect. To be conscientious, to clarify your ideas, *engage in a dialogue with yourself*. Discipline yourself to write comments as you go along; do not wait until your program is finished. This will be too late. Mathematicians have long understood this: they carefully *write up* up proofs before believing them to be true.

- If your program is false, or if you must return to it after six months, you will be happy to find explanations which indicate how the program was conceived.

It is annoying that technology which would facilitate good documentation has not appeared. The text editor that comes with a compiler falls far short of a full featured word processor; it is so primitive and rustic that one could cry. To be really lucid, one often needs to write formulae or include a sketch. When will we have access to an editor in a compiler that is worthy of the name? I dream of a compiler where one can make comments appear or disappear with a simple click, as in hypertext.

1.7. Descartes' Discourse on the Method

Descartes' *Discours de la méthode* was published in 1637. This fascinating text anticipates modern programming methods!

As the multiplicity of laws often furnishes excuses for vices, so that a state is much better ruled when, not having but very few laws, these are very strictly observed; so, in place of the large number of precepts of which logic is composed, I believed that I would have enough with the following four, provided that I were to make a firm and constant resolution not to fail, even a single time, to observe them.

The first was never to accept anything as true that I did not evidently know to be such: that is to say, carefully to avoid precipitation and prejudice; and to include in my judgements nothing more than that which would present itself to my mind so clearly and so distinctly that I were to have no occasion to put it in doubt.

The second to divide each of the difficulties I would examine into as many parts as would be possible and as would be required in order to better resolve them.

The third, to conduct my thoughts in an orderly manner, by beginning with those objects the most simple and the most easy to know, in order to ascend little by little, as by degrees, to the knowledge of the most composite ones; and by supposing an order even among those which do not naturally precede one another.

And the last, everywhere to make enumerations so complete and reviews so general that I were assured of omitting nothing.[1]

[1] Translation by George Hefferman, *Discourse on the Method*, Univ. of Notre Dame Press, Notre Dame, 1994.

2. Review of Arithmetic

2.1. Euclidean Division

Let $a, b \in \mathbb{Z}$ be two integers with $b > 1$. Using the order relation, one can show that there exists a unique pair $(q, r) \in \mathbb{Z} \times \mathbb{N}$ such that

$$a = bq + r \quad \text{and} \quad 0 \le r < b.$$

The integer q is called the *quotient* and r the *remainder* upon euclidean division of a by b. Let $[x]$ denote the integer part of the real number x (this means that $[x]$ is the integer defined by the conditions $[x] \le x < [x] + 1$). Then we have $q = [a/b]$. One often wants small remainders to make an algorithm "converge." In this case, we use a variant of euclidean division with *centered remainder*:

$$a = bq + r, \quad -\tfrac{1}{2}b < r \le \tfrac{1}{2}b.$$

Here, again, the pair (q, r) is unique. If $\bar{q}$ and $\bar{r}$ are the quotient and remainder upon ordinary euclidean division, then it is clear that

$$r = \begin{cases} \bar{r} & \text{if } 2\bar{r} \le b, \\ \bar{r} - b & \text{if } 2\bar{r} > b; \end{cases} \qquad q = \begin{cases} \bar{q} & \text{if } 2\bar{r} \le b, \\ \bar{q} - 1 & \text{if } 2\bar{r} > b. \end{cases} \tag{2.1}$$

We say that d *divides* n, written $d \mid n$, if $n = dq$ for some $q \in \mathbb{Z}$. In particular, any number divides 0!

Exercise 1 *(Solution at the end of the chapter)*

Removing all multiples of 2 and 3 from $\mathbb{N}$ gives the sequence:

$$u_1 = 5,\ u_2 = 7,\ u_3 = 11,\ u_4 = 13,\ u_5 = 17,\ u_6 = 19,\ u_7 = 23,\ \ldots$$

Show that this sequence satisfies a first order recurrence relation. Generalize to the sequence of integers not divisible by 2, 3, or 5. Occasionally we can use this sequence to speed up an algorithm (for example, the algorithm that finds the smallest divisor of a given number).

2.2. Numeration Systems

No mathematician has ever seen an integer! When we read "1994", we read a *word* whose letters are called *numerals*. A very simple theorem asserts that there is a bijection between these words and $\mathbb{N}$.

Theorem 2.2.1 (Base b numeration). *Let $b > 1$ be a fixed integer. To every integer $x > 0$, one can associate a unique decomposition of the form*

$$x = x_n b^n + x_{n-1} b^{n-1} + \ldots + b x_1 + x_0 \tag{2.2}$$

where each x_i satisfies the condition $0 \leq x_i < b$ and where $x_n > 0$.

Proof. Suppose that the decomposition exists. Upon putting, $q_0 = x_n b^{n-1} + x_{n-1} b^{n-2} + \cdots + x_1$, we immediately obtain

$$x = q_0 b + x_0, \quad 0 \leq x_0 < b.$$

In other words, q_0 and x_0 are the quotient and remainder upon euclidean division of x by b, and are, therefore, uniquely determined. Reasoning by induction, we see that x_1 is the remainder upon division of q_0 by b, and so on.

On the other hand, we can exhibit such a decomposition by dividing x and its successive quotients by b. The algorithm terminates because the quotients are a strictly decreasing sequence of integers bounded below by zero. □

The integer b is called the *base* of the numeration system. If b is not too big, we can associate to each integer in the interval $[0, b-1]$ a typographical symbol:

- When b is the number "ten", the symbols are the arabic numerals 0, 1, 2, 3, 4, 5, 6, 7, 8, 9.
- When b is smaller than "ten", one chooses the corresponding subset of the arabic numerals.
- When b is the number "sixteen", it is traditional to use the arabic numerals and the first letters of the alphabet: 0, 1, 2, …, 9, A, B, C, D, E, F.

Theorem 2.2.1 allows us to associate to each integer x the *word* $\overline{x_n \cdots x_0}$ comprised of the numerals corresponding to the x_i (the overline indicates that numerals are juxtaposed, not multiplied).

This way of representing numbers is not the only one. We still use traces of numeration systems that have been used since antiquity (for example, to describe subdivisions of angles and time).

Theorem 2.2.2 (Numeration with multiple bases). *Let $(b_i)_{i \geq 1}$ be an infinite sequence of integers greater than 1. Then every integer $x > 0$ can be written uniquely in the form*

$$x = x_0 + x_1 b_1 + x_2 b_1 b_2 + \cdots + x_n b_1 \cdots b_n, \tag{2.3}$$

where each x_i satisfies the condition $0 \leq x_i < b_{i+1}$.

The proof is an immediate generalization of the preceding.

Exercise 2 *(Solution at the end of the chapter)*

Let P and Q be two polynomials with integral coefficients. Show that if $b > 1$ is sufficiently large, then $P(b) = Q(b)$ implies that $P = Q$.

2.3. Prime Numbers

The definition is often misstated or misunderstood.

Definition 2.3.1. *An integer $p \in \mathbb{Z}$ is called a prime number if it satisfies the following two properties:*

- *it is different from ± 1;*
- *its only divisors are $-p, -1, 1, p$.*

Contrary to widespread opinion, the numbers -1 and $+1$ *are not* prime numbers.[1] (Algebraists call them *units.*)

We have known since antiquity how to prove (using the sequence $n!+1$) that the set of prime numbers is infinite. We shall often use the ordered sequence $(p_i)_{i \geq 1}$ of prime numbers:

$$p_1 = 2,\ \ p_2 = 3,\ \ p_3 = 5,\ \ p_4 = 7,\ \ p_5 = 11,\ \text{ etc.}$$

Theorem 2.3.1 (The least divisor function). *Let $n > 1$ be any integer and let* LD(n) *be the least integer greater than* 1 *which divides n. Then*

(a) LD(n) *is a prime number;*
(b) *if n is not a prime number, then* $(\mathrm{LD}(n))^2 \leq n$.

Proof. We first note that LD(n) always exists: the integer $d = n$ is greater than 1 and divides n, so that the set of divisors of n which are greater than 1 is not empty and, thus, possesses a smallest element.

(a) If $p = \mathrm{LD}(n)$ is not prime, then we can write $p = p'p''$ with $1 < p' < p$. Since p' divides n, we obtain a contradiction.

(b) If n is not prime and $p = \mathrm{LD}(n)$, then $n = pn'$ with $n' > 1$. By definition of LD(n), we have $p \leq n'$ which implies that $p^2 \leq pn' \leq n$. □

We now recall *Bertrand's postulate*. This is an arithmetic result which we shall sometimes need and whose proof is somewhat technical[2] without being very difficult. This was conjectured by Bertrand in 1845 and proved by Tchebycheff in 1850.

---

[1] If 1 were considered to be a prime number, we would lose the uniqueness of the decomposition into primes.

[2] An elementary proof is given in *An Introduction to the Theory of Numbers*, by G.H. Hardy and E.M. Wright, Oxford Science Publications, 5th edition (1979), pp. 343–344

Theorem 2.3.2 (Bertrand's postulate). *Let $n \geq 1$ be an integer. There always exists a prime number p satisfying $n < p \leq 2n$.*

Corollary 2.3.1.

- *Let $(p_i)_{i \geq 1}$ be the increasing sequence of prime numbers. For every i, one has $p_{i+1} < 2p_i$.*
- *Let p be any prime number. There always exists a prime number q satisfying $p < q < p^2$.*

Proof. The first assertion follows from Bertrand's postulate upon putting $n = p_i$; the second follows upon remarking that $2n \leq n^2$ when $n \geq 2$. □

Exercise 3

(Solution at the end of the chapter.) Consider the doubly infinite table, called the *Sundaram sieve* (1934), whose rows and columns are the following infinite progressions:

$$\begin{array}{ccccccccl}
4 & 7 & 10 & 13 & 16 & \cdots & \leftarrow & \text{difference } 3 \\
7 & 12 & 17 & 22 & 27 & \cdots & \leftarrow & \text{difference } 5 \\
10 & 17 & 24 & 31 & 38 & \cdots & \leftarrow & \text{difference } 7 \\
13 & 22 & 31 & 40 & 49 & \cdots & \leftarrow & \text{difference } 9 \\
16 & 27 & 38 & 49 & 60 & \cdots & \leftarrow & \text{difference } 11 \\
\vdots & \vdots & \vdots & \vdots & \vdots & & &
\end{array}$$

Show that $2n + 1$ is prime if and only if n does not appear in the table above.

2.3.1. The number of primes smaller than a given real number

Let x be a positive *real* number. A celebrated arithmetic function is

$$\begin{aligned} \pi(x) &= \textit{number of primes} \leq x \\ &= \textit{largest index } i \textit{ such that } p_i \leq n. \end{aligned}$$

The table below displays some of its values:

x	10	20	30	40	50	60	70	80	90
$\pi(x)$	4	8	10	12	15	17	19	22	24
x	100	200	300	400	500	600	700	800	900
$\pi(x)$	25	46	62	78	95	109	125	139	154
x	1000	2000	3000	4000	5000	6000	7000	8000	9000
$\pi(x)$	168	303	430	550	669	783	900	1007	1117

This function has fascinated mathematicians for centuries. Here are some fundamental results concerning it which are difficult to prove [Hardy and Wright, *op. cit.*].

Theorem 2.3.3.

- $\pi(n) < 2\log 2 \cdot \dfrac{n}{\log n}$ *for every integer* $n \geq 2$.
- *As* x *tends to infinity,* $\pi(x) \sim \dfrac{x}{\log x}$.
- *There exist constants* $A, B > 0$ *such that for every integer* $n \geq 2$,

$$A\, n \log n < p_n < B\, n \log n.$$

2.4. The Greatest Common Divisor

It is easy to show using euclidean division that every additive subgroup of $\mathbb{Z}$ has the form $d\mathbb{Z}$. Moreover, if we require that $d \geq 0$, then the subgroup uniquely determines d.

Let $a, b \in \mathbb{Z}$ be two integers and let $a\mathbb{Z}+b\mathbb{Z}$ be the set of linear combinations of a and b with integral coefficients. These combinations form an additive subgroup of $\mathbb{Z}$. We define the *greatest common divisor* (GCD) of a and b to be the unique integer $d \geq 0$ satisfying:

$$d\mathbb{Z} = a\mathbb{Z} + b\mathbb{Z}.$$

Here are several immediate consequences of the definition:

$$\mathrm{GCD}(a, b) = \mathrm{GCD}(\pm a, \pm b) \tag{2.4}$$

$$\mathrm{GCD}(a, b) = \mathrm{GCD}(b, a) \tag{2.5}$$

$$\mathrm{GCD}(a, 0) = |a| \tag{2.6}$$

$$\mathrm{GCD}(a + \lambda b, b) = \mathrm{GCD}(a, b) \tag{2.7}$$

$$\mathrm{GCD}(\lambda a, \lambda b) = |\lambda|\, \mathrm{GCD}(a, b) \tag{2.8}$$

Since a and b and $d = \mathrm{GCD}(a, b)$ are elements of $a\mathbb{Z} + b\mathbb{Z}$, both a and b are multiples of d. That is, the GCD divides both a and b:

$$a = da', \quad b = db', \quad \text{if} \quad d = \mathrm{GCD}(a, b). \tag{2.9}$$

When a is not zero, the equality $a = da'$ shows that the GCD is greater than 0, and similarly if $b \neq 0$. On the other hand, the GCD is zero when $a = b = 0$:

$$\mathrm{GCD}(a, b) = 0 \iff a = b = 0.$$

2.4.1. The Bézout Theorem

Another consequence of the fact that a, b and $\mathrm{GCD}(a, b)$ belong to the subgroup $a\mathbb{Z} + b\mathbb{Z}$ is the following.

Theorem 2.4.1 (Bézout theorem – first version). *Let $a, b \in \mathbb{Z}$ be any two integers. There exist $u, v \in \mathbb{Z}$ such that*

$$au + bv = \mathrm{GCD}(a, b) \tag{2.10}$$

Corollary 2.4.1. *If $a \neq 0$ or $b \neq 0$, then $d = \mathrm{GCD}(a, b)$ is the greatest integer greater than or equal to 1 which simultaneously divides a and b.*

Proof. If δ is an integer which divides a and b, then δ divides $au + bv = d$. Since d is not zero, it follows that $|\delta| \leq d$. □

Thus, the GCD merits its name when $a \neq 0$ or $b \neq 0$; but doesn't when $a = b = 0$, since any integer divides a and b.

Definition 2.4.1. *Two numbers a and b are called relatively prime (or coprime or strangers) when their* GCD *is* 1.

• The classical terminology "relatively prime" is unfortunate because beginners often confound the assertions "p is prime" and "p and q are relatively prime".

• Observe that 0 is not relatively prime to 0 because $\mathrm{GCD}(0, 0) = 0$. So, if a and b are relatively prime, at least one of them is nonzero.

When $a \neq 0$ or $b \neq 0$, we can divide (2.10) by $d = \mathrm{GCD}(a, b) > 0$ and use formula (2.8). This gives us

$$a = da', \quad b = db', \quad \mathrm{GCD}(a', b') = 1 \ \text{ if } \ d = \mathrm{GCD}(a, b) > 0. \tag{2.11}$$

Theorem 2.4.2 (Bézout theorem – second version). *Two integers a and b are relatively prime if there exist $u, v \in \mathbb{Z}$ such that*

$$au + bv = 1. \tag{2.12}$$

Proof. The first version of Bézout's theorem shows that u and v exist when a and b are relatively prime. Conversely, if $au + bv = 1$ then the number 1 belongs to the subgroup $a\mathbb{Z} + b\mathbb{Z}$. This shows that $a\mathbb{Z} + b\mathbb{Z} = \mathbb{Z}$; that is, $d = 1$. □

Remark 2.4.1. We shall see in Chapter 8 (Blankinship's algorithm) how to effectively calculate u and v.

2.4.2. Gauss's Lemma

Theorem 2.4.3 (Gauss's lemma). *Let a, b be any two integers and suppose that d divides their product ab. If d is relatively prime to a, then d divides b.*

Proof. Applying Bézout's theorem to the pair (a, b) shows that $au + dv = 1$. Multiplying both sides by b gives $abu + bdv = b$. But then d divides abu because it divides ab, and it clearly divides bdv. Thus, it must divide their sum b. □

Application: Bézout's equation

Suppose that we want to solve for all integers x and y satisfying the equation:

$$ax + by = 1, \quad (a, b \geq 1 \text{ relatively prime}) \tag{2.13}$$

Bézout's theorem guarantees that $u, v \in \mathbb{Z}$ exist such that

$$au + bv = 1. \tag{2.14}$$

In other words, (u, v) is a particular solution of the equation. Suppose that (x, y) is another solution. Subtracting equation (2.14) from (2.13) gives

$$a(x - u) = b(v - y),$$

which shows, in particular, that a divides $b(v - y)$. Applying Gauss's lemma shows that $v - y = ak$ for an appropriate k. One shows that b divides $(x - u)$ and obtains finally that

$$x = u + bk, \quad y = v - ak, \quad k \in \mathbb{Z}.$$

The converse is immediate.

2.5. Congruences

Let $n > 1$ be any integer. We say that a and b are *congruent modulo n*, and write $a \equiv b \pmod{n}$ if $a - b$ is divisible by n; that is, if there exists an integer $k \in \mathbb{Z}$ such that $a - b = kn$.

Congruence is an equivalence relation on $\mathbb{Z}$. We let $\mathbb{Z}_n$ denote the quotient set. It contains n elements which we identify with the integers in the interval $[0, n-1]$ when working with classical euclidean division and with the integers between $-\frac{1}{2}n$ and $\frac{1}{2}n$ when working with centered remainders.

We denote the class of x by $\overline{x}$. However, when the notation becomes too cumbersome, we systematically omit the line over x. In this case, the context will make clear what is meant.

Congruence mod n is compatible with addition and multiplication:

$$a \equiv b \text{ and } a' \equiv b' \implies \begin{cases} a + a' \equiv b + b', \\ aa' \equiv bb'. \end{cases}$$

As a result, addition and multiplication on $\mathbb{Z}$ carry over to the quotient set $\mathbb{Z}_n$ which is then naturally endowed with the structure of a commutative ring.

For beginners

We often use the locution "to lift an equality from $\mathbb{Z}_n$ to $\mathbb{Z}$". This means that if we are given an equation $\alpha = \beta$ between two classes in $\mathbb{Z}_n$, we choose representatives of these classes, that is, integers a and b with $\overline{a} = \alpha$ and $\overline{b} = \beta$, and obtain an equality $a = b + kn$ in $\mathbb{Z}$. Since we systematically omit the congruence symbol and the overbars above representatives, the sentence

"the congruence $a \equiv b \pmod{n}$ lifts to the equality $a = b + kn$"

translates into the lapidary (and, to a beginner, somewhat puzzling) expression

"lifting $a = b$, we obtain $a = b + kn$."

One becomes rapidly accustomed to this sort of intellectual yoga which depends very strongly on the context.

Proposition 2.5.1. *The units of $\mathbb{Z}_n$ are the classes of integers relatively prime to n.*

Proof. Let $\varepsilon \in \mathbb{Z}_n$ be such that there exists $\varepsilon' \in \mathbb{Z}_n$ with $\varepsilon\varepsilon' = 1$ (notice the absence of overbars, despite the fact that we are in $\mathbb{Z}_n$). Lifting this equality to $\mathbb{Z}$ gives $\varepsilon\varepsilon' + kn = 1$ which means that ε and n are relatively prime. The converse is immediate. □

Corollary 2.5.1. *The ring $\mathbb{Z}_p$ is a field if and only if p is a prime number.*

Proof. If $\mathbb{Z}_p$ is a field, all elements which are not equal to zero are invertible. Lifting to $\mathbb{Z}$, we see that all integers between 1 and $p - 1$ are relatively prime to p. Thus, p s prime. The converse is immediate. □

2.6. The Chinese Remainder Theorem

This theorem was known to Chinese mathematicians in the first century our era – it allowed them to solve certain problems involving conjunctions of stars.

Theorem 2.6.1 (Chinese remainder theorem – weak version). *If $n > 1$ and $m > 1$ are relatively prime integers, the system of congruences*

$$x \equiv \alpha \pmod{n}, \quad x \equiv \beta \pmod{m},$$

possesses a unique solution $x \in [\![0, nm - 1]\!]$.

Proof. It is clear that the map

$$\Phi : \mathbb{Z}_{nm} \longrightarrow \mathbb{Z}_n \times \mathbb{Z}_m$$

defined by $x \pmod{nm} \mapsto (x \pmod{n}, x \pmod{m})$ is a ring homomorphism (that is, it is compatible with addition and multiplication). Let x be an element in the kernel of Φ. If we lift to $\mathbb{Z}$, the conditions $x \equiv 0 \pmod{n}$ and $x \equiv 0$ mod m) show that x is simultaneously divisible by n and m. Since n and m are relatively prime, x must be divisible by nm. Descending to $\mathbb{Z}_{nm}$, we obtain that $x = 0$, which means that Φ is injective.

The sets $\mathbb{Z}_{nm}$ and $\mathbb{Z}_n \times \mathbb{Z}_m$ both have nm elements. Thus, ϕ is surjecive and the given system has a unique solution in $\mathbb{Z}_{nm}$, which lifts to infinitely many solutions in $\mathbb{Z}$. □

We actually learn a little more from the proof: if x_0 is a particular solution, the other solutions are of the form $x = x_0 + knm$ with $k \in \mathbb{Z}$.

Exercise 4

Let $u, v \in \mathbb{Z}$ be such that $au + bv = 1$. Show that the map $\mathbb{Z}_n \times \mathbb{Z}_m \to \mathbb{Z}_{nm}$ defined by $(r, s) \mapsto mvr + nvs$ is the isomorphism Φ^{-1}. Check *directly* that Φ^{-1} is compatible with multiplication, a fact that is not at all evident at first glance.

Theorem 2.6.2 (Chinese remainder theorem – strong version).
If $n_1, \ldots, n_\ell > 1$ are pairwise relatively prime integers, the system of congruences

$$x \equiv \alpha_1 \pmod{n_1}, \quad \ldots \quad, x \equiv \alpha_\ell \pmod{n_\ell}, \tag{2.15}$$

possesses an infinite number of solutions. If x_0 is a particular solution, the other solutions are the numbers $x = x_0 + kn_1 \cdots n_\ell$ with $k \in \mathbb{Z}$.

Proof. We do not present the usual proof found in algebra books. Instead, we give a more natural proof[3] which leads to a program which is very easy to write. We expand x in a base of truncated multiples of the numbers $n_1, \ldots, n_\ell$ (see 2.2.2):

$$\begin{cases} x = x_0 + x_1 n_1 + x_2 n_1 n_2 + \cdots + x_\ell n_1 n_2 \cdots n_\ell \,, \\ 0 \le x_0 < n_1, \quad \ldots \quad, \; 0 \le x_{\ell-1} < n_\ell \quad \text{and} \quad x_\ell \in \mathbb{Z}. \end{cases} \tag{2.16}$$

In order not to obscure the proof, let us suppose henceforth that $\ell = 3$, in which case we write

$$x = x_0 + x_1 n_1 + x_2 n_1 n_2 + x_3 n_1 n_2 n_3.$$

- By combining (2.16) with the first congruence (2.15) , we already get:

$$x_0 \equiv \alpha_1 \pmod{n_1}.$$

This shows that x_0 exists and is unique. If we identify $\mathbb{Z}_n$ with $[\![0, n_1 - 1]\!]$ and if α_1 belongs to this interval, then $x_0 = \alpha_1$. Otherwise, $x_0 = \alpha_1 \bmod n_1$.

- Knowing the value of x_0, we can then determine that of x_1 from the second congruence in (2.15):

$$x_0 + n_1 x_1 \equiv \alpha_2 \pmod{n_2}.$$

[3] H. Garner, *The Residue Number System*, IRE Trans, EC8 (1959), pp. 140–147.

We know that n_1 is invertible in $\mathbb{Z}_{n_2}$ because n_1 and n_2 are relatively prime. As a result the first degree equation has a single unique solution $x_1 \in [\![0, n_2 - 1]\!]$.

• Knowing now the values of x_0 and x_1, we use the third congruence in (2.15) to obtain

$$x_0 + x_1 n_1 + x_2 n_1 n_2 \equiv \alpha \pmod{n_3}.$$

The integers n_1 and n_2 define two elements of $\mathbb{Z}_{n_3}^*$ since both are coprime to to n_3; their product is therefore invertible in $\mathbb{Z}_{n_3}^*$. As a result, the first degree equation in x_2 defines a unique integer $x_2 \in [\![0, n_3 - 1]\!]$.

Conversely, if x_0, x_1, x_2 are defined by the preceding equations, all integers of the form

$$x = x_0 + x_1 n_1 + x_2 n_1 n_2 + x_3 n_1 n_2 n_3,$$

with $x_3 \in \mathbb{Z}$ arbitrary are solutions of the system (2.15). □

Example 2.6.1. Suppose that we want to solve the system

$$x \equiv 2 \pmod 5, \quad x \equiv 4 \pmod 6, \quad x \equiv -1 \pmod{49}.$$

We seek an x of the form

$$x = x_0 + 5x_1 + 30x_2 + 1470x_3.$$

We solve the congruences one after another:

- The first congruence immediately gives $x_0 = 2$.
- The equation $2 + 5x_1 = 4$ in $\mathbb{Z}_6$ has $x_1 = 4$ as its single solution.
- Finally, the equation $22 + 30x_2 = -1$ has the solution $x_2 = 27$ in $\mathbb{Z}_{49}$.

Thus, the solutions of the system are the numbers

$$x = 832 + 1470x_3 \quad \text{for all} \quad x_3 \in \mathbb{Z}.$$

2.7. The Euler phi Function

The *Euler phi function* $\varphi : \mathbb{N}^* \to \mathbb{N}^*$ is defined by the formula:

$$\begin{aligned}\varphi(n) &= \operatorname{Card} \mathbb{Z}_n^* \\ &= \textit{number of integers } 1 \le k \le n \textit{ such that } \operatorname{GCD}(k, n) = 1\end{aligned}$$

The first few values of $\varphi(n)$ are:

n	1	2	3	4	5	6	7	8	9	10	11	12	13	14
$\varphi(n)$	1	1	2	2	4	2	6	4	6	4	10	4	12	6

If a and b are two relatively prime integers, the ring isomorphism $\mathbb{Z}_{ab} \simeq \mathbb{Z}_a \times \mathbb{Z}_b$ induces a isomorphism of multiplicative groups $\mathbb{Z}_{ab}^* \simeq \mathbb{Z}_a^* \times \mathbb{Z}_b^*$. Consequently, the Euler phi function is *multiplicative* which means that:

$$\forall a \in \mathbb{Z},\ \forall b \in \mathbb{Z}, \quad \mathrm{GCD}(a, b) = 1 \Rightarrow \varphi(ab) = \varphi(a)\varphi(b).$$

This property reduces the calculation of $\varphi(n)$ to that of $\varphi(p^\alpha)$ for p prime and $\alpha \geq 1$. To evaluate $\varphi(p^\alpha)$ we first seek integers $h \in [1, p^\alpha]$ which are not relatively prime to p. Since any such integer is necessarily divisible by p, it is of the form

$$h = pq, \quad 1 \leq q \leq p^{\alpha-1}.$$

Thus, there are $p^{\alpha-1}$ integers h which are relatively prime to p. This gives

$$\varphi(p^\alpha) = p^\alpha - p^{\alpha-1} = p^\alpha\Big(1 - \frac{1}{p}\Big).$$

If $n = p_1^{\alpha_1} \cdots p_r^{\alpha_r}$ is the decomposition of n into distinct prime factors, we obtain the formula:

$$\varphi(n) = (p^{\alpha_1} - p^{\alpha_1-1}) \cdots (p^{\alpha_r} - p^{\alpha_1-r}) = n\Big(1 - \frac{1}{p_1}\Big) \cdots \Big(1 - \frac{1}{p_r}\Big).$$

2.8. The Theorems of Fermat and Euler

Theorem 2.8.1 (The little Fermat theorem). *If a is any integer and p any prime number, then $a^p \equiv a$ modulo p. If, moreover, a is not divisible by p, then $a^{p-1} \equiv 1 \pmod{p}$.*

Proof. Transported to $\mathbb{Z}_p$, this assertion becomes: "For all $a \in \mathbb{Z}_p$, one has $a^p = a$ and if $a \neq 0$, then $a^{p-1} = 1$."

- This is evident if $a = 0$.
- If $a \neq 0$, consider the map $x \mapsto ax$ of $\mathbb{Z}_p$ to itself. Since $\mathbb{Z}_p$ is a field, this map is a bijection. That is, we have $ax \neq 0$ if $x \neq 0$. If we let $x_1, \ldots, x_{p-i}$ denote the elements of $\mathbb{Z}_p^*$, we have

$$\mathbb{Z}_p^* = \{x_1, \ldots, x_{p-1}\} = \{ax_1, \ldots, ax_{p-1}\}.$$

Multiplying all elements of $\mathbb{Z}_p^*$ together, we obtain

$$x_1 \cdots x_{p-1} = a^{p-1} x_1 \cdots x_{p-1}.$$

The result now follows upon dividing both sides by $x_1 \cdots x_{p-1} \neq 0$. □

This result generalises immediately using the Euler phi function (it suffices to copy the proof above).

Theorem 2.8.2 (Euler). *If a is any integer relatively prime to n, then*

$$a^{\varphi(n)} \equiv 1 \pmod{n}.$$

2.9. Wilson's Theorem

In the proof of Fermat's theorem, the product $P = (p-1)!$ of all elements of $\mathbb{Z}_p^*$ plays a role. Let us specify its value.

Theorem 2.9.1 (Wilson). *An integer $p > 1$ is prime if and only if*

$$(p-1)! \equiv -1 \pmod{p}.$$

Proof. First suppose that p is a prime: we must show that the product $P = x_1 \cdots x_{p-1}$ of all elements of $\mathbb{Z}_p^*$ is equal to -1. Since $\mathbb{Z}_p$ is a field, we know that each element $x_i \in \mathbb{Z}_p^*$ possesses an inverse $x_i^{-1} \in \mathbb{Z}_p^*$ such that $x_i x_i^{-1} = 1$. By associating to each x_i its inverse, we obtain 1 in the product P as many times as there are pairs (x_i, x_i^{-1}) such that $x_i \neq x_i^{-1}$. Regrouping, we see that P is equal to the product of "orphans" in $\mathbb{Z}_p^*$; that is, P equals the product of $x \in \mathbb{Z}_p^*$ such that $x = x^{-1}$. But then $x^2 = 1$. That is, $(x-1)(x+1) = 0$. Thus the orphans are precisely the elements 1 and -1, which shows that $P = 1 \times -1 = -1$.

To show the converse, suppose that p is not prime. The p possesses a divisor d such that $1 < d < p$. If $(p-1)!$ were congruent to -1 modulo p, since d appears in the factorial, we would have $0 \equiv -1 \pmod{d}$ which is absurd. □

Corollary 2.9.1. *Let p be an odd prime. The equation*

$$x^2 + 1 = 0$$

possess roots in $\mathbb{Z}_p$ if and only if $p \equiv 1 \pmod{4}$. In this case, the roots are $x' = \ell!$ and $x'' = -\ell!$ where $\ell = \frac{1}{2}(p-1)$.

Proof. An odd number is of the form $4n+1$ or $4n+3$.

- We first consider the case $p = 2\ell + 1$ with $\ell = 2n$. Identify $\mathbb{Z}_p^*$ with the set $\{-\ell, \ldots, -1, 1, \ldots, \ell\}$ using centered remainders. Then, Wilson's theorem can be written:

$$-\ell \times \cdots \times -1 \times 1 \cdots \times \ell = (-1)^\ell (\ell\,!)^2 = -1.$$

Since $\ell = 2n$ is an even number, we see that $x = \ell!$ is a root of the equation $x^2+1 = 0$ in $\mathbb{Z}_p$. The other root is clearly $-\ell!$ since $x^2+1 = (x+\ell!)(x-\ell!)$.

- Now consider the case $p = 4n+3$. Suppose that there exists $x \in \mathbb{Z}_p$ such that $x^2+1 = 0$. That is, $x^2 = -1$. This allows us to write:

$$x^{p-1} = x^{4n+2} = (x^2)^{2n} x^2 = x^2 = -1.$$

On the other hand, by Fermat's theorem we have $x^{p-1} = 1$. Thus, $-1 = 1$ in $\mathbb{Z}_p$ which is absurd. □

Remark

Calculating a factorial is not an efficient way of finding the roots of the equation $x^2 + 1 = 0$ in $\mathbb{Z}_p$. We shall return to this subject later (Chapter 8).

Exercise 5 *(Solution at the end of the chapter)*

An old mathematical fantasy is that of finding a "simple formula" (in a sense to be made precise) or a "function" whose values are only prime numbers.

• Prove (using Taylor's formula) that a polynomial with integer coefficients cannot take only prime values.

• Here is another try which seems promising *a priori*. Let $x, y \in \mathbb{N}^*$. Put $T(x, y) = x(y + 1) - (y! + 1)$ and

$$f(x, y) = \begin{cases} 2 & \text{if } |T| \geq 1, \\ y + 1 & \text{if not.} \end{cases}$$

Show that the function $f : \mathbb{N}^* \times \mathbb{N}^* \to \mathbb{N}$ takes its values in the set of prime numbers and that each odd prime number is obtained exactly once. Write a Pascal program which displays the values $f(x, y)$ for $1 \leq x, y \leq 100$. What conclusion do you draw from this experience?

2.10. Quadratic Residues

An element $a \in \mathbb{Z}_n$ is said to be a *quadratic residue* modulo n if it is a square in $\mathbb{Z}_n$; that is, if there exists an element $x \in \mathbb{Z}_n$ such that $a = x^2$.

Being a square depends on the ring in which one works: a complex number is always a square; a real number is a square if and only if it is positive or zero. The following result sharpens Fermat's theorem and gives a very simple criterion for an element of $\mathbb{Z}_p$ to be a square when p is a prime number.

Theorem 2.10.1 (Euler). *Let $p = 2\ell + 1$ be an odd prime. For every element $x \neq 0$ in $\mathbb{Z}_p^*$, one has:*

$$x^\ell = \begin{cases} +1 & \text{if } x \text{ is a quadratic residue,} \\ -1 & \text{if } x \text{ is not a quadratic residue.} \end{cases}$$

Proof. Fermat's theorem shows that $(x^\ell)^2 = x^{p-1} = 1$. Since we are in a field, it follows that $x^\ell = \pm 1$.

Consider the map $\kappa : x \mapsto x^2$ of $\mathbb{Z}_p^*$ to itself. We have $x^2 = y^2$ if and only if $(x + y)(x - y) = 0$; that is, if and only if $y = \pm x$ since we are working in a field. This reasoning shows that every nonzero quadratic residue has exactly two preimages under κ. The number of quadratic residues $\neq 0$ must therefore equal $\frac{1}{2}(p - 1)$ and the number of nonresidues is also equal to $\frac{1}{2}(p - 1)$.

Let $x = a^2$ be a quadratic residue. Fermat's theorem shows that $x^\ell = a^{2\ell} = 1$, thereby showing that the roots of the polynomial $X^\ell - 1$ are nonzero quadratic residues (and that these are the only ones). If x is not a quadratic residue, we therefore have $x^\ell \neq 1$, which implies that $x^\ell = -1$. □

2.11. Prime Number and Sum of Two Squares

Let a be an integer of the form $4n + 3$. Since $x^2 + y^2 \equiv 0, 1, 2 \pmod 4$, one sees that immediately that a cannot be a sum of two squares.

Theorem 2.11.1 (Fermat). *Every prime number of the form* $4n + 1$ *is a sum of two squares.*

Proof. Consider the set of triples[4]

$$S = \{(x, y, z) \in \mathbb{N}^3 : x^2 + 4yz = p\}.$$

This set is nonempty because it contains the triple $(1, n, 1)$.

Consider now the map $\psi : \mathbb{N}^3 \to \mathbb{N}^3$ defined by :

$$\psi(x, y, z) = \begin{cases} (x + 2z, z, y - x - z) & \text{if } x < y - z, \\ (2y - x, y, x - y + z) & \text{if } y - z < x < 2y, \\ (x - 2y, x - y + z, y) & \text{if } x > 2y. \end{cases}$$

A simple verification shows that $\Psi(S) \subset S$ and that the restriction of Ψ to S is an involution that possess a fixed point that is unique. It follows first that the cardinality of S is odd, and thus that *every* involution possesses at least one fixed point. Considering now the involution $(x, y, z) \mapsto (x, z, y)$, we see that it possesses a fixed point (a, b, b) which gives $a^2 + 4b^2 = p$. □

Remarks

- This theorem, which is a little jewel of static mathematics, gives absolutely no indication of how we might explicitly find a and b. We will see this in Chapter 8 when studying Euler's proof.

- We will prove in Chapter 9 that if (a, b) is a particular solution of the equation $p = x^2 + y^2$, then all the other solutions are $(\pm a, \pm b)$ and $(\pm b, \pm a)$. Thus, the pair (a, b) is unique in a sense that one can easily make precise.

- One should not imagine that the condition $n \equiv 1 \pmod 4$ guarantees that n is a sum of two squares. A counterexample is 21. More generally, one has the following result.

[4] Following D. Zagier, *A one-sentence proof that every prime* $p \equiv 1 \bmod 4$ *is a sum of two squares* American Mathematical Monthly, 97 (1990), p. 144.

Theorem 2.11.2 (Jacobi, 1829). *Let $n \geq 1$ be an integer and $\Delta(n)$ the number of pairs $(x, y) \in \mathbb{Z}^2$ which are solutions of the equation $x^2 + y^2 = n$. If $d_1(n)$ is the number of divisors of n which are congruent to 1 modulo 4 and $d_3(n)$ is the number of divisors of n which are congruent to 3 modulo 4, then*

$$\Delta(n) = 4\big[d_1(n) - d_3(n)\big].$$

(The odd divisors are the only ones which occur in this decomposition.)

For enterprising readers, the proof is carried out in two steps: one begins by checking that the formula holds when n is a power of a prime number. Then one proves that the functions Δ and $d_1 - d_3$ are multiplicative.

2.12. The Moebius Function

Let $f : \mathbb{N}^* \to M$ be any function with values in an abelian group M. We write M additively. To f we can associate a function $\varphi : \mathbb{N}^* \to M$ defined by

$$\varphi(n) = \sum_{d \mid n} f(d). \tag{2.17}$$

For the first few values of n we have

$$\begin{aligned}
\varphi(1) &= f(1), \\
\varphi(2) &= f(1) + f(2), \\
\varphi(3) &= f(1) + f(3), \\
\varphi(4) &= f(1) + f(2) + f(4), \\
\varphi(5) &= f(1) + f(5), \\
\varphi(6) &= f(1) + f(2) + f(3) + f(6), \text{ etc.}
\end{aligned}$$

Conversely, we might ask whether it is possible to reconstruct f if we know φ. If we view the $f(k)$ as unknowns, the equations above make up a system of linear equations with integral coefficients whose matrix is triangular with coefficients on the diagonal equal to 1. This is a Cramer system which means that there is a unique function f. The inverse matrix T^{-1} has integral coefficients because $\det T = 1$. The explicit solution of this system is remarkable.

Theorem 2.12.1 (The Moebius inversion formula). *For every integer $n \geq 1$, one has*

$$f(n) = \sum_{d \mid n} \mu\Big(\frac{n}{d}\Big)\, \varphi(d) = \sum_{d \mid n} \mu(d)\, \varphi\Big(\frac{n}{d}\Big), \tag{2.18}$$

where $\mu : \mathbb{N}^* \to \{-1, 0, 1\}$ *is the Moebius function*

$$\mu(n) = \begin{cases} 1 & \textit{if } n = 1, \\ 0 & \textit{if } n \textit{ is divisible by the square of a prime number,} \\ (-1)^k & \textit{if } n \textit{ is the product of k distinct prime numbers.} \end{cases}$$

The first few values of the Moebius function are:

n	1	2	3	4	5	6	7	8	9	10	11	12	13	14
$\mu(n)$	1	−1	−1	0	−1	1	−1	0	0	1	−1	0	−1	1

Before proving the inversion formula (2.18), we give some immediate properties of the Moebius function that we will need in what follows.

Lemma 2.12.1. *If a and b are relatively prime, then $\mu(ab) = \mu(a)\mu(b)$. For all $n > 1$, one has* $\sum_{d\,|\,n} \mu(d) = 0$.

Proof. The first assertion is an immediate consequence of the definition. To establish the second, consider a prime divisor p of n. We can write $n = p^\alpha m$ with $\alpha \geq 1$ and m not divisible by p.

- If $m = 1$, we have

$$\sum_{d\,|\,n} \mu(d) = \mu(1) + \mu(p) + \mu(p^2) + \cdots = \mu(1) + \mu(p) = 0.$$

- If $m > 1$, the divisors δ of n are of the form $d, pd, p^2d, \ldots, p^\alpha d$ where d is a divisor of m. Thus, we can write

$$\begin{aligned} \textstyle\sum_{\delta\,|\,n} \mu(\delta) &= \textstyle\sum_{d\,|\,m} \mu(d) + \sum_{d\,|\,m} \mu(pd) + \sum_{d\,|\,m} \mu(p^2d) + \cdots \\ &= \textstyle\sum_{d\,|\,m} \mu(d) + \sum_{d\,|\,m} \mu(pd) \\ &= \textstyle\sum_{d\,|\,m} \mu(d) + \mu(p) \sum_{d\,|\,m} \mu(d) \\ &= 0. \end{aligned}$$

□

Proof of formula (2.18). We begin with the sum

$$S = \sum_{d\,|\,n} \mu(n/d)\varphi(d) = \sum_{d\,|\,n} \mu(n/d) \sum_{\delta\,|\,d} f(\delta).$$

If d divides n and if δ divides d, we have $n = d\delta m$. Therefore, we can rearrange the sum:

$$S = \sum_{\delta m\,|\,n} f(\delta)\mu(m) = \sum_{\delta\,|\,n} f(\delta) \sum_{m\,|\,n\delta^{-1}} \mu(m).$$

By the lemma, the sum of the $\mu(m)$ is zero if $n\delta^{-1} > 1$ and is equal to 1 if $n\delta^{-1} = 1$, that is, if $\delta < n$ or $\delta = n$. We are left with $S = F(n)$.

Remarks

• If we write the group M multiplicatively, formulas (2.17) and (2.18) become

$$\varphi(n) = \prod_{d \mid n} f(d) \Longrightarrow f(n) = \prod_{d \mid n} \varphi(d)^{\mu(n/d)}. \tag{2.19}$$

This version will be very useful when we deal with cyclotomic polynomials in Chapter 10.

• The Moebius function appears in many fascinating formulas. For example, if $s > 1$, one has

$$\zeta(s) = \sum_{1}^{\infty} \frac{1}{n^s} \Longrightarrow \frac{1}{\zeta(s)} = \sum_{1}^{\infty} \frac{\mu(n)}{n^s}.$$

The proof of this formula is easy. Because both series are absolutely convergent for $s > 1$ we have the right to rearrange their product in the following way:

$$\Big(\sum_{1}^{\infty} \frac{1}{p^s}\Big)\Big(\sum_{1}^{\infty} \frac{\mu(q)}{q^s}\Big) = \sum_{p,q \geq 1} \frac{\mu(q)}{(pq)^s} = \sum_{n \geq 1} \frac{1}{n^s}\Big(\sum_{q \mid n} \mu(q)\Big) = 1.$$

2.13. The Fibonacci Numbers

Fibonacci (*filius Bonacci*, 1180–1228), also known under the name of Leonardo of Pisa, introduced the sequence which immortalized him while studying the growth of rabbits on a desert island. This harmless sequence, of vital importance in computer science, is defined by the initial conditions

$$F_0 = 0, \quad F_1 = 1$$

and the second order recurrence relation

$$F_n = F_{n-1} + F_{n-2} \qquad n \geq 2.$$

The first few Fibonacci numbers are therefore

n	0	1	2	3	4	5	6	7	8	9	10	11	12
F_n	0	1	1	2	3	5	8	13	21	34	55	89	144

The roots of the characteristic equation $X^2 = X + 1$ are the *golden numbers* $\gamma = \frac{1}{2}(1 + \sqrt{5})$ and $\delta = \frac{1}{2}(1 - \sqrt{5})$. Because the sequence satisfies the same recurrence relation as the F_n, we obtain *Binet's formula* (which is not of much interest when working over the integers):

$$F_n = \frac{1}{\sqrt{5}}\Big\{\Big(\frac{1}{2}(1 + \sqrt{5})\Big)^n - \Big(\frac{1}{2}(1 - \sqrt{5})\Big)^n\Big\}.$$

Exercise 6

Show (by induction) that $F_n > \gamma^{n-2}$ for $n \geq 3$.

2.14. Reasoning by Induction

We are going to recall and elaborate the basic principles of the technique of mathematical induction which we use frequently (especially when dealing with questions involving recurrence).

Theorem 2.14.1 (Principle of weak induction). *Suppose that $\mathbb{D}$ is a subset of $\mathbb{N}$ with the following two properties*

(a) *the integer* 0 *is in* $\mathbb{D}$;
(b) *anytime that n is in $\mathbb{D}$, one can show that $n+1$ is in $\mathbb{D}$.*

Under these conditions, $\mathbb{D} = \mathbb{N}$.

Proof. Suppose that $\mathbb{D} \neq \mathbb{N}$. Then $\mathbb{D}^c = \mathbb{N} - \mathbb{D}$ is not empty and therefore has a least element μ. Condition (i) implies that $\mu > 0$. So we can consider the integer $\mu' = \mu - 1 \geq 0$. The definition of μ shows that μ' belongs to $\mathbb{D}$. From condition (ii), it follows that $\mu' + 1 = \mu$ belongs to $\mathbb{D}$ which is a contradiction. □

Theorem 2.14.2 (Principle of strong induction). *Let $\mathbb{D}$ be a subset of $\mathbb{N}$ with the following two properties:*

(a) *the integer* 0 *belongs to* $\mathbb{D}$;
(b) *any time that the interval $[\![0, n]\!]$ is contained in $\mathbb{D}$, one can demonstrate that $n+1$ belongs to $\mathbb{D}$.*

Under these conditions, $\mathbb{D} = \mathbb{N}$.

Proof. It suffices to copy the preceding proof replacing the sentence "the definition of μ shows that μ' belongs to $\mathbb{D}$" by "the definition of μ shows that the integers $< n$ are in $\mathbb{D}$, which gives the inclusion $[\![0, \mu']\!] \subset \mathbb{D}$". □

Experience shows nevertheless that these two principles of proof do not suffice because one is often obliged to induct on $\mathbb{N}^2$ or on sets which are much more baroque.

Definition 2.14.1. *One says that an ordered nonempty set E is well-ordered if every nonempty subset $\mathbb{D} \subset E$ possesses a least element; that is, an element $\mu \in \mathbb{D}$ less than or equal to all other elements of $\mathbb{D}$.*

Let us establish some immediate consequences of this definition.

• A well-ordered set is *totally ordered.* In fact, two elements x, y of E are always comparable because $\mathbb{D} = \{x, y\}$ possesses a least element which therefore must be comparable to the others.

• A well-ordered set always possesses a least element since $\mathbb{D} = E$ is a nonempty subset of E. Unless explicitly stated otherwise, we will always denote the least element by 0.

• The smallest element μ of a nonempty subset $\mathbb{D}$ is unique. In what follows we let $\mu = \min \mathbb{D}$ denote the smallest element of $\mathbb{D}$.

• In an well-ordered set which is *not bounded above*, every element possesses a *successor* x'. This is defined as the unique element possessing the following properties:

(a) $x < x'$;

(b) there is no element between x and x' (in other words, the inequalities $x \leq y \leq x'$ imply that $y = x$ or $y = x'$).

Proof. Since E is not bounded above, the set $\mathbb{D} - [0, x]$ is not empty. The successor of x is then $x' = \min \mathbb{D}$ (this is immediate). □

Examples

• The set $\mathbb{N}$ is well-ordered.

• A nonempty interval of $\mathbb{N}$ is well-ordered – this shows that there exist well-ordered finite sets.

• There are many other simple sets which are well-ordered as the following result shows.

Theorem 2.14.3. *For every integer $k \geq 2$, the lexicographic order on $\mathbb{N}^k$ is a well-ordering.*

Proof. To simplify, we are going to prove the theorem in the case when $k = 3$. Let $\mathbb{D}$ be a nonempty set of triples $(x, y, z) \in \mathbb{N}^3$. Consider the set of first coordinates of elements of $\mathbb{D}$. Since this is a subset of $\mathbb{N}$, we know that it has a least element:

$$\xi = \min \{x \in \mathbb{N} : \ (x, y, z) \in \mathbb{D}\}.$$

Consider in turn the elements

$$\eta = \min \{y \in \mathbb{N}; (\xi, y, z) \in \mathbb{D}\} \quad \zeta = \min \{z \in \mathbb{N}; (\xi, \eta, z) \in \mathbb{D}\}$$

(note the presence of ξ in the definition of η and of ξ, η in the definition of ζ). It is clear that (ξ, η, ζ) belongs to $\mathbb{D}$. To show that it is the smallest element, suppose that there exists $(x, y, z) \in \mathbb{D}$ and that $(x, y, z) \leq (\xi, \eta, \zeta)$ (the ordering being lexicographic). Then we have $x \leq \xi$, which implies that $x = \xi$ in view of the definition of ξ. This implies that $y \leq \eta$ and, thus, that $y = \eta$ in view of the definition of η. Similarly, one shows that $z = \zeta$. □

This result is very interesting. First of all, it allows us to reason by induction on $\mathbb{N}^2$ or $\mathbb{N}^3$. But it also shows that we should be wary of extrapolating from the set of integers: *in a well-ordered set, some elements may fail to have a predecessor!*

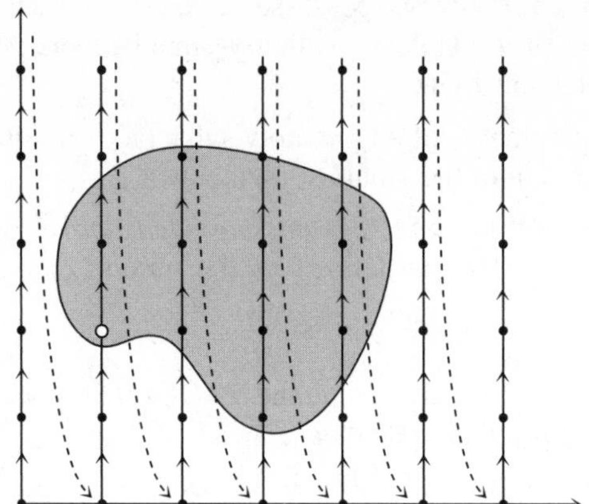

Fig. 2.1.

In $\mathbb{N}^2$ for example, the couple (1,0) has no predecessor (see Fig. 2.1). The smallest element of the set $\mathbb{D}$ of points in the grey region is the white point $(1, 2)$ Each element of $\mathbb{N}^2$ possesses a successor, but the points on the horizontal axis do not have a predecessor.

Exercise 7

Show that in $\mathbb{N}^2$ with the lexicographic ordering, the points with no predecessor are the points on the horizontal axis. To understand the structure of the order on $\mathbb{N}^2$ a little better, consider the map $\varphi : \mathbb{N}^2 \to \mathbb{R}$ defined by

$$\varphi(x, y) = x + 1 - \frac{1}{y + 1}.$$

What does $\varphi(\mathbb{N})$ look like? Show that this is a strictly increasing bijection between $\mathbb{N}^2$ and its image.

Theorem 2.14.4 (Principle of transfinite induction). *Let E be a well-ordered set and $\mathbb{D}$ a subset with the following two properties:*

(a) *the set $\mathbb{D}$ contains the least element* 0 *of E;*

(b) *each time that one has $[0, x) \subset \mathbb{D}$, one can show that x belongs to $\mathbb{D}$.*

Under these conditions, one has $\mathbb{D} = E$.

Proof. We adapt the preceding proofs, sorting them out so as to make no reference to predecessors. Suppose for a moment that $\mathbb{D}$ is not the set E. Then $\mathbb{D}^c = E - \mathbb{D}$ is not empty and has a least element $\mu > 0$ after (i). From the definition of μ and (ii), one sees that the elements of E which satisfy the inequality $x < \mu$ are in $\mathbb{D}$ (this is the way we eliminate recourse to the predecessor of μ). From the induction hypothesis, one concludes that μ is an element of $\mathbb{D}$, which is a contradiction. □

Example

Here is a typical example of transfinite induction. Let $u, v\colon \mathbb{N} \to \mathbb{N}$ and $w : \mathbb{N}^2 \to \mathbb{N}$ be any three functions. Consider the function $f : \mathbb{N}^2 \to \mathbb{N}$ defined by

$$f(x, y) = \begin{cases} u(x) & \text{if } y = 0, \\ v(y) & \text{if } x = 0, \\ f\big(x-1, w(x, y)\big) + f(x, y-1) & \text{if } x, y > 0. \end{cases}$$

(We suppose that $u(0) = v(0)$ for this definition to be coherent.)

We call $\mathbb{D}$ the domain of definition of this function. We are going to demonstrate that it is equal to $\mathbb{N}^2$ with the aid of transfinite induction.

- First of all, $f(0, 0)$ exists.
- Now let (x, y) be an element of $\mathbb{N}^2$ and suppose that $f(u, v)$ exists for all pairs (u, v) satisfying the condition $(u, v) < (x, y)$ with respect to the lexicographic order. We must prove that $f(x, y)$ exists.
  - ▹ This is evident if x or y is zero.
  - ▹ If x and y are not zero, since we have

$$\big(x-1, u(x, y)\big) < (x, y) \quad \text{and} \quad (x, y-1) < (x, y)$$

for the lexicographic order, the induction hypotheses guarantees the existence of the numbers $f(x-1, u(x, y))$ and $f(x, y-1)$. Therefore,

$$f(x, y) = f\big(x-1, u(x, y)\big) + f(x, y-1)$$

is well-defined.

We will see another application of this principle in Chapter 12.

2.15. Solutions of the Exercises

Exercise 1

If we calculate the differences $u_{i+1} - u_i$, we see that we get a periodic sequence:

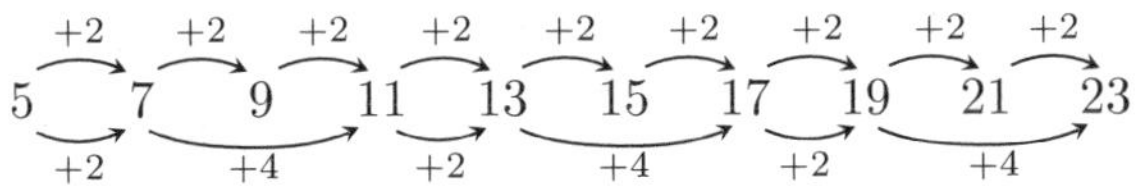

+2 +2 +2 +2 +2 +2 +2 +2 +2
5 7 9 11 13 15 17 19 21 23
+2 +4 +2 +4 +2 +4

More formally, every integer is of the form $6q + r$ with $r = 0, \ldots, 5$. If we eliminate the multiples of 2 and 3, the only possible values of r are $r = 1$ or $r = 5$. The terms of the sequence (u_n) are thus

$$\ldots, 6q+1,\ 6q+5,\ 6(q+1)+1,\ 6(q+1)+5,\ 6(q+2)+1, \ldots$$

As a result, we go from u_i to u_{i+1} by alternately adding 2 or 4, which gives:

$$u_{i+1} = \begin{cases} u_i + 2 & \text{if } i \equiv 1 \pmod 2, \\ u_i + 4 & \text{if } i \equiv 0 \pmod 2. \end{cases} \tag{2.20}$$

Exercise 2

Suppose first of all that the coefficients of P and Q are in $\mathbb{N}$ and strictly less than b. Then $P(b) = Q(b)$ expresses the fact that the numbers $P(b)$ and $Q(b)$ have the same expression in terms of the base b. Therefore the coefficients are equal. If the coefficients of P and Q have arbitrary sign, one can reduce to the preceding case by adding a suitable multiple of $X^n + X^{n-1} + \cdots + X + 1$ to both polynomials.

Exercise 3

The element in the x-th row and y-th column is:

$$N = 3x + 1 + (y - 1)(2x + 1) = 2xy + x + y.$$

- Let $2n + 1 = pq$ be a composite number. Since $2n + 1$ is odd, we can put $p = 2x + 1$ and $q = 2y + 1$ with $x, y \geq 1$. Then $n = 2xy + x + y$ figures in the table.
- Conversely, if $n = 2xy + x + y$ figures in the table, one has $2n + 1 = (2x + 1)(2y + 1)$ which shows that $2n + 1$ is composite.

Write a Pascal program which calculates and displays the Sundaram sieve up to n rows and n columns, and which uses it to obtain the corresponding prime numbers.

Exercise 5

Suppose that there exists a polynomial $F \in \mathbb{Q}[X]$ such that $F(x)$ is a prime number for all $x \in \mathbb{N}$ greater than or equal to x_0. Put $F = G/d$ with $G \in \mathbb{Z}[x]$ and $d \in \mathbb{N}^*$ and let $p = F(x_0)$. Applying the Taylor formula gives:

$$F(x_0 + phd) = F(x_0) + p\Big(hdF'(0)\Big) + p^2\Big(\frac{(hd)^2}{2\,!}F''(x_0)\Big) + \cdots + p^n\Big(\frac{(hd)^n}{n\,!}F^{(n)}(x_0)\Big).$$

Since $G^{(k)}(x_0)/k!$ is an integer for $k \geq 1$, so is $(hd)^k F^{(k)}(x_0)/k!$. Hence $F(x_0 + hp)$ is an integer multiple of p strictly greater than p if one chooses h sufficiently large. This is a contradiction.

Now let us see what happens with the second attempt.

- If $|T| > 1$, we have $f(x, y) = 2$ which is a prime number.

• If $T = 0$, we have $x(y + 1) = (y! + 1)$ which implies that $y! + 1 \equiv 0$ modulo $(y + 1)$. Wilson's theorem then assures us that $f(x, y) = y + 1$ is indeed a prime number.

Now, suppose that we want to solve the equation $f(x, y) = p$ when p is an odd prime. We have

$$T = 0 \implies y + 1 = p \implies x = \big[(p - 1)! + 1\big]/p,$$

and Wilson's theorem guarantees that x is an integer. There is a unique such pair. Conversely, computing $f(x, y)$ gives again the prime p.

We have a magnificent reformulation of Wilson's theorem and a "simple" function which only takes prime values. It is annoying that this function almost always takes the value 2 since T is almost never 0: for example, one has $T(x, 10) = 0$ iff $x = x_0 = 329{,}891$ which means that $f(x, 10) = 2$ for $x \neq x_0$ and $f(x_0, 10) = 11$. This is not an efficient way to generate prime numbers ...

3. An Algorithmic Description Language

An *algorithm* is a "recipe", a minute description of the operations that one must perform to obtain a desired result. In order to avoid any ambiguity, this description necessarily uses a very restricted language: two persons separated by thousands of kilometers must perform exactly the same sequence of operations.

The language that we are going to present to describe algorithms will be based on the computer language Pascal. However, we will stray from this language without apology when necessary, because an algorithm must be understood by a person, and human beings absolutely do not function like compilers!

The distinction between an algorithm and a program is important:

- An algorithm is a description, as clear and as vivid as possible, of a set of actions which do not depend on a given machine; it is intended for our brain.
- A program is a painstakingly precise, punctilious text intended for a compiler. It is written in a specific language. A program written in Lisp cannot be understood by a Pascal, or a C, or an Ada compiler.

An algorithm generally suppresses many details that one cannot ignore when dealing with a compiler. In a manner of speaking, an algorithm is the quintessence of a program. In contrast, a program is an *implementation* of one or more algorithms.

Like any living language, a computer language consists of words that are organised into sentences. The set of precise rules which govern how correct words and sentences can be formed is called the *syntax* of the language. Syntax is mechanical: it allows us to give orders to the compiler.

The word *semantics* refers to the meaning of the text that one writes. The meaning exists only for (and in) our brain. By its very definition, it is inaccessible to a machine which is just a set of switches and a clock.

If you are a mathematician, you will easily grasp the distinction: haven't you ever been able to repeat a proof, without understanding it?

3.1. Identifiers

Calculating involves manipulating variables, each of which must be given a name called an *identifier*. For ease of reading, mathematicians prefer to use identifiers consisting of a single letter: the abcissa x, time t, etc, even to the point of borrowing from several alphabets (roman, greek, hebrew, etc.).

As the complexity of the objects increases, however, they use more and more identifiers consisting of several letters (SO, PSL, Hom, End, etc.).

A mathematical proof uses relatively few identifiers. By contrast, a program can contain hundreds, or even thousands, of variables and uses words of arbitrary length as identifiers. The convention used is very natural: an identifier always begins with a letter (which can be upper or lower case); and is followed by (unaccented) letters or numerals.

For example, "x", "$x1$", "$x12$", "$x1y2z3$" and "*toto*" are identifiers. In contrast, "$1x$" and "*déjà*" are not.

Suppose that we wish to call an identifier *initial velocity*. We cannot leave the space between the two words, because a space is neither a letter nor a numeral. We could avoid the difficulty by writing *initialvelocity* or *InitialVelocity*.

In order to improve the readability of programs, Pascal, and many other computer languages, treat the symbol "_" (called the *underscore* or *underline* or *break* symbol, and obtained by pressing the key combination SHIFT – on a keyboard) as a letter. Consequently, "*initial_velocity*", "_1", "_x" and "x_" are identifiers.

For beginners

Choosing good identifiers is crucial: they should inform the reader of their meaning and are an important form of *self-documentation* of the program. It is very easy to render an algorithm or program illegible by an awkward, unitelligible, or bizarre choice of identifiers.

Consider, for instance, the statement:

$$distance := initial\_distance + speed * time.$$

You have, of course, the right to rebaptize these identifiers and replace *distance* everywhere in the program by *acceleration*, *initial_distance* by *speed* and *speed* by *initial_distance* to obtain the bizarre statement:

$$acceleration := speed + initial\_distance * time.$$

For a compiler, identifiers mean nothing: they are anonymous addresses and cannot mislead. But it should not surprise you if your brain is led astray in a sufficiently hostile context ...

3.2. Arithmetic Expressions

An *arithmetic expression* is a collection of identifiers, numbers, and symbols such as "$a - (b * (c + x/y) - \cos(t + 1))$" or "$a * x[2 * i + i + 1] + b$". There is nothing special to say concerning their construction or syntax.[1]

3.2.1. Numbers

An arithmetic expression can contain numbers: "$2 * x + 3.14$". Since one does not have access to the notation 10^n on a keyboard, the numbers 3.56×10^{12} and 1.7×10^{-4} are denoted in Pascal by "$3.56E12$" and "$1.7E - 4$".

3.2.2. Operations

The operations "$+, -, *, /$" are *left-associative*, which means, for example, that a computer will *evaluate* "$x/y/z/t$" as if it were written "$((x/y)/z)/t$".

Be careful! The operation "/" manufactures real numbers. When you type "4/2", you obtain the real number 2.000... in Pascal, not the integer 2 (in other words, the computer does not view $\mathbb{Z}$ as a subset of $\mathbb{R}$).

Let a, b be integers with $b > 0$ and let q, r be the quotient and remainder upon division of a by b:

$$a = bq + r, \quad 0 \le r < b.$$

The quotient q is denoted "a div b" and the remainder r by "$a$mod b". We thus have two internal operations div , mod : $\mathbb{Z} \times \mathbb{Z} \to \mathbb{Z}$ which have two peculiarities of which you should be aware:

- The operations " div " and "mod " have priority over addition and multiplication. This means that "$a + b$mod n" and "$a$mod $p * p$" are interpreted as if they were the expressions "$a + (b$mod $n)$" and "$(a$mod $p) * p$". Do not forget the parentheses if what you want is "$(a + b)$mod n" or "$a$mod $(p * p)$"!

- Be very careful: when b is negative, "a div b" and "$a$mod b" are not what a mathematician means by the quotient or remainder: there is often a shift. Experiment to find out what convention is being used.

For beginners

In mathematics, the product of two variables is generally denoted by concatenating the names of the variables: the product $x \times y$ is denoted by xy. In computer science, the use of the symbol "$*$" is indispensable because a program may very well contain the identifiers x, y and xy.

---

[1] You already learned about these constructions empirically. But what are the precise rules? A beginning of a response will be given in Chapter 13. A rigorous description of these rules uses the abstract concept of a *grammar* which we cannot take up here.

3.2.3. Arrays

In computer science, the indexed objects that mathematicians use (vectors, matrices, etc.) are called *arrays*. Because indexes are not available on a keyboard, they are placed in square brackets:

- instead of talking about a vector $x = (x_1, \ldots, x_n)$, a computer scientist considers an array $x[1..n]$ whose elements are $x[1], x[2], \ldots, x[n]$;
- a matrix is an array with two indexes of the form $A[1..n, 1..m]$; the element $A_{i,j}$ is written $A[i, j]$.

You can use arrays with three, four, etc. indexes.

An arithmetic expression can contain references to an array:

$$A[i, j] + x[i + 1] * y[2 * j + 2 * t[u, v + w] - 4].$$

It is possible to replace an index by an arithmetic expression if the value of the latter is a whole number.

3.2.4. Function calls and parentheses

An arithmetic expression can also contain *function calls*:

$$2 * x + \cos(3 * y * y + 0.5) / \log(A[i, j + 1] - \sin(t)).$$

In mathematics, one suppresses redundant parentheses whenever possible in order to reduce clutter. One writes, for example, "$\cos x$" instead of "$\cos(x)$" and, hence, "$y + \cos x$" rather than "$\cos x + y$". In contrast, one cannot forget the parentheses in a program: one systematically writes "$\cos(x)$".

Square brackets and braces (curly brackets) are not allowed: "$\log[x + \cos(y)]$" is incorrect, it is necessary to write "$\log(x + \cos(y))$": square brackets are reserved for arrays and curly brackets enclose comments.

3.3. Boolean Expressions

Boolean expressions are arithmetic expressions which take the values *true* or *false*. They are obtained as follows:

- by using a boolean variable which takes the values *true* or *false*;
- by comparing two arithmetic expressions: for example $x + y \neq z - t + u$ or $x + y * (z + x) \geq 3 - \cos(x + y)$;
- by combining boolean expressions using the logical operations "or", "and" and "not". For example,

$$\textbf{not } (x = y + 1) \textbf{ or } (a \geq b) \textbf{ and } (c + d > z + 1) \textbf{ or not } \mathit{finish}$$

In an arithmetic expression, multiplication has priority over addition and subtraction. In a similar manner, the operation "and" has priority over "or" as well as the operations "$+, -, *, /$". As a result of this convention, "$(a < b)$ or $((x = y)$ and $(u > v))$" has the same meaning as "$(a < b)$ or $(x = y)$ and $(u > v)$", thereby allowing us to suppress some parentheses.

For beginners

• The logical operations "and" and "or" do not at all corrrespond to the way we use them when we speak! We have a tendency to give the operation "or" an *exclusive value* (the computer scientist's "xor"). When we assert, for example: "This is butter *or* margarine!" we understand that it is *either* one *or* the other, but certainly not both. We also say "this property is true for $i < 10$ and $i > 20$" although the boolean expression "$(i < 10)$ and $(i > 20)$" is always false.

• Since the symbols $\leq$, $\geq$, and $\neq$ are not available on a keyboard, they are replaced in Pascal by the compound symbols <=, >=, and <> (without a space between the characters). Having said this, we will *continue* to use without further comment the classical mathematical symbols *in the programs and algorithms in this book*. Reading and understanding a program is not the same as typing the program!

• Do not forget to use parentheses *systematically* whenever the logical operations "not", "and", and "or" occur: parentheses are indispensable in Pascal. The operations "and" and "or" also have priority over arithmetic operations: a Pascal compiler reads "$a < b$ or $x = y$" as "$a < (b$ or $x) = y$" which is devoid of sense.

• The boolean expression "n mod $2 = 0$" allows one to test the parity of n: it is true if and only if n is even.

• Suppose that *finish* is a boolean variable (that is, a boolean expression reduced to an identifier). Do not write "*finish* = *true*" in your tests; simply use the identifier "*finish*". The effect will be the same since these two boolean expressions take the same values.

• If a and b are two integers, the boolean expressions "$a * b = 0$" and "$(a = 0)$ or $(b = 0)$" are mathematically equivalent. A programmer will systematically use the latter because it is stupid to use a multiplication which is much slower than a test.

3.4. Statements and their Syntax

We communicate with the help of sentences. In computer science, a *statement* corresponds to a sentence; a *program* is a sequence of statements. Since our goal is to content ourselves with a minimum of Pascal, we shall only use three types of statements: assignments, conditionals, and loops.

- An *assignment*:
  - ▷ ⟨*identifier*⟩ := ⟨*arithmetic expression*⟩
  - ▷ ⟨*element of an array*⟩ := ⟨*arithmetic expression*⟩
- A *conditional*, with or without "else":
  - ▷ **if** ⟨*boolean expression*⟩ **then** ⟨*statement*⟩ **else** ⟨*statement*⟩
  - ▷ **if** ⟨*boolean expression*⟩ **then** ⟨*statement*⟩
- A *loop*, of which there are three types:
  - ▷ **for** ⟨*assignment*⟩ **to** ⟨*arithmetic expression*⟩ **do** ⟨*statement*⟩
  - ▷ **for** ⟨*assignment*⟩ **downto** ⟨ *arithmetic expression*⟩ **do** ⟨*statement*⟩
  - ▷ **while** ⟨*boolean expression*⟩ **do** ⟨*statement*⟩
  - ▷ **repeat** ⟨*sequence of statements*⟩ **until** ⟨*boolean expression*⟩

The angle brackets ⟨ ⟩ indicate, for example, that ⟨*identifier*⟩ is to be replaced by an identifier. Sometimes we will use a "case of" statement which generalises and simplifies certain constructions made from "if then else". Consult your Pascal manual for details.

3.4.1. Assignments

These are the simplest statements to write:

$$x := a + b * c, \qquad A[i, j + k] := \log(x + y/x) + x * x + x + 1.$$

The symbol ":" which precedes the symbol "=" cannot be omitted. There is no space between these two symbols: in other words, it is necessary to consider ":=" as a new symbol, the *assignment symbol.*

3.4.2. Conditionals

Here are two simple examples.

if $(x = 1)$ **or** $(y > 0)$ **then** $y := a + \cos(x)$ **else** $y := a - \sin(x)$

if $x > u + v$ **then** $A[i, j] := x[i] + y[j]$

For beginners

It is necessary to distinguish carefully between tests and assignments:

if $x := 1$ **then** $y = u + v$	**if** $x = 1$ **then** $y := u + v$

The code on the left contains two syntax faults at the outset!

- "$x := 1$" is a statement which is forbidden after "if" (a statement does not have a value: what value can one give to an action?);
- "$y = u + v$" is a boolean expression and is forbidden after "then".

Computer scientists are more careful than mathematicians, for whom the meaning of the equals sign depends strongly on the context. (However, this situation is evolving: one encounters the assignment symbol ":=" more and more frequently in recent books and papers on mathematics.)

3.4.3. For loops

Consider the "for" loop:

$$\textsf{for } i := 1 \textsf{ to } n * n + 1 \textsf{ do } x[2 * i + 1] := a * i + b$$

• The variable i on the left of the assignment symbol is called the *control variable* of the loop. This name is reserved because a compiler uses it when it diagnoses an error. It is necessarily an variable of integer type; it cannot be an element of an array.

• The single statement that follows the "do" is called the *body of the loop.*

For beginners

From time to time, the following error

$$\textsf{for } i := 1 \textsf{ to } 10 \textsf{ do } xi := i * i$$

is made when trying to effect the assignments $x1 := 1 * 1$, $x2 := 2 * 2$, $x3 := 3 * 3$, etc. The compiler will refuse because it does not recognize the variable "xi". It is necessary to define an array $x[1..10]$ and write $x[i] := i * i$.

3.4.4. While loops

Here are two very simple examples:

$$\textsf{while } x > 0 \textsf{ do } x := x + 3$$

$$\textsf{while } (x > 0) \textsf{ and } (x \leq 10) \textsf{ do } x := x + 1$$

• The boolean expression is called the *exit test* of the loop.

• The single statement that follows the "do" is called the *body of the loop*.

3.4.5. Repeat loops

Here is an example, although we have not yet defined a *sequence* of statements:

$$\textsf{repeat } x := x + i \text{ ; } i := i + 1 \textsf{ until } x \geq 100$$

In a loop of this type, the *body of the loop* is formed by all the statements between "repeat" and "until".

For beginners

• In a "while" loop, one first encounters the exit test, then the body of the loop. In contrast, in a "repeat" loop, one encounters the body before the exit test.

• The body of a "while" loop contains only a single statement while the body of a "repeat" loop can contain as many as one wants. We will see a little later how one can handle this apparent asymmetry using a block of statements.

3.4.6. Sequences of statements

In general, a novel is made up of many sentences, each ending in a period. Similarly, an algorithm (or a program) contains many statements. A *sequence of statements* consists of a finite (nonempty) set of statements.

Unlike sentences in a novel which end in a period, in Pascal, statements in a sequence of statements are *separated* by a semicolon: that is, there must be an statement on *each side* of the semicolon. If the letter S designates a statement, a sequence of statements is as follows:

S_1 ; S_2 ; S_3 ← *correct*

S_1 ; S_2 ; S_3; ← *the last semicolon is incorrect.*

For beginners

Experience shows, alas, that adherence to this convention is not easy, because we spontaneously tend to follow a statement by a semicolon, a reflex inspired, no doubt, by the period that ends our sentences. Pascal compilers are tolerant, and accept, whenever possible, redundant semicolons and the empty statements that they evoke. We cite an instance a little later where this is not possible.

This said, there is nothing to be gained in maintaining that the syntax of Pascal is difficult; the placement of semicolons is *very simple*, contrary to what one sometimes reads. Teaching programming shows that a student who has difficulty with semicolons is one who *does not know by heart* the list of statements.[2]

3.4.7. Blocks of statements

It frequently happens that one must repeat several statements in a loop. For example, consider

while $x > 0$ **do**
  $sum := sum + 2 * x$;
  $x := x - 1$

[2] To know by heart is to be capable of responding in a *tenth of a second*, without reflection; as a reflex. Consequently, if you hesitate, if you have to mentally review all possible statements, then you must learn and relearn the list of statements: several minutes each day for a week will suffice.

where the vertical line indicates that we want to execute the two statements *as long as x is greater than* 0. But a "while" loop accepts only a single statement after the "do"; as a result, the Pascal compiler "sees" the following:

$$\textbf{while } x > 0 \textbf{ do } sum := sum + 2 * x \,;$$
$$x := x - 1$$

Thus, we have created an infinite loop which repeatedly adds $2 * x$ to the variable *sum*; the statement $x := x - 1$ is therefore *never* executed.

We need a mechanism for *grouping* statements, and making a sequence of statements into a single statement. A *block of statements*, which is then considered as a new statement, is a sequence of statements preceded by a "begin" and followed by an "end":

$$\underbrace{\textbf{begin } \langle \textit{sequence of instructions} \rangle \textbf{ end}}_{\text{a single instruction}}$$

Thus, the solution of our problem is as follows:

```
while x > 0 do begin
   | sum := sum + 2 * x ;
   | x := x - 1
end
```

For beginners

• We now understand better why a statement which follows the "do" in a "for" loop or a "while" loop is called the body of the loop; most of the time the body of a loop is a block of statements.

• If *S* is a statement, one can write "begin *S* end", but this does not accomplish any more than writing *S* alone. Thus, one uses a "begin end" starting with two or more statements.

• A "repeat until" by itself forms a statement block. Thus, there is no point in typing

$$\textbf{repeat begin } S_1 \,;\; S_2 \,; \ldots ;\; S_n \;\; \textbf{end until} \;\ldots$$

3.4.8. Complex statements

Now that we have learned how to write simple statements, we can assemble and nest them to obtain more and more complex statements. Consider the following two texts. The text on the left is a sequence of two statements; the one on the right contains a single statement. We note therefore that it is possible to write arbitrarily long statements.

```
for x := a + b to a * a + b + 1 do
    for y := c to a * c + 1 do
        for z := 1 to n * n + 1 do
            U[x, z, z] := x + y ;
repeat
| x := x + y + z
| y := y * x - z
| u := u + cos(u + v)
until x > t
```

```
for x := a + b to a * a + b + 1 do
        for z := 1 to n * n + 1 do
while z > 0 do begin
| z := x + y ;
| if z = 0
| then u := u + v
| else u := u - v ;
| z := x + y + z div 2
end
```

3.4.9. Layout on page and control of syntax

The layout of a program on a page is very important. If you neglect it, mastering the syntax and understanding the text becomes very difficult. Which would you prefer? To read a kilometer of text such as the following:

```
for x := a + b to a * a + b + 1 do while x > 0 do begin
z := x + y; repeat if z = 0 then u := u + v else begin
u := u - v ; v := v * v end until z < -1 ; y := y - 2 end
```

or to read the more structured text below?

```
for x := a + b to a * a + b + 1 do
while x > 0 do begin
| z := x + y ;
| repeat
| | if z = 0
| | then u := u + v
| | else begin u := u - v ;  v := v * v end
| until z < -1 ;
| y := y - 2
end
```

They produce, however, exactly the same effect. A compiler "sees" no difference between the two because, from its standpoint, passing from one line to another is nothing more than a single space; on the other hand, I challenge you to tell me rapidly whether the three lines of the first text above are syntactically correct, without some sort of preliminary layout.

Let us return to our kilometer of text and check its syntax.

- The text begins with a "for" loop which is the start of the statement:

```
for x := a + b to a * a + b + 1 do ...
```

This loop is syntactically correct if its body is a statement.

- The text which follows the "do" begins with a new loop

```
while x > 0 do begin ... end
```

and we are led to checking whether the body of the "while" loop is a legal statement, *i.e.* whether the text between the "begin" and "end" is a correctly written sequence of statements.

• The body of the "while" loop contains three statements

$$z := x + y\,;\ \textbf{repeat}\ \ldots\ \textbf{until}\ z < -1\,;\ y := y - 2$$

• We are thus led to checking whether the body of the "repeat ... until" loop is correct, which one verifies easily:

$$\textbf{if}\ z = 0\ \textbf{then}\ u := u + v\ \textbf{else begin}\ u := u - v;\ v := v * v\ \textbf{end}.$$

All that remains is to read a semicolon followed by a conditional. The text we started with is therefore syntactically correct and consists of two statements.

Syntactic analysis proceeds like peeling an onion: one inspects the outside layer first, and begins again with the inside layers. If the program is intelligently laid out, the analysis can be made at a glance. Vertical bars, combined with *indentation*, allow you to instantly see the extent of the different blocks of statements; the "if then else" conditions are laid out vertically whenever necessary.

For beginners

The ideal is to have one statement per line; however, when a statement is too long, it must run over a line. Don't be too rigid in the way you lay things out. Use space harmoniously; you should be guided by aesthetic considerations, that is, the comfort of the reader. It is quite all right, for example, to stack the three pieces "if", "then", "else" of a conditional vertically one on the other; however, when each is very short, it will be easier to read if it is all on one line.

Some individuals pass down to the line after each "begin", so that they can place the corresponding "end" vertically beneath; this is inconvenient as it wastes a line. Also the screen of a computer monitor is small! You are looking at a landscape through a keyhole ....

Here are two classic blunders that beginners make.

• Typing the whole text (or the reserved words) in capitals. This makes reading very painful. (Capitals were not designed for rapid reading, keep them for monuments.)

• Indenting your text too much. Don't – three spaces suffice on a screen. At the back of your eye, on the same axis as the lens, there is a tiny yellow spot very rich in nerves called the *fovea*. This region is responsible for the recognition of forms. Look directly in front of you: if someone approaches you from the side, you will realize that a person is approaching, but you will not be able to identify him or her because you are not facing the person and, consequently, his or her image will not fall on your fovea.

It is for this reason that the eye moves ceaselessly. At normal reading distance, our brain only recognizes the contents of a disk about ten centimeters in diameter. If the indentation is too large, it forces you to move your focus instead of being able to grasp everything at a single glance, and this makes syntactical analysis and comprehension much more difficult.

An important programming tip

First type "begin end"; then return and insert between "begin" and "end" the sequence of statements that is to become a block. Do the same with "repeat until" and with "case of end". If you adhere to this discipline, you will *never* have to worry about closing blocks; and you will not pass your time counting on your fingers how many "begin"'s you still have to close.

When you write out a program by hand, always follow a "begin" with a vertical line. This facilitates syntactic analysis; if you change a page, you will know exactly how many "begin"'s you have to close.

3.4.10. To what does the else belong?

When you nest "if then else" statements, the "else" always belongs to the closest "then". An intelligent layout (indentation and vertical lines) is very useful in facilitating comprehension and analysis. Thus you write:

```
if x > 0
then y := y + 1
else if x = 0
     then z := z - 1
     else if x < -1
          then u := u + 1
          else u := u - 1
```

It is sometimes necessary – but very, very rarely – to detach an "else" from the closest "then". This is done with a block, like this:

```
if x < a                                          if x ≥ a
then begin                                        then y := x div 2 + 1
| if x mod 2 = 0 then y := x div 2       ⟺       else if x mod 2 = 0
end                                                    then y := x div 2
else y := x div 2 + 1
```

3.4.11. Semicolons: some classical errors

Let's put ourselves in the place of a beginner who decides to simplify life once and for all by ending each statement with a semicolon. Since an assignment is a statement, he or she types

```
if x > 0
then x := a ;  ← incorrect semicolon!
else x := b
```

The compiler analyes the correct statement "if $x > 0$ then $x := a$". It then expects to find a statement after the first semicolon. But since an "else" can never begin a statement, it protests.

Here is another classic fault. At the left is a beginner's program; at the right is what the compiler "understands":

if $x > 0$	**if** $x > 0$
then $x := u$	**then** $x := u$
else if $x = 0$	**else if** $x = 0$
then $x := v$; $y := x * x$	**then** $x := v$;
else $x := w$	$y := x * x$ **else** $x := w$

The placement on the left suggests that it is necessary to simultaneously execute the two statements "$x := v$" and "$y := x * x$" when x is zero; however, the block "begin end" which would make these two statements into one is missing. As a result, the "else" appears in the middle of a legal arithmetic expression. The compiler has good reason to protest.

For beginners

We end with a problem that worries many novice programmers who have not learned (or understood) their definitions: when one nests blocks, is it necessary to put semicolons between the "end"s?

Recall that semicolons *separate* statements. Since an statement never begins with an "end", we note that the two first semicolons are questionable.

```
begin
| ...
| begin
| | ...
| | begin
| | | ...
| | end ;  ← superfluous semicolon but accepted
| end ;  ← superfluous semicolon but accepted
end ;  ← correct semicolon
```

On the contrary, the last semicolon is indispensable when the last "end" is followed by a statement. The reasoning becomes evident when the code is written on a single line:

```
begin ... begin ... begin ... end ; end ; end ;
```

3.5. The Semantics of Statements

Identifiers only exist for our intellectual comfort; a computer only recognizes *addresses*, which are whole munbers, in its memory (we will return to this subject in Chapter 6). The *contents* of the memory at the address corresponding to the identifier is the *value* of this identifier.

To better understand what this means, imagine a letter box: the name on it corresponds to the identifier and the letter that one slides into it corresponds to its value.

Each day, your letter box receives letters; in an analogous manner, the program (considered as a mailman) can modify the *value* of an identifier. In mathematics, a variable does not change its value during a proof. In contrast, the contents of a variable in a program can change thousands, or hundreds of thousands, of times in a second!

The analogy with a letter box breaks down however. When a program needs to transfer the value of a variable into a microprocessor, it "photocopies" the letter, it does not withdraw it! In other words, reading is not destructive.

3.5.1. *Assignments*

An assignment describes a *process*. In order to execute the statement

$$x := b + a * x$$

the computer first calculates the value of the arithmetic expression $b + a * x$ by *recopying* in a suitable order (here, a, x, b) the contents of the variables as it does its calculations. This done, it *overwrites* the value in the address corresponding to the variable x; this has the effect of *erasing* the previous contents. (Certain languages use the notation $x \leftarrow b + a * x$ to better indicate this.)

One *increments* the variable x by writing "$x := x + 1$" and *decrements* it by writing "$x := x - 1$".

For beginners

The type of a variable is very important: the assignment "$u := v$" is only possible if u and v have the same type. In case of error, the compiler will announce that there is a *type mismatch*. However, there is an exception: the assignments $\langle \textit{real} \rangle := \langle \textit{integer} \rangle$ are legal.

3.5.2. *Conditionals*

Consider the statement

$$\textbf{if } x = 0 \textbf{ then } y := y + 1$$

If the contents of the variable x is 0, y is incremented; in the contrary case, that is, if $x \neq 0$, nothing happens and the computer executes the next statement (if there is one).

The conditional with an "else" is also easy to grasp:

$$\textbf{if } x = 0 \textbf{ then } y := y + 1 \textbf{ else } z := z - 1$$

If the contents of x is zero, y is incremented and the program *skips* the rest of the statement to execute the next statement in the program: thus, the value of the variable z is not changed. On the contrary, if x is zero, the program *skips* the beginning of the conditional and decrements z: thus, the variable y does not change its value.

3.5.3. *First translations*

Before examining the semantics of loops, we first familiarize ourselves with our algorithmic description language by translating several common mathematical constructions into it.

1) To express whether or not x belongs to an interval, the result returned being a boolean, we can write

$$(a \leq x) \textbf{ and } (x \leq b)$$

if we know that $a \leq b$. But if we don't know this, we should be prudent and write:

$$(a \leq x) \textbf{ and } (x \leq b) \textbf{ or } (b \leq x) \textbf{ and } (x \leq a)$$

The priority of "and" over "or" ensures that the translation is not ambiguous.

Beginners should note that in order to reliably translate $x \in [a, b] \cup [c, d]$ it is advisable to first write "$(x \in [a, b])$ or $(x \in [c, d])$", then replace $x \in [a, b]$ and $x \in [c, d]$ by the appropriate code.

2) In mathematics, a comma frequently plays the role of "and". Thus, for example, one translates the condition $i < x < j,\ x \neq k$ by:

$$(i < x) \textbf{ and } (x < j) \textbf{ and } (x \neq k)$$

3) The classical notation

$$R = \begin{cases} R_1 & \text{if } \langle condition \rangle, \\ R_2 & \text{otherwise} \end{cases}$$

is translated simply by

$$\textbf{if } \langle condition \rangle \textbf{ then } R := R_1 \textbf{ else } R := R_2$$

4) Mathematicians often write:

$$R = \begin{cases} R_1 & \text{if } \langle condition_1 \rangle, \\ R_2 & \text{if } \langle condition_2 \rangle. \end{cases}$$

There are two legitimate translations: one, on the left, using one statement, the other, on the right, using two.

if $\langle condition_1 \rangle$ **then** $R := R_1$ **else if** $\langle condition_2 \rangle$ **then** $R := R_2$	**if** $\langle condition_1 \rangle$ **then** $R := R_1$; **if** $\langle condition_2 \rangle$ **then** $R := R_2$

What happens if both conditions are true? The solution on the left gives R the value R_1, while that on the right gives R the value R_2. This is not too serious: if the original mathematical assignment was coherent, then one has $R_1 = R_2$ when both conditions hold.

For beginners

Nevertheless, you should systematically use the solution on the left:

- it executes more rapidly (one test instead of two);
- the translation on the right is perilous, as we are going to see.

5) We can complicate the game (we suppose that i is an integer):

$$X = \begin{Bmatrix} a \text{ if } i > 0 \\ b \text{ if } i = 0 \\ c \text{ if } i = -1, -2 \\ d \text{ if not} \end{Bmatrix} \Longrightarrow \begin{cases} \textbf{if } i > 0 \\ \textbf{then } X := a \\ \textbf{else if } i = 0 \\ \textbf{then } X := b \\ \textbf{else if } (i = -1) \textbf{ or } (i = -2) \\ \textbf{then } X := c \\ \textbf{else } X := d \end{cases}$$

There is only one statement!

For beginnners

Using the option “else” is *indispensable* for automatically obtaining reliable code. If you don’t use it, you risk writing nonsense.

Consider, for example, the following translation proposed by a beginner, who refuses to use “else” imagining that it will somehow simplify life:

$X := d$;
if $i > 0$ **then** $X := a$;
if $i = 0$ **then** $X := b$;
if $(i = -1)$ **or** $(i = -2)$ **then** $X := c$

First of all, this translation begins with the "trick" $X := d$, which is not at all clear, especially for a beginner! Then, when $i > 0$, the algorithm still executes the tests $i = 0$, $i = -1$, and $i = -2$, which is idiotic.

To better see why this intellectual laziness is suicidal, consider the following example, patterned on the above:

$$X = \left\{ \begin{array}{ll} a & \text{if } X > 0, \\ b & \text{if } X = 0, \\ c & \text{if } X = -1, -2, \\ d & \text{otherwise.} \end{array} \right\} \Longrightarrow \left\{ \begin{array}{l} X := d\,; \\ \textbf{if } X > 0 \textbf{ then } X := a\,; \\ \textbf{if } X = 0 \textbf{ then } X := b\,; \\ \textbf{if } (X = -1) \textbf{ or } (X = -2) \textbf{ then } X := c \end{array} \right.$$

This translation is flagrantly false! Because X is modified at the outset, the tests that follow have nothing to do with the initial value of the variable X, but pertain instead to the values of d, a, b and c. A good translation is very natural and executes more rapidly:

```
if X > 0 then X := a
else if X = 0 then X := b
     else if (X = -1) or (X = -2) then X := c
          else X := d
```

3.5.4. The boustrophedon order

In some ancient languages the direction in which one reads changes from line to line; there is no "carriage return". One reads, for example, the first line from right to left, the second from left to right, the third from right to left, etc. This serpentine writing is called *boustrophedon writing*.

Consider the rectangle $\mathcal{R}$ of points with integer coordinates (Fig. 3.1) satisfying $0 \le x \le a$, $0 \le y \le b$ with $a \ge 0$ and $b \ge 0$. Inspired by the serpentine pattern of boustrephedon writing, we can endow $\mathcal{R}$ with a total order called the *boustrophedon order*. One traverses $\mathcal{R}$ in the increasing direction by

- leaving the origin (0,0) and moving along the line $y = 0$ towards the right until we get to $(a, 0)$;
- then climbing to the line $y = 1$ and moving along it to the left starting at $(a, 1)$ and continuing to the point $(0, 1)$, etc.

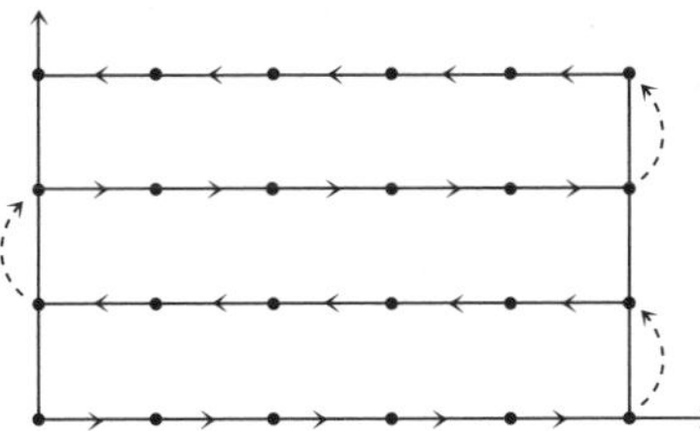

Fig. 3.1. *Boustrophedon order*

Thus, the smallest element of the rectangle $\mathcal{R}$ is the point $(0, 0)$; the largest is (a, b) if b is even and $(0, b)$ otherwise.

The resulting total order is given by:

$$(x, y) \underset{B}{<} (x', y') \quad \begin{cases} \text{if } \ x < x' & \text{and} \quad y = y' \equiv 0 \pmod 2, \\ \text{if } \ x > x' & \text{and} \quad y = y' \equiv 1 \pmod 2, \\ \text{if } \ y < y' \end{cases}$$

and the successor of an element (x, y), when it exists, is:

$$succ(x, y) = \begin{cases} (x + 1, y) & \text{if } x < a \text{ and } y \text{ even}, \\ (a, y + 1) & \text{if } x = a \text{ and } y \text{ even}, \\ (x - 1, y) & \text{if } x > 0 \text{ and } y \text{ odd}, \\ (0, y + 1) & \text{if } x = 0 \text{ and } y \text{ odd}. \end{cases}$$

To translate the above into code, we argue according to the parity of y. Since the largest element does not have a successor, we need a boolean variable which we call *exist*.

```
exist := true ;
if y mod 2 = 0   {y is even}
then if x < a
     then x := x + 1
     else if y < b   {and x = a}
          then y := y + 1
          else exist := false
else {now y is odd}
     if x > 0
     then x := x − 1
     else if y < b   {and x = 0}
          then y := y + 1
          else exist := false
```

- If *exist* is true, the new values of x and y are those of the successor of (x, y).
- If *exist* is false, the new values of x and y mean nothing.

Exercises 1

• Close this book and construct your own code to calculate the successor of (x, y).

• Define a boustrephedon order on $[\![0, a]\!] \times [\![0, b]\!] \times [\![0, c]\!]$ as follows: to go in the increasing direction, augment x when y is even, and diminish it if y is odd; similarly, augment y when z is even and diminish it otherwise. In this way, the parity of y modifies the order relation for x and the parity of z modifies it for y.

• Generalize to a product of n intervals.

3.5.5. The for loop

Let i, n, a be integer variables and consider the loop

$$\textbf{for } i := n + a \textbf{ to } n * n \textbf{ do } x := x + i$$

Here the *body* of the loop reduces to a single statement; or, to be more precise, we do not have a statement block. We call i the *control variable*.

To execute this loop:

• The progam evaluates *for once and for all* the bounds $min = n + a$ and $max = n * n$.

• If $min \leq max$, the variable i successively (and automatically) takes the values $min, min+1, \ldots, max$. Each time, the program executes the statement(s) in the body of the loop. What happens in our example is as if the program executed the sequence of statements:

$$x := x + a + 1; \quad x := x + a + 2; \ \ldots \ ; x := x + a * a.$$

(Exercise: how much is the value of x augmented in total?)

• If $min > max$ nothing happens: the program skips to the statement that follows (if it exists) and x does not change its value.

The downto variant

The loop

$$\textbf{for } i := n * n \textbf{ downto } n - a \textbf{ do } x := x + i$$

functions in a similar manner. In executing this loop,

• the program begins by evaluating once and for all the bounds $max = n * n$ and $min = n - a$ (note the inversion of the bounds).

• If $max < min$, the program does nothing and skips to the statement that follows the loop (if it exists).

• If $max \geq min$ the program successively gives the control variable the values $max, max - 1, \ldots, min$ and executes each time the statement(s) in the body of the loop.

For beginners

The language Pascal was conceived to teach good programming. Thus the “for” loop is *protected* in a manner so as to resist attempts to branch out of it.

- There is no point typing

```
for i := 1 to n do begin
| S := S + i ;
| if S ≥ 0 then n := 0
end
```

in the hopes of leaving the loop as soon as $S \geq 0$. Recall that the bounds $min = 1$ and $max = n$ are evaluated once and for all *before* the body of the loop is executed; since the program compares the value of i to the number max, the program is not able to take account of the modified value of n.

- There is no point trying to modify the value of the control variable in order to leave the loop prematurely by typing, for example,

```
for i := 1 to n do begin
| S := S + i ;
| if S ≥ 0 then i := n + i
end
```

We will see a little later (when we discuss the "while" loop) how to realise very simply what the attempts above unsuccessfully try to do.

Remark

Modern implementations of Pascal allow one to leave any type of loop using special statements (such as "leave", "break" or "exit", depending on the dialect used).

However, professional programmers are reluctant to use these statements without good reason. In general, when they modify a large program, they content themselves with examining the test β which controls the loop[3] without reading the body of the loop. If the body of the loop does not contain the statement "leave", one knows that the condition "not β" is true on exiting the loop; but this need not be the case if one leaves the loop by some other means. And rare is the programmer who signals this and carefully makes precise what condition is satisfied on leaving the loop in a different way.

Nevertheless, these statements are used very sparingly in certain circumstances when they hugely simplify the programming task.

3.5.6. The while loop

Consider for example the loop

```
i := a ;
while i ≤ b do begin
| x := x + i ; i := i + 1
end
```

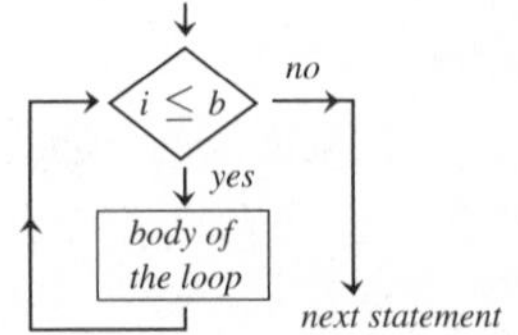

[3] Programmers must work as quickly as possible!

which means "repeat the body of the loop as long as $i \leq b$". To execute this loop, the program:

- Begins by evaluating the boolean expression $i \leq b$.
- If the value of the expression is true, it *penetrates* into the body of the loop and executes the statements found there.
- This done, it *returns* to before the test and repeats the same sequence of actions, as suggested by the arrow.
- When the boolean expression becomes false, it skips to the statement immediately following the body of the loop (if such exists): we say that it *leaves* the loop.

In our example, the program does not modify the value of x if $a > b$ since it does not penetrate into the interior of the loop. In contrast, when $a \leq b$, what occurs is as if program executed the statements

$$x := x + a;\ x := x + a + 1;\ \ldots\ ;\ x := x + b.$$

In a "while" loop, the test comes first. Thus it is entirely possible not to penetrate into the loop.

Example

If we wish to calculate the sum S of the even numbers less than some fixed number n, it suffices to use one of the following two loops:

```
S := 0 ; i := 1 ;
while 2 * i <= n do begin
| S := S + 2 * i ; i := i + 1
end
```

```
S := 0 ; i := 2 ;
while i <= n do begin
| S := S + 2 * i ; i := i + 2
end
```

Why is the loop on the right better?

For beginners

1) A "for" loop is an *abbreviation* of the following "while" loop:

```
for i := min to max do S   <==>   i := min ;
                                  while i <= max do begin
                                  | S ; i := i + 1
                                  end .
```

Similarly, the "downto" variant is translated using a "while" loop as follows:

```
for i := max downto min do S   <==>   i := max ;
                                      while i >= min do begin
                                      | S ; i := i - 1
                                      end
```

Recall that a "for" loop functions all alone: incrementation (or decrementation) of the control variable is *automatic*.

2) We can now explain how the "for" loop is protected. When a compiler encounters the loop

for $i := n + 1$ **to** $n * n *$ **do** S

it creates three special variables (inaccessible to the programmer) which we call here α, ω and κ. The compiler then translates the following loop into binary.

```
α := n + 1 ; ω := n * n ; κ := α ;
while κ ≤ ω do begin
    | S ; ← body of the original "for" loop
    | κ := κ + 1 ;
    | i := κ
end
```

3.5.7. The repeat loop

The statement "repeat until" means "repeat the body of the loop until the exit test becomes true." When the program encounters the two statements

```
i := a ;
repeat
    | x := x + i ; i := i + 1
until i > b
```

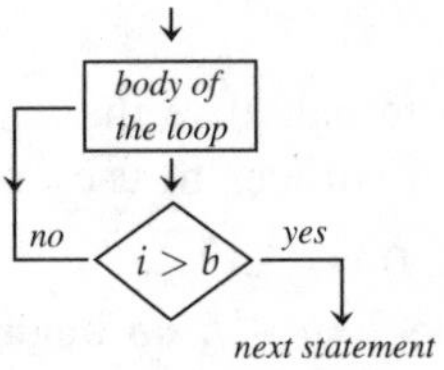

it first gives i the value a (the first statement), then penetrates *unconditionally* into the body of the loop where it executes the assignments "$x := x + i$" and "$i := i + 1$". Only then, does it compare *for the first time* the values of i and b. If $i > b$, the program leaves the loop and executes the statements that follow (if there are any); otherwise, it returns to beginning of the loop and repeats the same sequence of operations. If we are interested in the variable x, what happens is the same as if one executed the sequence of statements

$$\underbrace{x := x + a;}_{\text{always}}\ \underbrace{x := x + a + 1;\ \ldots\ ;\ x := x + b}_{\text{only if } a < b}.$$

For beginners

One always enters at least once into the body of a "repeat" loop. We have the equivalence :

$$\textbf{repeat } \alpha \textbf{ until } \beta \iff \begin{cases} \alpha\ ; \\ \textbf{while not } \beta \textbf{ do } \alpha \end{cases}$$

3.5.8. Embedded loops

Two embedded "for" loops can work minor miracles. Consider, for example, the loop

```
for i := 1 to a do                   {external loop}
for j := 1 to b do write(i, j)       {internal loop}
```

The body of the external loop "for i := 1 to a do ... " is the internal loop "for j := 1 to b do ...". The external loop successively gives i the values $1, 2, \ldots, a$. Each time that i takes a new value, the variable j sweeps out the integers in the interval $[\![1, b]\!]$. The result of these two loops is to write on the screen the pairs

$$\begin{array}{cccc} (1,1) & (1,2) & \ldots & (1,b) \\ (2,1) & (2,2) & \ldots & (2,b) \\ \vdots & \vdots & \vdots & \vdots \\ (a,1) & (a,2) & \ldots & (a,b) \end{array}$$

in the order in which we read them. Mathematically, writing the couples (i, j) one after the other defines a total order on the couples. In this case, the order is the *lexicographic order*. That is, one compares the first coordinates: if they are different, the couple with the larger first coordinate is the larger; if they are equal, one compares the second coordinate.

Remark

One cannot measure the difficulty of an algorithm by the number of embedded loops that it contains because one can translate embedded loops into single loops. In the last example, observe that Euclidean division sets up a bijection between integers $n = bi + j \in [\![0, ab-1]\!]$ and couples $(i, j) \in [\![1, a]\!] \times [\![1, b]\!]$:

```
for n := 0 to a * b - 1 do begin
  | i := n div b + 1 ;
  | j := n mod b + 1 ;
  | write(i, j)
end
```

3.6. Which Loop to Choose?

We need a loop any time that one deals with *repetition* of a given process. In order to select the right loop, keep in mind the following:

- Can the process be controlled by an integral variable which runs over an interval whose bounds are known in advance? If yes, use a "for" loop.
- Is the process to be effected $n \geq 0$ times? If yes, use a "while" loop.
- Is the process to be effected $n \geq 1$ times? If yes, use a "repeat" loop.

3.6.1. *Choosing a for loop*

Is the vector $x = (x_1, \ldots, x_n)$ zero? To answer, we examine $x_1, x_2, \ldots, x_n$ successively. This amounts to letting the index i run over the interval $[\![1, n]\!]$:

```
place := −1 ;
for i := 1 to n do
    if x[i] ≠ 0 then place := i
```

```
place := −1 ;
for i := n downto 1 do
    if x[i] ≠ 0 then place := i
```

The variable *place* remains equal to -1 when the vector x is zero (because it is necessary to take everything into consideration!). Otherwise it equals the largest index i such that $x_i \neq 0$ in the solution on the left, and the smallest such index in the right.

3.6.2. *Choosing a while loop*

Here are two examples which would be difficult to handle using a "for" loop:

```
S := 0 ; i := 0 ;
while i * i * i + i < N do begin
  | S := S + i ;
  | i := i + 1
end
```

```
x := abs(a) ; y := abs(b) ;
while (x > 0) and (y > 0) do
    if x ≥ y
    then x := x − y
    else y := y − x
```

The loop on the left is controlled by an integer which runs over an interval whose upper limit is not explicitly known. In the loop on the right, the pair (x, y) controls the process.

3.6.3. *Choosing a repeat loop*

To pick an integer n between 1 and 10, one uses a " repeat " loop (because there is at least one such):

```
repeat
  | readln(n)
until (1 ≤ n) and (n ≤ 10)
```

For beginners

In a "for" loop, the control variable is incremented (or decremented) automatically. By contrast, in a "repeat " or "while " loop, one needs a "motor". If you forget, you create an infinite loop ...

3.6.4. *Inspecting entrances and exits*

Each time that you write a loop, *stop and reread* what you have written and try to mentally execute the code. Carefully inspect the entrance and exit to a loop, for these are the places where one most often goes astray.

Suppose that for some integer $n \geq 1$, one wants to execute a sequence of statements,

$$process(1); process(2); \ldots ; process(n).$$

The best solution is, of course, the program

for $i := 1$ **to** n **do** $process(i)$

and one scarcely needs to simulate this because of the simplicity of the program.

Now, consider a solution which uses "while" loop. Beginners often write:

while $i < n$ **do** $process(i)$

Let us try to enter the loop. We must compare i and n. But the value of i does not exist.[4] Thus, the test will function in an unforseeable manner. Thus, we must *initialize the variable* i:

$i := 1$; **while** $i < n$ **do** $process(i)$

Now that the problem of entering the loop is settled, let us begin anew and try to execute this new code. We leave $i = 1$ and are authorized to enter the body of the loop which has us effect *process*(1). After this, we return to the entrance of the loop with the *same value* $i = 1$. We have just detected an infinite loop!

The diagnosis is simple: the loop does not contain a "motor".

For the sake of demonstration, we correct this in an exceedingly clumsy manner.

```
i := 1 ;
while i < n do begin
   | i := i + 1 ;
   | process(i)
end
```

Let us begin again our mental execution of the code: we enter the loop with the value $i = 1$ (recall that $n \geq 1$). The variable i is immediately incremented, then we execute *process*(2). We detect our first fault: we have forgotten to execute *process*(1) and risk a crash if *process*(2) needs to be preceded by *process*(1). We return again to the entrance of the loop, increment i, then execute *process*(3), etc.

To test the exit of the loop, suppose that i has the value $i = n - 1$. This authorizes us to re-enter the loop. The variable i takes the value n, we execute $process(n)$, and then return to the entrance to the loop. But since the boolean expression $i < n$ takes the value *false*, we leave the loop since we no longer have the right to enter. We exit the loop correctly.

[4] More precisely, the value exists, but it must be considered as aleatory – see the discussion on "litter" in Chapter 6.

This simulation shows that the initialization of the variable i is incorrect. We should have written $i := 0$ or changed the placement of the motor and modified the test.

```
i := 0 ;                          i := 0 ;
while i < n do begin              while i ≤ n do begin
|  i := i + 1 ;                   |  process(i) ;
|  process(i)                     |  i := i + 1
end                               end
```

Remark

It is quite legitimate to pass to a "repeat" loop here because the process is done $n \geq 1$ times. Passing from a "while" loop to a "repeat" loop is mechanical: it suffices to take the negation of the entrance test to the first loop to obtain the exit test for the second loop:

```
i := 0 ;                 i := 1 ;
repeat                   repeat
|  i := i + 1 ;          |  process(i) ;
|  process(i)            |  i := i + 1
until i ≥ n              until i > n
```

For beginners

This painstaking inspection should become a reflex: *never dispense with it.* You will detect lots of faults of the sort found above: non-initialized variables, incorrect initialization, poorly chosen loops, missing motors. The minute that you "lose" in inspection will save hours of debugging. Your choice.

3.6.5. Loops with accidents

Let $x[1 \cdots n]$ be any sequence of integers. The code that follows was intended to answer to the question: is the number a in this sequence?

```
for i := 1 to n do present := (a = x[i])
```

Alas, the code is faulty as the following counter-example shows: $x[1] = 1$, $x[2] = 2$, $x[3] = 3$ and $a = 2$. The variable *present* successively takes the values *false*, *true* and *false*. Here we must interrupt the "for" loop as soon as we detect the presence of the number a.

Knowing that it is not possible to interrupt a "for" loop in standard Pascal (we refuse here to allow ourselves to take refuge in the modern statements "exit" or "break"), we first transform the loop into a while loop:

```
i := 1 ;
while i ≤ n do begin
|  present := (a = x[i]) ;
|  i := i + 1
end
```

We can now insert the boolean *present* in the exit test to *interrupt* the loop at the appropriate moment. We also do not forget to initialize the boolean.

```
i := 1 ; present := false ;
while (i ≤ n) and not present do begin
| present := (a = x[i]) ;
| i := i + 1
end
```

A number of loops will handle the general case in which a process is terminated by (one or more) specific cases. When the situation is sufficiently complicated, it is preferable to use the general case; we will introduce exceptions afterwards.

3.6.6. Gaussian elimination

Suppose that we want to implement the Gaussian elimination algorithm on a square matrix of dimension $n > 1$ (perhaps we wish to invert the matrix or calculate its determinant). In order to do this, we successively process the columns $1, \ldots, n$. We deliberately stay at a relatively high level of generality by not detailing what is involved in processing a column. The constraints are:

- if the current column is not zero, we process it;
- if the current column is zero, we halt (because we know that the matrix is not invertible or that its determinant is zero).

First Approximation. Let us go down the wrong road in order that we may understand the right one. If the matrix is invertible, the loop

```
for k := 1 to n do process_column(k)
```

does the job perfectly. However, this solution is incorrect if the matrix is not invertible, because it does not respect the constraint “stop processing as soon as we encounter a column that is zero”.

Second Approximation. Thus, we must inquire before acting. To do this, suppose that we introduce a boolean function *zero_column* which takes the value *true* when the current column is zero and modify the preceding loop.

```
for k := 1 to n do begin
| if zero_column(k)
| then «interruption»
| else process_column(k)
end
```

Third Approximation. Since interrupting a “for” loop is not allowed in standard Pascal, we transform it to a “while” loop by introducing a boolean which manages the interruption (and we don’t forget the motor!).

```
k := 1 ; finish := false ;
while (k ≤ n) and not finish do begin
| if zero_column(k)
| then finish := true
| else process_column(k) ;
| k := k + 1
end
```

Remarks

1) Another solution is

```
k := 1 ; finish := false ;
while not finish do begin
| if zero_column(k) or (k > n)
| then finish := true
| else process_column(k) ;
| k := k + 1
end
```

2) We could have used a "repeat" loop since we at least have to explore the first column, if only to determine whether it is zero and we have to interrupt the processing right away.

```
k := 0 ; finish := false ;
repeat
| k := k + 1 ;
| if zero_column(k)
| then finish := true
| else process_column(k) ;
until (k ≥ n) or finish
```

```
k := 0 ; finish := false ;
repeat
| if zero_column(k)
| then finish := true
| else process_column(k) ;
| k := k + 1 ;
until (k > n) or finish
```

For beginners

To set up a delicate loop, proceed by successive approximations and ruthlessly criticize your own code. First set up the *external shell* of your loop, and then fill in the body of the loop.

```
k := 1 ; finish := false ;
while (k ≤ n) not finish do begin
| ... ← part to fill in eventually
| k := k + 1
end
```

3.6.7. How to grab data

Suppose that we want to write a program that repeats the following sequence a variable number of times:

- choose two integers a and b;
- process the data and display the results (for example, a curve that depends on the parameters a and b).

Suppose moreover that the modules for processing and displaying are reliable only if a and b are both > 0. Thus we require that the program terminate as soon as one of the integers a or b is ≤ 0.

Here are two solutions typical of beginners. The solution on the left functions correctly. Nevertheless, repeating the statement "*choose*(a, b)" is a blunder arising from the wrong choice of loop.

```
choose(a, b) ;
while (a > 0) and (b > 0) do begin
  | process(a, b) ;
  | choose(a, b) ;
end
```

```
repeat
  | choose(a, b) ;
  | process(a, b) ;
until (a ≤ 0) or (b ≤ 0)
```

The solution on the right, although it does not have this defect, is dangerous! To leave the program, we could for example enter $a = 0$, $b = -3$. The program, however, performs processing with incorrect values of the parameters: we risk an infinite loop or a crash. We might unwittingly provoke this catastrophe the moment the values are read.[5]

After this avalanche of criticism, our beginner decides to protect him or herself with a test:

```
repeat
  | choose(a, b) ;
  | if (a ≥ 0) and (b ≥ 0) then process(a, b)
until (a ≤ 0) or (b ≤ 0)
```

"This will work for sure!" our beginner says. This is true, but the code has an esthetic defect: when we want to stop, the program first evaluates the boolean expression "$(a > 0)$ and $(b > 0)$", then its negation "$(a \leq 0)$ or $(b \leq 0)$" which is superfluous. Here are two more elegant solutions which use a boolean variable to control the loop:

```
finished := false ;
repeat
  | choose(a, b) ;
  | if (a > 0) and (b > 0)
  | then process(a, b)
  | else finished := true
until finished
```

```
repeat
  | choose(a, b) ;
  | begin_again := (a > 0) and (b > 0) ;
  | if begin_again then process(a, b)
until not begin_again
```

This code is still not satisfatory because it is not ergonomic! When we want to leave the program we first respond $a = 0$ when it prompts us. But

[5] Another proverb: "Even the first time, it is necessary to know how to protect oneself..."

this does not stop it from asking *subsequently* for the value of b (imagine the exasperation[6] of a user who had to pointlessly enter the values of ten variables instead of two). It is necessary to dissociate the prompt for a from that of b:

```
finish := false ;
repeat
| choose(a) ;
| if a = 0
| then finish := true
| else begin
| | choose(b) ;
| | if b = 0 then finish := true else process(a, b)
| end
until finish
```

Laziness on the part of the programmer is no excuse. Never forget that it is the program that must adapt to human beings.

For beginners

From this discussion, you should especially retain the two schemas

```
finish := false ;            | repeat
repeat                       | | ...
| ...                        | | begin_again := ⟨boolean expression⟩ ;
| if condition               | | ...
| then finish := true        | until not begin_again
| else ...                   |
until finish                 |
```

which you will often have occasion to use.

Exercise 2

Imagine another solution using a procedure *choose*$(a, finish)$.

[6] I sometimes find student complaints on exams such as "Too long! Not enough time!" when they invoke a procedure such as *choose*(a, b, c, u, v, w) which contains the same code six times in succession. What is to stop them from defining a procedure *choose*(x) with a single argument, then writing: *choose*(a); ...; *choose*(w)? Let us make this into a proverb: "You have forgotten a procedure if you are writing the same code more than three times!"

4. How to Create an Algorithm

Do you remember how you learned to write proofs? It took several years. First you were presented with simple models which you learned by heart, then imitated. These became more and more complex, until one day you discovered that you could do it on your own.

This apprenticeship resembles the way an infant learns a language: he or she listens, reproduces sounds, words, simple sentences, changes a word here and there. The length and complexity of the sentences increase over time and the child winds up capable of coherent discourse.

To learn to write a program, we will follow the same path: contemplate and understand simple models, *learn them by heart*, modify them lightly, etc. First of all you will write little programs by copying then modifying[1] those given in the text or in other books. Since you have already undergone an apprenticeship in writing proofs, your progress will be very rapid.

You should, however, not be under any illusions. Writing algorithms is also difficult, often more difficult than writing proofs. A ten line algorithm can take many hours[2] of effort.

We will present and use three methods:

- *manipulation and enrichment* of existing code (for example, transforming "for" loops into "while" loops);
- *use of recurrent sequences*, which allow us to reduce to static thought when a problem becomes truly delicate;
- *deferral of code writing*, in order to deal with one difficulty at a time.

As we shall see in the examples that follow, these three methods are not independent and tend to interact with one another.

4.1. The Trace of an Algorithm

To obtain the *trace* of an algorithm, you assign reasonable values to the inputs and "run it by hand". That is, you execute the statements one by one as a

[1] Recopying then reconstructing is a very effective way of learning by heart.

[2] We are talking here about serious algorithms – in practice, 90 % of programs consist of trivial algorithms.

computer would do. Consider the algorithm

```
min := 1 + b div a ;  max := (2 * b) div a ;
for x1 := min to max do begin
  | a1 := x1 * a - b ;  b1 := b * x1 ;
  | if b1 mod a1 = 0 then x2 := b1 div a1
end
```

If $a = 2$ and $b = 9$, then $min = 5$, $max = 9$, which gives the trace:

x_1	a_1	b_1	x_2
5	1	45	45
6	3	54	18
7	5	63	–
8	7	72	–
9	9	81	9

The dash represents a value that has not changed. The layout on the page is important: present your calculations in tabular form, as in the example.

For beginners

This technique is the best way to familiarize yourself with the *sequential* thought foreign to most mathematicians who are more familiar with static thought. Do not kid yourself: *step-by-step simulation of the functioning of a computer is of capital importance*. Dedicate a number of hours to this activity and practice it systematically: it will become second nature to you!

4.2. First Method: Recycling Known Code

It often happens that a problem resembles one that has already been solved. Then, you can recycle old code.

4.2.1. Postage stamps

Let $1 < a < b < c$ be three relatively prime integers that we imagine to be the price in cents of three postage stamps. One can show that there exists[3] a threshold $\chi = \chi(a, b, c) \leq (a-1)(c-1)$ above which the equation

$$ax + by + cz = n$$

[3] This result generalizes to n stamps such that $\mathrm{GCD}(a_1, \ldots, a_n) = 1$. When $n = 2$, it is easy to prove that the threshold is $\chi(a, b) = (a-1)(b-1)$. No formula is known for $n \geq 3$, but there are very effective algorithms for determining the threshold $\chi(a_1, \ldots, a_n)$. If we suppose that $1 < a_1 < \cdots < a_n$, one can prove the inequality $\chi(a_1, \ldots, a_n) \leq (a_1 - 1)(a_n - 1)$.

admits at least[4] one solution $(x, y, z) \in \mathbb{N}^3$. In other words every amount $n \geq \chi$ of postage is *realizable* with our three stamps. In contrast, we cannot supply the *exact* postage if $n = \chi - 1$.

When $a = 5$, $b = 6$ and $c = 16$, the first realizable amounts are:

x	y	z	n	x	y	z	n	x	y	z	n
1	0	0	5	1	2	0	17	5	0	0	25
0	1	0	6	0	3	0	18	2	0	1	26
2	0	0	10	4	0	0	20	1	1	1	27
1	1	0	11	1	0	1	21	0	2	1	28
0	2	0	12	0	1	1	22	1	4	0	29
3	0	0	15	1	3	0	23	0	5	0	30
0	0	1	16	0	4	0	24	3	0	1	31

When n_0 is realizable, so is $n = n_0 + ka$ for $k \geq 0$. As a result, if $n_0, n_0 + 1, \ldots, n_0 + a - 1$ are realizable, so is every amount $n \geq n_0$. This remark, and an examination of the table above shows that we have

$$\chi(5, 6, 16) = 20.$$

4.2.2. How to determine whether a postage is realizable

If x, y, z are solutions, we have $0 \leq x \leq n/a$ and two similar inequalities involving y and z. Since our goal is only to acquaint ourselves with the problem, we employ brute force and test all possible triples (x, y, z). To do this, we recycle three nested loops:

```
realizable := false ;
for x := 0 to n div a do
for y := 0 to n div b do
for z := 0 to n div c do
     if a * x + b * y + c * z = n
     then realizable := true
```

This code functions very well, but there is no reason to continue to test other triples (x, y, z) after we have found a solution. This brings up the problem of interrupting a loop. We apply our method: that is, we replace "for" loops with "while" loops and put motors "$x := x + 1$", etc. at the head of the loops.

---

[4] Uniqueness is of no interest because the equation under consideration always has solutions $(x, y, z) \in \mathbb{Z}^3$ of which many will be ≥ 0 as soon as n is sufficiently large

```
realizable := false ; x := -1 ;
na := n div a ; nb := n div b ; nc := n div c ;
while not realizable and (x < na) do begin
| x := x + 1 ; y := -1 ;
| while not realizable and (y < nb) do begin
| | y := y + 1 ; z := -1 ;
| | while not realizable and (z < nc) do begin                 (4.1)
| | | z := z + 1 ;
| | | if a * x + b * y + c * z = n then realizable := true
| | end
| end
end ;
if realizable then writeln(x, y, z)
```

As long as *realizable* remains false, the three loops test the triples (x, y, z) in lexicographic order. When *realizable* becomes true for the first time, the three loops are interrupted one after the other *without* x, y, z *changing value* because the motors are at the head of the loops. Verifying the correctness of the result is then very easy.

4.2.3. Calculating the threshold value

It is clear that $n < a$ is not realizable (recall that $a < b < c$). In order to find χ, we successively examine $n = a, a + 1, a + 2, \ldots$ and stop when we detect a consecutive realizable postages.

realizable	×	×				×	×	×			×	×	×	×		×	×	×	×	×
n	5	6	7	8	9	10	11	12	13	14	15	16	17	18	19	20	21	22	23	24
num_succ	1	2	0	0	0	1	2	3	0	0	1	2	3	4	0	1	2	3	4	5

Since we do not have any code to recycle, we experiment. A few tries will show that generating the third line above will allow us to determine χ. Call *num_successive* the value of an integer on the last line; the value of the next integer on the same line is calculated according to the rule:

```
if realizable
then num_successive := num_successive + 1                     (4.2)
else num_successive := 0
```

The entire line is obtained by repeating this operation; it terminates when *num_successive* takes the value a. It is most natural to use a "repeat" loop here because the number of attempts is greater than or equal to $a > 1$.

```
n := a - 1 ;
repeat
| n := n + 1 ;                                                (4.3)
| «calculate num_successive»
until num_successive = a
```

It remains to *assemble* our fragments of code by inserting (4.1) and (4.2) into (4.3) to obtain the definitive code:

```
n := a − 1 ; num_successive := 0 ;
repeat
| n := n + 1 ;
| «insert here the code (4.1) which defines realizable»
| if realizable
| then num_successive := num_successive + 1
| else num_successive := 0
until num_successive ≥ a ;
χ := n − a + 1
```

Remark

We have just used two techniques for rewriting code:

- we have refined a trivial code (three nested loops) and adapted it to our needs;
- we have assembled fragments of code.

Read carefully the warning at the end of Section 4: it is necessary to use the second technique with moderation to avoid writing incomprehensible code.

Exercise 1 *(Solution at the end of the chapter)*

A celebrated theorem of Lagrange states that any integer is a sum of four squares (Chapter 8). This result is best possible in the sense that there exist integers which are not a sum of three squares. Write an algorithm that finds integers $n \in [\![0, 2000]\!]$ which are not sums of three squares.

To verify the algorithm that you have just created, we avail ourselves of the following result.

Theorem 4.2.1 (Gauss). *An integer n is not a sum of three squares if and only if it is of the form $n = 4^k(8q + 7)$.*

For example, here are the numbers ≤ 311 which are not sums of three squares:

7	15	23	28	31	39	47	55	60	63
71	79	87	92	95	103	111	112	119	124
127	135	143	151	156	159	167	175	183	188
191	199	207	215	220	223	231	239	240	247
252	255	263	271	279	284	287	295	303	311

Let $\mathbb{E}$ be the set of integers of the form $n = 4^k(8q+7)$. How are we going to be able to write the elements $\leq N$ of $\mathbb{E}$? Let S denote the arithmetic sequence

$\{8q + 7, q \geq 0\}$ and notice that

$$\mathbb{E} = S \cup 4\mathbb{E} \tag{4.4}$$

Put $\mathbb{E}_N = \mathbb{E} \cap [\![0, N]\!]$ and apply (4.4) repeatedly to get

$$\mathbb{E}_N \subset \mathbb{E} = S \cup 4S \cup 4^2S \cup \ldots \cup 4^nS \cup 4^{n+1}\mathbb{E}. \tag{4.5}$$

Choose $N = 2000$. Since $4^5 = 1024$, we conclude from (4.5) that all integers in $4^5\mathbb{E}$ are greater than or equal to $4^5 \times 7 = 7168$, which gives the inclusion

$$\mathbb{E}_N \subset S \cup 4S \cup 4^2S \cup 4^3S \cup 4^4S. \tag{4.6}$$

Suppose that we have already listed the first few elements of $\mathbb{E}_N$. Let x_ℓ denote the smallest number in $4^\ell S$ which has not yet been listed. The inclusion relation (4.6) shows that the next number we should list is

$$x = \min\{x_0, x_1, x_2, x_3, x_4\}.$$

To easily find the value of x we retain the values of the auxiliary variables $x_0, x_1, \ldots, x_4$. When we write x_ℓ, we replace it by its successor $x_\ell + 8 \cdot 4^\ell$ in $4^\ell S$. Thus, the desired code is

```
x0 := 7 ; x1 := 4 * x0 ; x2 := 4 * x1 ;
x3 := 4 * x2 ; x4 := 4 * x3 ;
repeat
  | x := min(x0, x1, x2, x3, x4) ; write(x) ;
  | if x = x0 then x0 := x0 + 8 ;
  | if x = x1 then x1 := x1 + 32 ;
  | if x = x2 then x2 := x2 + 128 ;
  | if x = x3 then x3 := x3 + 512 ;
  | if x = x4 then x4 := x4 + 2048 ;
until x > 2000
```

Exercise 2 *(Solution at the end of the chapter)*

Let $\mathbb{E}$ be the set of integers which are a sum of two squares. Using the cover of $\mathbb{E}_N = \mathbb{E} \cap [\![0, N]\!]$ by the sets $C_\ell = \{\ell^2 + x^2 : x \geq k\}$, write an algorithm which lists the elements of $\mathbb{E}$ in increasing order. Do the same with sums of cubes.

4.3. Second Method: Using Sequences

For a mathematician, the value of a variable is immutable. In contrast, the variables in a program often change value during its execution. Imagine a program that calculates $\int_0^1 f(x)\,dx$ by dividing the interval [0,1] into 10^3 subintervals. Then the variable x would take a thousand values, and it is inconceivable to tie up a thousand places in memory for a single variable. It is for this reason that a variable x in a program represents an *adddress* in the memory of the computer, the value of x corresponds to the *contents* of this address (we will return to this in Chapter 6). Thus the variable x is a dynamic object which we cannot manipulate as we would in a proof, where everything is static. Happily, one can reconcile mathematics and computer science very simply by introducing time.[5] If we let x_t denote the contents of the variable x at the instant t, then we obtain a number which does not change. From this point of view, we can associate to each variable x in a program the sequence (x_t) of succcessive values[6] taken by x:

$$\textit{identifier } x \textit{ in a program} \rightleftarrows \textit{mathematical sequence } (x_t)$$

An algorithm carries out a sequence of operations and stops when it reads the final result. As a result, to write an algorithm, most of the time it suffices to ask yourself what are *the sequences whose last term must be calculated*. It happens, but very rarely, that the desired solution consists in calculating *all* terms in a sequence. In good cases (the ones that we can handle ...), the value of a sequence at instant $t+1$ can be obtained relatively simply from the value at the instant t (if not, one does not have an algorithm). In other words, we can write a first order recurrence relation

$$x_{t+1} = f(x_t) \tag{4.7}$$

Once we have the recurrence, the algorithm is not much further. It suffices to replace (4.7) by the assignment:

$$x := f(x) \tag{4.8}$$

Example

Let n be an integer ≥ 1 and suppose that we want to calculate the sum

$$S = \sum_{i=1}^{n} u_i \,.$$

Here the notation means that S is the *last term* of the recurrent sequence:

$$S_0 = 0, \quad S_1 = S_0 + u_1, \quad S_2 = S_1 + u_2, \quad \ldots \; , \quad S_n = S_{n-1} + u_n .$$

We begin by replacing the three dots (which are the rustic loops that mathematicians use) by a "for" loop

$$\begin{cases} S_0 := 0 \,; \\ \textbf{for } i := 1 \textbf{ to } n \textbf{ do } S_i := S_{i-1} + u_i \end{cases} \tag{4.9}$$

[5] Time here is not clock time, but conceptual time resulting from mental subdivision of the task.

[6] This is the idea of a stroboscope.

Note that we are still dealing with mathematics here; we have only improved the presentation of the sequence (S_i) by using the more precise language borrowed from computer science.

To transform (4.9) into an algorithm, we can consider S_i as the contents of the memory S at the instant i. Having made this choice, we suppress the time index i, which *automatically* gives us the algorithm:

$$S := 0\ ;\ \textbf{for}\ i := 1\ \textbf{to}\ n\ \textbf{do}\ S := S + u[i] \qquad (4.10)$$

The *last* value of S contains the desired sum.

Remark

We cannot suppress the index i in the u_i because $u_1, \ldots, u_n$ are not the successive contents of the variable u, but the *data* that existed before our fantasizing about time.

4.3.1. Creation of a simple algorithm

The underlying idea is very simple:

Build up to the algorithm by starting with the trace that you imagine.

In other words, you need to know the algorithm that you are looking for.

This seems paradoxical, but experience shows that this method succeeds very often. Proceed in steps:

1) Try to obtain the result you want using a sequence of calculations. Do not be preoccupied by rigor, but let your imagination roam. Experiment with simple, but not stupid, examples. Present your calculations in tabular form, as if it were a trace.

2) When you are sufficiently at ease with your "recipe", systematize the methods by becoming a mathematician. That is, introduce sequences and indexes. Precisely define the objects that you are manipulating (this helps comprehension enormously). Do not yet introduce loops because one can only do one thing well at a time;[7] content yourself instead with the three dots "..." of the mathematician. Try to handle the general case. The introduction of indices will usually result in one or more first order recurrences.

3) When you are at ease with the mathematical description, refine and replace the mathematician's three dots "..." by the appropriate loops. Note that you are still in the realm of mathematics, but it is expressed in a more modern language.

[7] One should not climb stairs while chewing gum.

4) Choose a time index in each recurrence; replace the recurrence $x_{i+1} = f(x_i)$ by the assignment $x := f(x)$ and replace equalities that are not tests by assignments.
5) Check the algorithm obtained by executing several traces. Eventually you will want to prove it. (The technique will be presented at the end of the chapter.)

4.3.2. The exponential series

Let x be a real number and $N \geq 0$ an integer. We want to calculate

$$S_N(x) = \sum_{k=0}^{N} \frac{x^k}{k!}.$$

As we have already remarked, the number S_N is the *last term* of the recurrent sequence:

$$S_0 = 1,\ S_1 = S_0 + \frac{x}{1!},\ S_2 = S_1 + \frac{x^2}{2!},\ \ldots,\ S_N = S_{N-1} + \frac{x^N}{N!}.$$

Since we can't type x^k and $k!$ directly into our program, we *name* the objects that inconvenience us by introducing the auxiliary sequences $P_k = x^k$ and $F_k = k!$ and then transform them into recurrent sequences:

$$P_0 = 1,\ P_k = x * P_{k-1}\,; \quad F_0 = 1,\ F_k = k * F_{k-1}.$$

We now present the calculation of $S_N(x)$ as a trace:

$$\begin{array}{lll}
P_0 = 1 & F_0 = 1 & S_0 = 1 \\
P_1 = x * P_0 & F_1 = 1 * F_0 & S_1 = S_0 + P_1/F_1 \\
P_2 = x * P_1 & F_1 = 2 * F_1 & S_2 = S_1 + P_2/F_2 \\
\vdots & \vdots & \vdots \\
P_N = x * P_{N-1} & F_N = N * F_{N-1} & S_N = S_{N-1} + P_N/F_N.
\end{array}$$

We can *condense* this trace using a "for" loop (this description remains correct when N is zero):

$$\begin{array}{l}
P_0 := 1\ ;\ F_0 := 1\ ;\ S_0 := 1\ ; \\
\textbf{for}\ k := 1\ \textbf{to}\ N\ \textbf{do begin} \\
\quad P_k = x * P_{k-1}\ ; \\
\quad F_k = k * F_{k-1}\ ; \\
\quad S_k = S_{k-1} + P_k/F_k \\
\textbf{end}
\end{array} \tag{4.11}$$

Note that we are still in the domain of mathematics: (4.11) is nothing but a more modern preentation of (4.10).

Now, let the index k be the time in (4.11). If we suppress it and replace the equalities by assignments, we obtain without effort the algorithm

```
P := 1 ; F := 1 ; S := 1 ;
for k := 1 to N do begin
|P = x * P ; F = k * F ; S = S + P/F
end
```

Remark

The initialization is delicate. If you start from the table:

$$
\begin{array}{lll}
S_0 = 1 & P_1 = x & F_1 = 1 \\
S_1 = S_0 + P_1/F_1 & P_2 = x * P_1 & F_1 = 2 * F_1 \\
S_2 = S_1 + P_2/F_2 & P_3 = x * P_2 & F_3 = 3 * F_2 \\
\vdots & \vdots & \vdots \\
S_{N-1} = S_{N-2} + P_{N-1}/F_{N-1} & P_N = x * P_{N-1} & F_N = N * F_{N-1}
\end{array}
$$

you wind up with a much clumsier algorithm because you are obliged to repeat the statement "$S := S + P/F$" outside the loop

```
S := 1 ; P := x ; F := 1 ;
for k := 1 to N − 1 do begin
|S = S + P/F ; P = x * P ; F = k * F ;
end ;
S = S + P/F
```

Think of this each time that you encounter a schema like that on the right; choose the one on the left instead.

Good temporal breaks	Bad temporal breaks
(1 2) (3 4) (5 6) (7 8) (9 10)	(1) (2 3) (4 5) (6 7) (8 9) (10)

Exercise 3

- Improve this algorithm by supposing that $x = a/b$ is a rational number. The result $S_N = Num_N/Den_N$ must be a rational number (that is, a pair of integers).
- Write an algorithm which calculates the sum

$$S_N = \sum_{i=0}^{N} \binom{N}{i} x^i$$

using first order recurrences for the sequences $\binom{N}{i}$ and x^i.

• The Bernoulli numbers $(B_n)_{n\geq 1}$ are defined as follows:

$$B_1 = \frac{1}{6}, \quad B_n = \frac{1}{2n+1}\left[n - \frac{1}{2} - \sum_{i=1}^{n-1} \binom{2n+1}{2i} B_i\right] \quad \text{if} \quad n \geq 2.$$

Write an algorithm which calculates the first N such numbers, the calculations taking place over the field of rational numbers. (The first Bernoulli numbers are $B_1 = \frac{1}{6}$, $B_2 = -\frac{1}{30}$, $B_3 = \frac{1}{42}$, $B_4 = -\frac{1}{30}$, $B_5 = \frac{5}{66}$.)

4.3.3. Decomposition into prime factors

Let $n > 1$ be a given number. To decompose n into prime factors, we all know the following method: look for a prime divisor, divide, then begin again with the quotient. When $n = 60{,}900$, this gives

60,900	2	1,015	5
30,450	2	203	7
15,225	3	29	29
5,075	5	1	← *stop*

We enrich this presentation by passing to trace mode; that is, by introducing identifiers and indices (the exchange of the columns n and d is to facilitate presentation of the algorithm).

	$n_0 = 60{,}900$	$d_4 = 5$	$n_4 = 1{,}015$
$d_1 = 2$	$n_1 = 30{,}450$	$d_5 = 5$	$n_5 = 203$
$d_2 = 2$	$n_2 = 15{,}225$	$d_6 = 7$	$n_6 = 29$
$d_3 = 3$	$n_3 = 5{,}075$	$d_7 = 29$	$n_7 = 1$ ← *stop*

Now that we have sequences, we write down first order recurrences which indicate how to pass from one line to the next.

• This is simple for the sequence (n_i) because $n_i = n_{i-1}/d_i$.

• In contrast, we cannot find a recurrence relation[8] for the sequence (d_i). In order to get around this obstacle, two solutions present themselves.

The first consists in supposing that we have access to an array $p[1\,..\,N]$ which contains the sequence of prime numbers: $p[1] = 2$, $p[2] = 3$, $p[3] = 5, \ldots$. This "static" solution, which beginners often propose, is not very appetizing because it raises more problems than it solves: which value should we give N? how should we fill out our table? how should we choose the next prime divisor (that is, how can we choose the index k_i in the formula $d_i = p[k_i]$)?

[8] If you could, you would become as famous as Euclid because letting n be the product of the N first prime numbers would then give a recurrence for prime numbers.

Since we can't advance further, we turn to the proverb "tomorrow things will be better, ...", and *avoid solving* the problem right away which is easily done by *giving a name* to the smallest divisor which we do not know. In fact, we already have a name since we have seen in Chapter 2 that the $\mathrm{LD}(n) = \{$least divisor > 1 of $n\}$ is always a prime number. Let us again resume our trace:

$$\begin{array}{ll} & n_0 = \textit{given integer} > 1; \\ d_1 = \mathrm{LD}(n_0); & n_1 = n_0/d_1; \\ d_2 = \mathrm{LD}(n_1); & n_2 = n_1/d_2; \\ \ldots\quad ; & \ldots\quad ; \\ d_k = \mathrm{LD}(n_{k-1}); & n_k = n_{k-1}/d_k; \\ & \textit{stop because } n_k = 1. \end{array}$$

The three dots indicate the presence of a loop? Which type? The index of d_t is tempting because it takes the values $1, 2, 3, \ldots$. But since do not know the value of k in advance, we cannot use a "for" loop. Knowing that we must always seek *at least once* the LD of n, (if only to learn whether n is prime), we settle on a "repeat" loop:

```
n0 := given integer > 1 ; t := 1 ;               n := given integer > 1 ;
repeat                                           repeat
 | dt := LD(n(t-1)) ; write(dt) ;        ==>      | d := LD(n) ; write(d) ;
 | nt := n(t-1) div dt ;                          | n := n div d
 | t := t + 1                                    until n = 1
until nt = 1
```

The left is a mathematical description of the sequences (n_t) and (d_t) which manages the time index t "by hand" using the motor "$t := t+1$". The index t and the useless statement "$t := t + 1$" drop out on the right when we pass to the algorithm by suppressing time.

Exercise 4

Improve this algorithm to take account of repeated prime factors: if p^3 divides n, it is stupid to call the function LD three times in a row.

Exercise 5

Let $n = p_1^{\alpha_1} \cdots p_k^{\alpha_k}$ be the decomposition of n into prime factors. Modify the algorithm so that it *stores* the prime divisors and their exponents as vectors $p[1\,..\,N]$ and $\alpha[1\,..\,N]$.

4.3.4. The least divisor function

We briefly sketch how to calculate the value of the least divisor function. Suppose, for example, that we want to find the LD of 323. For lack of other ideas, we divide $n = 323$ by 2, 3, 4, ... until we get a divisor. To determine if d divides n, we turn naturally to the remainder $r = n \bmod d$ which gives the two lines:

d	2	3	4	5	6	7	8	9	10	11	12	13	14	15	16	17
r	1	2	3	3	5	1	3	8	3	4	11	11	1	8	3	0

Naming the two sequences, we have:

$d_1 = 2$	$d_2 = 3$	$d_3 = 4$	$d_4 = 5$	...	$d_{16} = 17$
$r_1 = 1$	$r_2 = 2$	$r_3 = 3$	$r_4 = 3$	...	$r_{16} = 0$

Having arrived at this stage, the temptation not to seek a recurrence relation is enormous because we have:

$$d_i = i + 1 \quad \text{and} \quad r_i = n \bmod d_i \,.$$

But this strategy leads to a dead end because this option gives rise to the following incomplete code

```
for d := 2 to ??? do r := n mod d
```

On the other hand, if we introduce the recurrence $d_i := d_{i-1} + 1$, we immediately obtain an explicit loop which we polish slightly

```
d := 1 ;                           d := 1 ;
repeat                             repeat
| d := d + 1 ;          ⟹         | d := d + 1
| r := n mod d                     until n mod d = 0
until r = 0
```

Remark

In Chapter 8 we present some more sophisticated algorithms for calculating this vital arithmetic function.

4.3.5. Cardinality of an intersection

Let $a_1 < a_2 < \cdots < a_n$ and $b_1 < b_2 < \cdots < b_m$ be two strictly increasing sequences of integers. We want to find the number of integers common to both sets; that is, the cardinality of the set $\{a_1, \ldots, a_n\} \cap \{b_1, \ldots, b_m\}$.

Consider the particular case of the sequences $2, 3, 5, 9$ and $3, 4, 5, 8, 9, 10$.

a	2	3		5		9	
b		3	4	5	8	9	10
num		1		2		3	

We compare $a_1 = 2$ and $b_1 = 3$. Since $a_1 < b_1$, we know that our intersection is equal to $\{3, 5, 9\} \cap \{3, 4, 5, 8, 9, 10\}$. We now compare $a_2 = 3$ and $b_1 = 3$. This reduces the problem to calculating the cardinality of the set $\{5, 9\} \cap \{4, 5, 8, 9, 10\}$. We stop when we compare a_n and b_m. The variable num contains the number of common elements that we have found.

$$
\begin{array}{l|lll}
t=1 & a_1 = 2 & b_1 = 3 & num_1 = 0 \\
t=2 & a_2 = 3 & b_1 = 3 & num_2 = 1 \\
t=3 & a_3 = 5 & b_2 = 4 & num_3 = 1 \\
t=4 & a_3 = 5 & b_3 = 5 & num_4 = 2 \\
t=5 & a_4 = 9 & b_4 = 8 & num_5 = 2 \\
t=6 & a_4 = 9 & b_5 = 9 & num_6 = 3 \\
t=7 & a_4 = 9 & b_6 = 10 & num_7 = 3
\end{array} \tag{4.12}
$$

When we examine the table (4.12), we see that row t does not contain a_t, b_t and num_t, but rather the sequences a_{i_t}, b_{j_t} and num_t whose indices are themselves sequences (this phenomenon is very frequent). Once the the biggest obstacle is overcome (by introducing the auxiliary sequences i_t and j_t), the passage from one line to the next is child's play:

$$
S(t) = \begin{cases}
\textbf{if } a_{i_t} = b_{j_t} \textbf{ then begin} \\
\quad | \, i_{t+1} = i_t + 1 \; ; \; j_{t+1} = j_t + 1 \; ; \; num_{t+1} = num_t + 1 \\
\textbf{end} \\
\textbf{else if } a_{i_t} < b_{j_t} \textbf{ then begin} \\
\quad | \, i_{t+1} = i_t + 1 \; ; \; j_{t+1} = j_t \; ; \; num_{t+1} = num_t \\
\textbf{end} \\
\textbf{else begin} \\
\quad | \, i_{t+1} = i_t \; ; \; j_{t+1} = j_t + 1 \; ; \; num_{t+1} = num_t \\
\textbf{end}
\end{cases} \tag{4.13}
$$

The first order recurrences require initial conditions:

$$i_1 = 1, \quad j_1 = 1, \quad num_1 = 0.$$

Since we stop as soon as $i_t > n$ or $j_t > m$, we can *condense* (4.12) as:

$$
\begin{cases}
i_1 = 1; \; j_1 = 1; \; num_1 = 0; \\
S(1); \; S(2); \; \ldots \; ; \; S(t); \\
\textit{stop as soon as } i_t > n \textit{ or } j_t > m
\end{cases} \tag{4.14}
$$

Now we refine the above by introducing a loop. Since we do not know the last value of k in advance and since it is always necessary to make at least one comparison, we choose a "repeat" loop.

$$
\begin{array}{l}
i_1 := 1 \;;\; j_1 := 1 \;;\; num_1 := 0 \;;\; t := 1 \;; \\
\textbf{repeat} \\
\quad S(t) \;; \\
\quad t := t + 1 \\
\textbf{until } (i_t > n) \textbf{ or } (j_t > m)
\end{array}
\tag{4.15}
$$

Substituting (4.13) into (4.15) gives

$$
\begin{array}{l}
i_1 := 1 \;;\; j_1 := 1 \;;\; num_1 := 0 \;;\; t := 1 \;; \\
\textbf{repeat} \\
\quad \textbf{if } a_{i_t} = b_{j_t} \textbf{ then} \\
\quad \textbf{begin } i_{t+1} = i_t + 1 \;;\; j_{t+1} = j_t + 1 \;;\; num_{t+1} = num_t + 1 \textbf{ end} \\
\quad \textbf{else if } a_{i_t} < b_{j_t} \textbf{ then} \\
\quad \textbf{begin } i_{t+1} = i_t + 1 \;;\; j_{t+1} = j_t \;;\; num_{t+1} = num_t \textbf{ end} \\
\quad \textbf{else} \\
\quad \textbf{begin } i_{t+1} = i_t \;;\; j_{t+1} = j_t + 1 \;;\; num_{t+1} = num_t \textbf{ end} \;; \\
\quad t := t + 1 \\
\textbf{until } (i_t > n) \textbf{ or } (j_t > m)
\end{array}
$$

We suppress the indices t and $t + 1$ and replace the equality signs which are not tests by assignments:

$$
\begin{array}{l}
i := 1 \;;\; j := 1 \;;\; num := 0 \;; \\
\textbf{repeat} \\
\quad \textbf{if } a[i] = b[j] \textbf{ then begin } i := i + 1 \;;\; j := j + 1 \;;\; num := num + 1 \textbf{ end} \\
\quad \textbf{else if } a[i] < b[j] \textbf{ then begin } i := i + 1 \;;\; j := j \;;\; num := num \textbf{ end} \\
\quad \textbf{else begin } i := i \;;\; j := j + 1 \;;\; num := num \textbf{ end} \;; \\
\quad t := t + 1 \\
\textbf{until } (i > n) \textbf{ or } (j > m)
\end{array}
$$

In this draft, we see that there are superfluous statements such as $num := num$. We suppress them together with the now superfluous "begin end" blocks they had necessitated. After tidying up a little, the final algorithm is then:

$$
\begin{array}{l}
i := 1 \;;\; j := 1 \;;\; num := 0 \;; \\
\textbf{repeat} \\
\quad \textbf{if } a[i] = b[j] \\
\quad \textbf{then begin } i := i + 1 \;;\; j := j + 1 \;;\; num := num + 1 \textbf{ end} \\
\quad \textbf{else if } a[i] < b[j] \\
\qquad \textbf{then } i := i + 1 \\
\qquad \textbf{else } j := j + 1 \;; \\
\textbf{until } (i > n) \textbf{ or } (j > m)
\end{array}
$$

When the algorithm terminates, the value of the variable num is the cardinality of the intersection.

4.3.6. The CORDIC Algorithm

The CORDIC (COordinate Rotation on a DIgital Computer) algorithm was published in 1959 by J. Volder.

We want to compute the value of the function $\tan\alpha$ of an prescribed angle $\alpha \in \left]0, \frac{1}{4}\pi\right[$. The starting idea is simple: suppose that we have a sequence

$$\alpha_1 = \pi_1, \ \alpha_2 = \pi_1 + \pi_2, \ \alpha_3 = \pi_1 + \pi_2 + \pi_3, \ \ldots$$

of progressively finer approximations of α. The continuity of the tangent function implies that $\tan\alpha_1, \tan\alpha_2, \tan\alpha_3, \ldots$ are better and better approximations to $\tan\alpha$. It suffices to express α_{t+1} using α_t and π_{t+1} to have first order recurrences appear. But how can we find the approximations α_t?

How to "weigh" a real number. To weigh an object using a scale with two platforms, we put the object to be weighed on one platform, the left, say, and add weights on the right, beginning with the heavier ones. The platform on the right remains lower that the one on the right – it is only at the very end that the two platforms come into equilibrium.

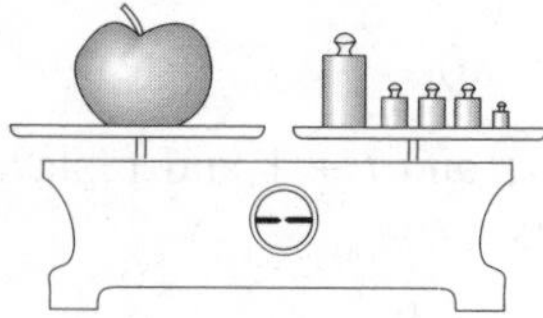

Fig. 4.1.

We suppose that we have a decreasing sequence $(\pi_n)_{n\geq 0}$ of weights tending zero. We suppose, moreover, that for each integer n, we have as many weights in the category π_n as we desire. Imagine, now, a weighing in which we successively put the weights $\pi_0, \pi_0, \pi_1, \pi_2, \pi_2, \pi_2$ on the right hand platform:

$$\begin{aligned}
\alpha_1 &= \pi_0 \\
\alpha_2 &= \pi_0 + \pi_0 \\
\alpha_3 &= \pi_0 + \pi_0 + \pi_1 \\
\alpha_4 &= \pi_0 + \pi_0 + \pi_1 + \pi_2 \\
\alpha_5 &= \pi_0 + \pi_0 + \pi_1 + \pi_2 + \pi_2 \\
\alpha_6 &= \pi_0 + \pi_0 + \pi_1 + \pi_2 + \pi_2 + \pi_2
\end{aligned}$$

At the instant t, we put a weight on the right hand platform if it does not make the platform descend; otherwise, we simply change the category of weights. If we deem that each of these two actions is performed in a unit of time, we will have introduced two sequences:

• The sequence (α_t) represents the sum of the weights found on the right hand platform at the instant t. This is an increasing sequence whose value necessarily approaches α.

• The second sequence is more subtle: at an instant t, we do not necessarily put the weight π_t on the right hand platform. Thus it is necessary to introduce the sequence (k_t) of indices of the weights π_{k_t} placed on the platform at time t.

These sequences specify the table above and oblige us to renumber the α's:

$$
\begin{array}{l|ll}
 & k_0 = 0 & \alpha_0 = 0 \\
t = 1 & k_1 = k_0 & \alpha_1 = \alpha_0 + \pi_{k_1} \\
t = 2 & k_2 = k_1 & \alpha_2 = \alpha_1 + \pi_{k_2} \\
t = 3 & k_3 = k_2 + 1 & \alpha_3 = \alpha_2 \\
t = 4 & k_4 = k_3 & \alpha_4 = \alpha_3 + \pi_{k_4} \\
t = 5 & k_5 = k_4 + 1 & \alpha_5 = \alpha_4 \\
t = 6 & k_6 = k_5 & \alpha_6 = \alpha_5 + \pi_{k_6} \\
t = 7 & k_7 = k_5 & \alpha_7 = \alpha_6 + \pi_{k_7} \\
t = 8 & k_8 = k_5 & \alpha_8 = \alpha_7 + \pi_{k_8}
\end{array}
$$

We now seek to understand how to pass from one line to the next:

• If $\alpha - \alpha_t \geq \pi_{k_t}$, at instant $t+1$ we place the weight π_{k_t} on the right hand platform, so that:

$$\alpha_{t+1} = \alpha_t + \pi_{k_t}, \quad k_{t+1} = k_t.$$

• If $\alpha - \alpha_t < \pi_{k_t}$, we change the category of weights at the instant $t+1$ and, as agreed, put *nothing* on the right hand platform:

$$\alpha_{t+1} = \alpha_t, \quad k_{t+1} = k_t + 1.$$

To simplify, we put:

$$S(t) = \begin{cases} \textbf{if } \alpha - \alpha_t \geq \pi_{k_t} \\ \textbf{then begin } \alpha_{t+1} = \alpha_t + \pi_{k_t} \; ; \; k_{t+1} = k_t \textbf{ end} \\ \textbf{else begin } \alpha_{t+1} = \alpha_t \; ; \; k_{t+1} = k_t + 1 \textbf{ end} \end{cases}$$

The weighing terminates when the approximation t is sufficiently good, by which we mean that $\alpha - \alpha_t < \varepsilon$, where $\varepsilon > 0$ is given in advance. Thus, we can write our weighing succinctly as:

$$\begin{array}{l} \alpha_0 := 0 \; ; \; k_0 := 0 \; ; \\ S(0) \; ; \; S(1) \; ; \; \ldots \; ; \; S(t) \\ \textit{stop when } \alpha - \alpha_t < \varepsilon \end{array} \tag{4.16}$$

Knowing that we have to place at least one weight on the right hand platform and that we do not know in advance the length of the weighing, we choose a "repeat" loop to replace the three dots in (4.19). As in the preceeding example, we are obliged to explicitly introduce time because a loop of this type needs a motor:

$$\begin{array}{l} \alpha_0 := 0\ ;\ k_0 := 0\ ;\ t := 0\ ; \\ \textbf{repeat} \\ \quad S(t)\ ; \\ \quad t := t+1 \\ \textbf{until}\ \alpha - \alpha_t < \varepsilon \end{array} \tag{4.17}$$

We emphasize once more: we are still in the domain of mathematics, but profiting from a more precise language.

Now let us turn to the calculation of the $\tan k$. The equality $\alpha_{t+1} = \alpha_t + \pi_{k_t}$ shows that

$$\tan \alpha_{t+1} = \frac{\tan \alpha_t + \tan \pi_{k_t}}{1 - \tan \alpha_t \cdot \tan \pi_{k_t}}.$$

If we use this formula, we must perform a division each time that t changes value. Since division is a long and complicated operation, Volder suggested calculating the numerator and denominator of $\tan t$ separately, thereby making do with a single division at the end of the loop. In addition, Volder proposed choosing[9] weights

$$\pi_k = \text{Arctan}\, 10^{-k}.$$

Put

$$\tan \alpha_t = \frac{Num_t}{Den_t}, \qquad \tan \pi_k = \frac{1}{10^k}.$$

Since

$$\tan \alpha_{t+1} = \frac{Num_t + Den_t * 10^{-k_t}}{Den_t - Num_t * 10^{-k_t}},$$

we can choose the recurrences

$$\begin{cases} Num_{t+1} = Num_t + Den_t * 10^{-k_t}, \\ Den_{t+1} = Den_t - Num_t * 10^{-k_t}. \end{cases} \tag{4.18}$$

Remarks

• For reasons of numerical stability, it is preferable to use multiplication by 10^{-k_t} rather than multiplication by 10^{k_t}.

• The reason for choosing $\pi_k = \text{Arctg}\, 10^{-k}$ is clear: one passes from Num_t to Num_{t+1} by adding to Num_t a number obtained from Den_t by a *truncation* which supresses the last k digits, an instantaneous operation on a computer.

[9] In binary, one clearly should choose the weights $\pi_k = \text{Arctan}\, 2^{-k}$.

• In the era when Volder created the CORDIC algorithm, memory was expensive. Since the difference between $\pi_k = \text{Arctg}\,10^{-k}$ and 10^{-k} rapidly becomes negligible, one can store the first few π_k in memory and replace the rest by 10^{-k} at the price of lengthening the code with several tests.

If we incorporate (4.19) into the loop (4.18) we obtain the following *mathematical* description of the weighing algorithm:

```
α₀ = 0 ; k₀ = 0 ; t := 0 ; Num₀ = 0 ; Den₀ = 1 ;
repeat
|  if α − α_t ≥ π_{k_t}
|  then begin
|  |  α_{t+1} = α_t + π_{k_t} ; k_{t+1} = k_t ;
|  |  Num_{t+1} = Num_t + Den_t * 10^{-k_t} ;
|  |  Den_{t+1} = Den_t − Num_t * 10^{-k_t}
|  end
|  else begin
|  α_{t+1} = α_t ; k_{t+1} = k_t + 1 end ;
|  t := t + 1
until α − α_t < ε
```

Having arrived at this stage, we see that it is preferable to replace t by $\beta = t$, which represents what remains to weigh.

We abandon mathematics by supressing the t and $t+1$ in the indices, and by replacing the equalities by assignments. But it is necessary to be prudent and retain the value of Num_t (by introduction of an auxiliary variable *temp*) because we need this number to calculate Den_{t+1}:

```
β := α ;  k := 0 ;  Num := 0 ;  Den := 1 ;
repeat
|  if β ≥ π_k then begin
|  |  β := β − π_k ; k := k ;
|  |  temp := Num ;
|  |  Num = Num + Den * 10^{-k} ;
|  |  Den := Den − temp * 10^{-k}
|  end
|  else begin β := β ; k := k + 1 end
until β < ε ;
{one has tan α ≃ Num/Den at the end of the loop}
```

This code contains the superfluous statements $k := k$ and $\beta := \beta$, which we suppress. The definitive CORDIC algorithm is then:

```
β := α ;  k := 0 ;  Num := 0 ;  Den := 1 ;
repeat
|  if β ≥ π_k then begin
|  |  β := β − π_k ;
```

```
        temp := Num ;
        Num = Num + Den * 10^-k ;
        Den := Den - temp * 10^-k
      end
      else k := k + 1
until β < ε ;
{one has tan α ≃ Num/Den at the end of the loop}
```

Remark

This algorithm has been enthusiastically studied since its creation; improvements of it were still being published in 1994! It has been generalised to all elementary functions. For more information, see J.-M. Muller, *Arithmétique des ordinateurs*, Masson, 1989.

4.4. Third Method: Defered Writing

An algorithm is often an assemblage of delicate elementary algorithms carried out with the aid of sequences. But when an algorithm contains many loops, it becomes very difficult to assemble *all* the sequences and recurrences. Let us try, for example, to calculate the number of divisors of an integer $n = p_1^{\alpha_1} p_2^{\alpha_2} \cdots p_k^{\alpha_k} > 1$ with the aid of the formula

$$d(n) = (\alpha_1 + 1)(\alpha_2 + 1) \cdots (\alpha_k + 1).$$

If we employ the factorization algorithm that we have detailed, we obtain the array in Table 4.1. This array is very complex because we see two different calculations superimposed, each with their own rythm:

- the calculation of the divisors p_i;
- for each p_i, the calculation of the corresponding exponent i.

We draw the following lesson from this example: *one does not contemplate a landscape (algorithm) with a microscope (a sequence).* In other words, we should avoid nesting one sequence inside another. To do this, we mask undesirable sequences by sentences that we *subsequently* transform into code.

It follows that developing a complex algorithm is like *peeling* an onion: we take care of the external layer (the first loop) by introducing a sequence if necessary. The other loops (if there are any) are masked by sentences.

We then make precise the sentences or statements which remain following the same method and taking care to never introduce more than one loop at a time.

An attentive reader will have noticed that we already used this technique without explicitly calling attention to it when we grafted the caluclation of the tangent into the weighing algorithm to end the CORDIC algorithm.

	n_0	$=$ *prescribed integer* > 1	$d_0 = 1$
$p_1 = \mathrm{LD}(n_0)$	n_1	$= n_0/p_1$	1
$p_1 = \mathrm{LD}(n_1)$	n_2	$= n_1/p_1$	2
$\vdots$		$\vdots$	$\vdots$
$p_1 = \mathrm{LD}(n_{\alpha_1-1})$	n_{α_1}	$= n_{\alpha_1-1}/p_1$	α_1
$p_2 = \mathrm{LD}(n_{\alpha_1})$	n_{α_1+1}	$= n_{\alpha_1}/p_2$	1 $d_1 = (\alpha_1 + 1)d_0$
$p_2 = \mathrm{LD}(n_{\alpha_1+1})$	n_{α_1+2}	$= n_{\alpha_1+1}/p_2$	2
$\vdots$		$\vdots$	$\vdots$
$p_2 = \mathrm{LD}(n_{\alpha_1+\alpha_2-1})$	$n_{\alpha_1+\alpha_2}$	$= n_{\alpha_1+\alpha_2-1}/p_2$	α_2
$p_3 = \mathrm{LD}(n_{\alpha_1+\alpha_2})$	$n_{\alpha_1+\alpha_2+1}$	$= n_{\alpha_1+\alpha_2}/p_3$	1 $d_2 = (\alpha_2 + 1)d_1$
$p_3 = \mathrm{LD}(n_{\alpha_1+\alpha_2+1})$	$n_{\alpha_1+\alpha_2+2}$	$= n_{\alpha_1+\alpha_2+1}/p_3$	2
$\vdots$		$\vdots$	$\vdots$
$p_3 = \mathrm{LD}(n_{\alpha_1+\alpha_2+\alpha_3-1})$	$n_{\alpha_1+\alpha_2+\alpha_3}$	$= n_{\alpha_1+\alpha_2+\alpha_3-1}/p_3$	α_3
$p_4 = \mathrm{LD}(n_{\alpha_1+\alpha_2+\alpha-3})$	$n_{\alpha_1+\alpha_2+\alpha_3+1}$	$= n_{\alpha_1+\alpha_2+\alpha_3}/p_4$	1 $d_3 = (\alpha_3 + 1)d_2$
$\vdots$		$\vdots$	$\vdots$

Table 4.1. *Trace of the calculation of $d(n)$.*

Let us put this advice into practice by calculating $d(n)$.

• We put in place the external loop by recycling the factorization algorithm which we know well by now:

```
n := prescribed integer > 1 ; d := 1 ;
repeat
| p := LD(n) ;
| «calculate α and n := n/p^α knowing p» ;
| d := d * (1 + α)
until n = 1
```

• We develop the internal loop to "calculate α and $n := n/p^\alpha$ knowing p":

```
α := 0 ;
while n mod p = 0 do begin
| α := α + 1 ; n := n div p
end
```

• We nest the two loops to obtain the definitive version:

```
n := prescribed integer > 1 ; d := 1 ;
repeat
```

```
| p := LD(n) ;
| α := 0 ;
| while n mod p = 0 do begin
| | α := α + 1 ;
| | n := n div p
| end ;
| d := d * (1 + α)
until n = 1
```

4.4.1. Calculating two bizarre functions

Let $n = p_1^{\alpha_1} \cdots p_k^{\alpha_k}$ be the decomposition of $n > 1$ into prime factors. Put

$$|\alpha| = \begin{cases} \pi_1 + \cdots + \pi_\ell & \text{if } \alpha = \pi_1^{\lambda_1} \cdots \pi_\ell^{\lambda_\ell} > 1, \\ 0 & \text{if } \alpha = 1; \end{cases}$$

$$\|\alpha\| = \begin{cases} \lambda_1\pi_1 + \cdots + \lambda_\ell\pi_\ell & \text{if } \alpha = \pi_1^{\lambda_1} \cdots \pi_\ell^{\lambda_\ell} > 1, \\ 0 & \text{if } \alpha = 1. \end{cases}$$

The values $|1| = \|1\| = 0$ are not gratuitous: they are the result of general conventions regarding "empty" sums. So, for example, we have:

$$|6| = 2 + 3 = 5, \quad |72| = |2^3 \times 3^2| = 2 + 3 = 5;$$
$$\|6\| = 2 + 3 = 5, \quad \|72\| = \|2^3 \times 3^2\| = 2 + 2 + 2 + 3 + 3 = 14.$$

We wish to calulate the functions

$$\Psi(n) = \prod \big(1 + |\alpha_i|\big) \quad \text{and} \quad \Lambda(n) = \prod \big(1 + \|\alpha_i\|\big).$$

Here are some values:

n	$\Psi(n)$	$\Lambda(n)$
$58320 = 2^4 \times 3^6 \times 5$	18	30
$67500 = 2^2 \times 3^3 \times 5^4$	36	60
$600000 = 2^6 \times 3 \times 5^5$	36	36
$8890560 = 2^6 \times 3^4 \times 5 \times 7^3$	72	120

Consider first the computation of the function Ψ. We first set up the main loop, masking the most difficult parts (that is, the other loops) with sentences:

```
Ψ := 1 ;
while n > 1 do begin
|p := LD(n) ;
|«calculate α and n := n/p^α knowing p» ;
|«calculate S := |α| knowing p» ;
|Ψ := (1 + S) * Ψ
end
```

• The prime factor p being chosen, the calculation of the exponent α and the division of n by p^α are easy to write:

```
α := 0 ;
while n mod p = 0 do begin
|p := LD(n) ;
|n := n div p ;
|α := α + 1
end
```

• The calculation of $S = |\alpha|$ is added naturally to the interior of the decomposition of α into prime factors. Since we know a prime divisor p of α, we add α to S and then "purge" α from its prime factor:

```
S := 0 ;
while α > 1 do begin
|p := LD(α) ;
|S := S + p ;
|repeat α := α div p until α mod p > 0
end
```

It remains to assemble the fragments of the code. The final code (Figure 4.2) contains four loops. If we replace the calculation of $S = |\alpha|$ by the simpler calculation of $T = \|\alpha\|$ we obtain code which calculates $\Lambda(n)$.

4.4.2. Storage of the first N prime numbers

This last example is rather difficult. We do not, as a result, recommend it for beginners; wait until you are at ease before beginning! Suppose that we know that

$$p_1 = 2, \quad p_2 = 3, \quad p_3 = 5, \quad p_4 = 7.$$

We want to store the first $N \geq 4$ prime numbers in an array: we search through the odd integers beginning with 9. To determine if n is prime, we divide it by odd prime numbers smaller than $\sqrt{n}$. If n is divisible by none of the numbers, we know that it is prime and store it. Otherwise, we end the tests and move to the next odd integer (see the trace of this algorithm in Figure 4.3).

```
Ψ := 1 ;
while n > 1 do begin
  p := LD(n) ; α := 0 ;
  while n mod p = 0 do begin
    p := LD(n) ;
    n := n div p ;
    α := α + 1
  end ;
  S := 0 ;
  while α > 1 do begin
    p := LD(α) ; S := S + p ;
    repeat α := α div p until α mod p > 0
  end ;
  Ψ := (1 + S) * Ψ
end
```

Calculation of $\Psi(n)$

```
Λ := 1 ;
while n > 1 do begin
  p := LD(n) ; α := 0 ;
  while n mod p = 0 do begin
    p := LD(n) ;
    n := n div p ;
    α := α + 1
  end ;
  T := 0 ;
  while α > 1 do begin
    p := LD(α) ; T := T + p
    α := α div p ;
  end ;
  Λ := (1 + T) * Λ
end
```

Calculation of $\Lambda(n)$

Fig. 4.2.

- The first draft of our algorithm is very natural:

```
p[1] = 2 ; p[2] = 3 ; p[3] = 5 ; p[4] = 7 ; n := 7 ;
for ℓ := 5 to N do begin
  «n := the smallest prime number > n» ;
  p[ℓ] := n
end
```

- The code that follows finds the prime number which follows n:

```
repeat
  n := n + 2 ;
  prime := «n is a prime number» ;
until prime
```

- To calculate the boolean *prime*, we respect the constraints: we divide n by the prime numbers $\leq \sqrt{n}$, which requires that we introduce the index k of the smallest prime numbers $> \sqrt{n}$.

```
«find the smallest k such that p_k^2 > n» ;
i := 2 ; prime := true ;
while (i < k) and prime do
    if n mod p[i] = 0
    then prime := false
    else i := i + 1
```

square	ℓ	n	*attempts*	*primes*
	1	2		$p_1 = 2$
	2	3		$p_2 = 3$
	3	5		$p_3 = 5$
	4	7		$p_4 = 7$
25	5	9	3	
		11	3	$p_5 = 11$
	6	13	3	$p_6 = 13$
	7	15	3	
		17	3	$p_7 = 17$
	8	19	3	$p_8 = 19$
	9	21	3	
		23	3	$p_9 = 23$
49	10	25	3, 5	
		27	3	
		29	3, 5	$p_{10} = 29$
	11	31	3, 5	$p_{11} = 31$
	12	33	3	
		35	3, 5	
		37	3, 5	$p_{12} = 37$
		39	3	
	13	41	3, 5	$p_{13} = 41$
	14	43	3, 5	$p_{14} = 43$
	15	45	3, 5	
		47	3, 5	$p_{15} = 47$
121	16	49	3, 5, 7	
		51	3	
		53	3, 5, 7	$p_{16} = 53$
	17	55	3, 5	
		57	3	
		59	3, 5, 7	$p_{17} = 59$
	18	61	3, 5, 7	$p_{18} = 61$
	19	63	3	
		65	3, 5	

Fig. 4.3. Search and storage of the first 18 prime numbers. The column *attempts* enumerates the prime numbers used to obtain a response: the current integer n is prime or composite.

• The determination of k requires that we introduce an auxiliary variable $square = p[k]^2$. If $n < p_k^2$ and $n + 2 \geq p_{k+1}^2$, then $n < p_k^2 + 2 \leq p_{k+1}^2$:

```
if n ≥ square then begin
| k := k + 1 ;  square := p[k]²
end
```

• It remains only to assemble the fragments of code (Fig. 4.4). But this can't be done too mechanically because the variables k and *square* make reference to an *old value*, so that the initialization must be imposed outside the principal loop.

```
p[1] = 2 ;  p[2] = 3 ;  p[3] = 5 ;  p[4] = 7 ;
n := 7 ;  k := 3 ;  square := 25 ;
for ℓ := 5 to N do begin
| repeat
| | n := n + 2 ;
| | if n ≥ square then begin k := k + 1 ;  square := p[k]² end ;
| | i := 2 ;  prime := true ;
| | while (i < k) and prime do
| |       if n mod p[i] = 0 then prime := false else i := i + 1
| until prime ;
| p[ℓ] := n
end
```

Fig. 4.4. Search and storage of the first N prime numbers

Remark

The method that we we have just practiced will be reprised and amply commented on in Chapter 7.

4.4.3. Last recommendations

We have just presented three techniques which frequently facilitate the writing and development of an algorithm.

• Experience shows that they tend to be used simultaneously. For example, the preceding section began by recycling the code (the first technique) for factoring an integer. We introduce recurrent sequences (the second technique) and provisionally mask the internal loops by sentences (the third technique).

• In closing, we remark on a trap which one should avoid. It is – *too easy* – to create a one or two page monster by nesting too many little alorithms. The result is *illegible*, hence difficult to control. If one of the subalgorithms is defective, we must throw out the whole chain of deductions because the results are so interwoven. The same holds if we want to modify the algorithm.

An algorithm must be, insofar as possible, a brief text. One does not really know very well what is going on beyond a dozen lines.... Divide your task into independent subalgorithms which can be treated as procedures or distinct functions when your code becomes too long.

4.5. How to Prove an Algorithm

To show that an algorithm is correct we must assure ourselves of the following:

- that it never crashes;
- that there are no infinite loops;
- that it always furnishes the desired result.

4.5.1. Crashes

The list is endless! In general, the program crashes when it tries an impossible operation. The most classical cases are the following.

- Division by zero or, more generally, calling a function or a procedure with an inappropriate value of the parameter (for example, calling *sqrt*(-1)).
- Attempting to access an object which does not exist, for example[10], the element $t[0]$ or $t[n+1]$ in the array $t[1\,..\,n]$.

4.5.2. Infinite loops

We clearly do not have to worry about "for" loops not ending. On the contrary, "while" and "repeat" loops are often a cause for worry: how can one be sure that they will terminate? The most celebrated loop is, without doubt, the "$3n+1$ loop":

$$\textbf{while } n > 1 \textbf{ do } n := T(n),$$

where $n > 1$ is a given integer and $T : \mathbb{N} \to \mathbb{N}$ is the function

$$T(n) = \begin{cases} \frac{1}{2}n & \text{if } n \text{ is even,} \\ \frac{1}{2}(3n+1) & \text{if } n \text{ is odd.} \end{cases}$$

It is still not known[11] whether this loop terminates for any value of n!

Most of the time, one shows that a loop terminates by using the technique of *infinite descent*. This technique was developed by Fermat and is based on the following innocuous remark:

[10] The R+ option (*range error*) on the compiler allows one to detect this type of error.

[11] To get an idea of the irregularity of the values of n calculated by this loop, show by induction that one has $T^k(2^k n - 1) = 3^k n - 1$ for all $k, n > 0$.

There does not exist a strictly decreasing infinite sequence in $\mathbb{N}$.

We try to find an $\mathbb{N}$-valued function whose values, when calculated at the moment of the exit test of the loop, are strictly decreasing. If we can find such a function, it is clear that there cannot be an infinite loop. Here are two examples.

4.5.3. Calculating the GCD of two numbers

Let a and b be two integers > 0. Consider the loop

while $(a > 0)$ **and** $(b > 0)$ **do** $\leftarrow$ *evaluate $f(a, b)$ here*
  if $a \geq b$ **then** $a := a - b$ **else** $b := b - a$

Here, it is clear that the function $f(a, b) = a + b \geq 0$ is strictly decreasing on each passage through the loop. As result, the loop must terminate after finite time.

4.5.4. A more complicated example

Suppose that we are given integers $t_1, \ldots, t_n > 0$ and that $L \subset [\![1, n]\!]$ is a nonempty set of integers to start. Consider the algorithm

while $L \neq \emptyset$ **do begin** $\leftarrow$ *evaluate the function g here*
  | «*withdraw any element ℓ from L* » ;
  | «*calculate $t > 0$ and $k \in [\![1, n]\!]$ using ℓ (it doesn't matter how)*» ;
  | **if** $t_k > t$ **then begin** $t_k := t$; «*add k to L*» **end**
end

Here the function Card L does not suffice because, when one withdraws an element ℓ, one can add k to L which does not decrease Card L. One must also use the t_i, which leads very naturally to the function

$$g = \operatorname{Card} L + \sum_{i=1}^{n} t_i \, .$$

We check what happens to its value on passing through the loop. Let L' and $t'_1, \ldots, t'_n$ denote the new values of the parameters after passage through the loop.

- If one does not add k to L, we know that the t_i are intact. Thus, $\operatorname{Card} L' = \operatorname{Card} L - 1$ and $t_1^i = t_i$ which shows that the value of g diminishes by 1.
- If one adds k to L, then $t'_k < t_k$ and $t'_j = t_j$ for $j \neq k$. Consequently, $\operatorname{Card} L' = \operatorname{Card} L$ and $\sum t'_i \leq \sum t_i - 1$ which shows that the value of g diminishes by at least 1.

As a result, the loop terminates after a finite time, which is not evident at first glance.

Remark

When $\mathbb{N}$ does not suffice, one can use the technique of infinite descent on any well-ordered set (for example, $\mathbb{N}^k$ with the lexicographic order).

4.5.5. The validity of a result furnished by a loop

We use a variant of reasoning by induction.

Definition 4.5.1. *Suppose that f is a function that we evaluate at the same time as the exit test from a loop. If we can show that these values are equal to those taken by the function just before the entry into the loop, we will say that f is an invariant of the loop.*

We demonstrate the use of loop invariants using two simple examples.

Calculating a sum

We want to calculate the sum $S = \sum_{i=1}^{n} u_i$. Consider the two loops:

	$S := 0$; $i := 0$;		$S := 0$; $i := 1$;
evaluate ⟶ *the invariant at this point*	**while** $i < n$ **do begin**	⟶	**while** $i \leq n$ **do begin**
	$i := i + 1$;		$S := S + u_i$;
	$S := S + u_i$		$i := i + 1$
	end		**end**

An invariant of the left hand loop is the function

$$f = S - \sum_{k=1}^{i} u_k.$$

In effect, when we present ourselves for the first time at the entry of the loop, we have $S = i = 0$, whence $f = 0$, in view of the usual conventions regarding sums on an empty set of indices. Now, suppose that at a given moment we have $f = 0$, and consider the new value f' of f when we come to the exit test after having executed the body of the loop. If S, i are the old values and S', i' the new, we clearly have $S' = S + u_{i+1}$ and $i' = i + 1$ and, hence, the implication

$$S = \sum_{k=1}^{i} u_k \implies S' = \sum_{k=1}^{i'} u_k.$$

Knowing that f continues to take the value 0, we still have $f = 0$ when we leave the loop. At this moment, $i = n$ which gives $S = \sum_{k=1}^{n} u_k$, and the algorithm calculates the sum correctly.

Similarly, one shows that the function $g = S - \sum_{k=1}^{i-1} u_k$ is an invariant of the right hand loop which always equals 0.

Remark

Recall that a "for" loop is an abbreviation for a "while" loop similar to the right hand loop in which the control variable is incremented at the end of the loop. Thus, the invariant should be evaluated at the entrance to the loop.

Calculation of the GCD of two numbers

We want to calculate the greatest common divisor of $a, b > 0$ using the algorithm

```
«evaluate the invariant just before entry to the loop»
repeat
| if a ≥ b then a := a − b else b := b − a
until (a = 0) or (b = 0) ;  ← evaluate the invariant here
if b = 0 then GCD := a else GCD := b
```

Since we are dealing with a "repeat" loop, which moves the test, we must also calculate the value of the invariant just before the entrance to the loop.

Consider the function $f = \text{GCD}(a, b)$. Its properties show immediately that it is an invariant. Consequently, if a_ω and b_ω are the final values taken by the variables a and b, we have

$$\text{GCD}(a_\alpha, b_\alpha) = \text{GCD}(a_\omega, b_\omega) = \begin{cases} a_\omega & \text{if } b_\omega = 0, \\ b_\omega & \text{if not.} \end{cases}$$

Thus, the algorithm correctly calculates the GCD.

4.6. Solutions of the Exercises

Exercise 1

In view of the symmetry of the unknowns, we can content ourselves with listing the solutions of the equation

$$n = x^2 + y^2 + z^2, \quad 0 \leq x \leq y \leq z.$$

We certainly have $x \leq (\frac{1}{3}n)^{1/2}$ and $y \leq (\frac{1}{2}n)^{1/2}$. We do not seek to majorize z because we can do better: as soon as $x^2 + y^2 + z^2$ exceeds n, it is useless to continue to increment z; thus we interrupt the loop which handles z using an "interrupter" *next_y*.

```
for n := 0 to N do begin
| n_x := n div 3 ; x_max := 0 ;
| while sqr(x_max + 1) ≤ n_x do x_max := x_max + 1 ;
| n_y := n div 2 ; y_max := 0 ;
| while sqr(y_max + 1) ≤ n_y do y_max := y_max + 1 ;
```

```
|sum_three_squares := false ; x := −1 ;
|while not sum_three_squares and (x ≤ x_max) do begin
||x := x + 1 ; y := x − 1 ;
||while not sum_three_squares and (y ≤ y_max) do begin
|||y := y + 1 ; z := y − 1 ; next_y := false ;
|||while not sum_three_squares and not next_y do begin
||||z := z + 1 ;
||||S := sqr(x) + sqr(y) + sqr(z) ;
||||if S = n then sum_three_squares := true else
||||if S > n then next_y := true ;
|||end
||end
|end ;
|if sum_three_squares then writeln(x, y, z)
end
```

Exercise 2

Let $\mathbb{E}_N$ be the set of integers $n \leq N$ which are of the form $n = x^2 + y^2$. Cover $\mathbb{E}_N$ with the subsets $C_\ell = \{\ell^2 + y^2; y \geq \ell\}$. If we choose the index *k_max* so that $N < \min C_{k\_max+1}$, we can be certain that the covering is *finite*:

$$\mathbb{E}_N \subset C_0 \cup \ldots \cup C_{k\_max}.$$

We are thus reduced to the preceding technique. Because there are many sets C_ℓ, we use an array instead of independent variables to store the integers that have still not been written.

If we define $y[\ell] \geq \ell$ by the property "the least element not already written from C_ℓ is $x_\ell = \ell^2 + y[\ell]^2$", then it is clear that the element that we should write is $x = \min(x_0, \ldots, x_{k\_max})$. Once x is written, we need to remember to increment $y[\ell]$ each time that $x = x_\ell$.

The algorithm then consists simply of bringing to light $y[0 \ldots k\_max]$; the function $\ell = where\_is(x)$ returns an index ℓ which satisfies $x \in C_\ell$. Since ℓ is not always unique, the auxiliary variable *old_x* avoids repeated listings of the same number.

```
old_x := −1 ; x := 0 ; ℓ := 0 ; k_max := 0 ;
while 2 * sqr(k_max + 1) ≤ N do k_max := k_max + 1 ;
for k := 0 to k_max do y[k] := k ;
while x ≤ N do begin
|ℓ := where_is(x) ;
|if old_x < x then begin writeln(x, ℓ, y[ℓ]) ; old_x := x end ;
|y[ℓ] := y[ℓ] + 1 ;
|x := min(y, k_max) {i.e. the minimum of k² + y_k², k = 0, ..., k_max}
end
```

One proceeds similarly with the sum of the two cubes.

5. Algorithms and Classical Constructions

You must learn the material in this chapter by heart, because we will encounter it frequently in many different forms.

5.1. Exchanging the Contents of Two Variables

To exchange the contents of two variables, a beginner will often suggest $y := x$; $x := y$. This is incorrect because the value of y is destroyed by the assignment $y := x$. Therefore, we safeguard the value in a temporary variable:

$$temp := x \ ; \ \ x := y \ ; \ \ y := temp$$

Along the same lines, suppose that we try to encode a planar iteration as follows:

$$(x, y) := \big(u(x, y), v(x, y)\big).$$

If we naively write $x := u(x, y)$; $y := v(x, y)$ we make the same error because the value of x used in the satement $y := v(x, y)$ is destroyed by the first assignment. Once again, we need to safeguard the value of x in an auxiliary variable:

$$temp := x \ ; \ \ x := u(temp, y) \ ; \ \ y := v(temp, y)$$

(We have already encountered this difficulty in dissecting the CORDIC algo-rhm in Chapter 4.)

Exercise 1 *(Solution at the end of the chapter)*

Let A be a 2×2 matrix. Translate the statement

$$\begin{pmatrix} x \\ y \end{pmatrix} := A \begin{pmatrix} x \\ y \end{pmatrix}$$

into code.

5.2. Diverse Sums

5.2.1. A very important convention

Apprentice mathematicians often protest when one writes the formula

$$S = a_0 + a_1 M_0 + a_2 M_0 M_1 + \cdots + a_n M_0 M_1 \cdots M_{n-1}$$

in the condensed form

$$S = \sum_{k=0}^{n} a_k \prod_{i=0}^{k-1} M_i .$$

They will argue that $\prod_{i=0}^{k-1}$ does not mean anything when $k = 0$ and that you should write

$$S_k = a_0 + \sum_{k=1}^{n} a_k \prod_{i=0}^{k-1} M_i .$$

The initial way of writing the sum is nevertheless correct thanks to a very useful convention which is vital to assimilate. To explain it, suppose that I is a *finite* set. Then the associativity and commutativity of addition and multiplication allow one to define the symbols

$$\sum_{i\in I} u_i \quad \text{and} \quad \prod_{i\in I} u_i .$$

Consider the partition $I = I_1 \cup I_2$ of I into two disjoint subsets. We can write

$$\sum_{i\in I} u_i = \sum_{i\in I_1} u_i + \sum_{i\in I_2} u_i \qquad (I_1 \cap I_2 = \emptyset).$$

When $I = I_1$ and $I_2 = \emptyset$, this formula becomes

$$\sum_{i\in I} u_i = \sum_{i\in I} u_i + \sum_{i\in\emptyset} u_i .$$

If we wish to avoid interminable (and parasitic) discussions in our proofs, it is necessary to accept "sums or products over the empty set" and to adopt the conventions

$$\sum_{i\in\emptyset} u_i = 0 \quad \text{and} \quad \prod_{i\in\emptyset} u_i = 1.$$

When we write $\prod_{i=0}^{k-1} u_i$, we understand $\prod_{i\in I} u_i$ where

$$I = \{ i \in \mathbb{Z};\ \ i \geq 0 \ \text{ and } i \leq k-1 \}.$$

When $k = 0$, the index set I is empty, so that the formula $\prod_{i=0}^{-1} u_i = 1$ with which we began this discussion is correct!

It is important to respect the conventions employed by mathematicians and to adapt one's code as a consequence. The code on the left, which calculates

$$S = \sum_{i=1}^{n} u_i \quad \text{and} \quad P = \prod_{i=1}^{n} u_i,$$

respects these conventions, while that on the right does not.

$S := 0$; $\leftarrow$ *good*	$S := u[1]$; $\leftarrow$ *very bad*
for $i := 1$ **to** n **do** $S := S + u[i]$;	**for** $i := 2$ **to** n **do** $S := S + u[i]$;
$P := 1$;	$P := u[1]$;
for $i := 1$ **to** n **do** $P := P * u[i]$	**for** $i := 2$ **to** n **do** $P := P * u[i]$
Correct code	*Incorrect code*

5.2.2. Double sums

To calculate the double sum

$$S := \sum_{i=1}^{p} \sum_{j=1}^{q} u_{i,j},$$

we can run through the matrix line after line starting with the first (the so-called *television scanning* order, after the way the pixels on a television screen are refreshed). Since the control variable is the couple (i, j), a translation with the aid of a first order recurrence must have the form

$$S_{(i',j')} := S_{(i,j)} + u_{i',j'},$$

where (i', j') is the successor of (i, j) with respect to the television scanning order; that is, the *lexicographic order*. But this order is easily realized by nesting two "for" loops:

for $i := 1$ **to** n **do**
    for $j := 1$ **to** n **do** ...

If we take the precaution of beginning with $S = 0$ in order that the cases $p = 0$ and $q = 0$ are correctly treated, the algorithm is

$S := 0$;
for $i := 1$ **to** n **do**
    for $j := 1$ **to** n **do**
        $S := S + u[i, j]$

We will generalize this to run through any product set using the lexicographic order.

Exercise 2

Calculate this sum using the boustrophedon order.

5.2.3. Sums with exceptions

Suppose we want to calculate the sum with an "exception":

$$S = \sum_{\substack{0 \le i \le n \\ i \ne k}} u_i \qquad (k \text{ is an integer}).$$

• *First solution*: We run through the integers in the interval $[0, n]$ being careful not to add a_k:

$$S_i = \begin{cases} S_{i-1} + u_i & \text{if } i \neq k, \\ S_{i-1} & \text{otherwise} \end{cases} \quad \Longrightarrow \quad \begin{cases} S := 0 \,; \\ \textbf{for } i := 0 \textbf{ to } n \textbf{ do} \\ \textbf{if } i \neq n \textbf{ then } S := S + u[i] \end{cases}$$

This solution is very clear, hence very certain.

• *Second solution*: We can calculate the sum of all the u_i, then cut out u_k when it is necessary:

```
S := 0 ;
for i := 0 to n do S := S + u[i] ;
if (0 ≤ k) and (k ≤ n) then S := S − u[k]
```

This solution is more doubtful. When one works over the integers, the computer will give the correct result; on the contrary, over the reals $a + x - a$ is only *approximately* equal to x. We have increased the speed (there is only one test), but at the cost of precision.

• *Third solution*: We set u_k equal to zero, then calculate the sum of all the u_i. But we must remember to *re-establish* the initial value of the u_k, because the code that we write could very well wind up being inserted in a program which might perhaps need u_k:

```
if (0 ≤ k) and (k ≤ n) then begin temp := u[k] ;  u[k] := 0 end
S := 0 ;
for i := 0 to n do S := S + u[i] ;
if (0 ≤ k) and (k ≤ n) then u[k] := temp
```

This solution is more rapid than the first (there are fewer tests). But it is long and uses a trick which is not within reach of a beginner.

• *Fourth solution*: We calculate the subsums corresponding to the intervals $[\![0, k-1]\!]$ and $[\![k+1, n]\!]$:

```
S := 0 ;
if (0 ≤ k) and (k ≤ n) then begin
  | for i := 0 to k − 1 do S := S + u[i] ;
  | for i := k + 1 to n do S := S + u[i] ;
end
else for i := 0 to n do S := S + u[i]
```

This is correct, but too long to write.

For beginners

The best code is *always* the simplest: the first solution is preferable. When you program, never forget that the first priority is a code that functions the first time. If it is too slow, you can always improve the critical parts of the code. It serves no purpose to write rapid and clever code which is very difficult to debug. Which would you prefer: to refine an exact code to speed it up or make it more elegant or to lose hours (or perhaps days) debugging a program which obstinately refuses to function?

5.3. Searching for a Maximum

Let $x_1, \ldots, x_n$ be a vector of real numbers > 0. To find the largest of the x_i, it suffices to compare x_1 and x_2, and to retain the largest, then to compare this to x_3, etc. We can use this procedure to store the position of the maximum

```
max := x[1] ; place_max := 1 ;
for i := 2 to n do
    if x[i] > max then begin max := x[i] ; place_max := i end
```

The problem becomes more complicated when we need to find the maximum of a sub-family of the x_i satisfying a given property P. In fact, the preceding code is incorrect if $P(x_1)$ is not true.

• If we know that the x_i are greater than 0, for example, we give the maximum a value smaller than all the x_i (for example $max = 0$) and we begin with $i = 1$:

```
max := 0 ;
for i := 1 to n do
    if (x[i] > max) and P(x[i]) then
       begin max := x[i] ; place_max := i end
```

If *max* retains the initial value 0, then we know that the set of x_i which satisfy the property P is empty; in this case, *place_max* means nothing (the value is indefinite).

• If we do not have any hypotheses on the x_i, the safest course is to make a preliminary reconnaisance.

```
max_exists := false ;
for i := 1 to n do
if P(x[i]) then begin
| max_exists := true ;  max := x[i]
end ;
if max_exists then for i := 1 to n do
if (x[i] > max) and P(x[i]) then begin
| max := x[i] ; place_max := i
end
```

If you find this code too "industrial" and not very intelligent, reread the end of the preceding section ....

5.4. Solving a Triangular Cramer System

Consider the triangular Cramer linear system where the diagonal coefficients $a_{i,i}$ are all different from zero.

$$\begin{cases} a_{1,1}x_1 + a_{1,2}x_2 + \cdots + a_{1,n}x_n = b_1, \\ \qquad\quad a_{2,2}x_2 + \cdots + a_{2,n}x_n = b_2, \\ \qquad\qquad\qquad \ddots \qquad\quad \vdots \\ \qquad\qquad\qquad\qquad a_{n,n}x_n = b_n, \end{cases}$$

To solve this system, beginners tend to think in terms of *formulas* (that is, they think in a static manner). They write

$$x_n = \frac{b_n}{a_{n,n}}, \tag{5.1}$$

$$x_{n-1} = \frac{1}{a_{n-1,n-1}}(b_{n-1} - a_{n,n}x_n)$$

$$= \frac{1}{a_{n-1,n-1}}\left(b_{n-1} - a_{n-1,n}\frac{b_n}{a_{n,n}}\right), \tag{5.2}$$

$$x_{n-2} = \frac{1}{a_{n-2,n-2}}(b_{n-2} - a_{n-2,n-1}x_{n-1} - a_{n-2,n}x_n), \tag{5.3}$$

then give up, because x_{n-2} is much too complicated a function of the $a_{i,j}$ and the b_i. But do we really need a formula to create an algorithm?

Think dynamically: (5.1) allows us to calculate the *value* of x_n, then (5.2) that of x_{n-2}, etc. If we know the values of $x_n, \ldots, x_{\ell+1}$, we can find the *value* of x_ℓ thanks to the formula:

$$x_\ell = \frac{1}{a_{\ell,\ell}}(b_\ell - a_{\ell,\ell+1}x_{\ell+1} - \cdots - a_{\ell,n}x_n). \tag{5.4}$$

Thus, a first draft of an algorithm to solve the system might be:

```
x[n] := b[n]/a[n] ;
for ℓ := n − 1 downto 1 do
    «calculate x[ℓ] using(5.4)»
```

To obtain x_ℓ we calculate a sum:

```
temp := 0 ;
for j := ℓ + 1 to n do temp := temp + a[ℓ, j] ∗ x[j] ;
x[ℓ] := (b[ℓ] − temp)/a[ℓ, ℓ]
```

Inserting this fragment of code into the preceding code gives

```
x[n] := b[n]/a[n] ;
for ℓ := n − 1 downto 1 do begin
| temp := 0 ;
| for j := ℓ + 1 to n do temp := temp + a[ℓ, j] * x[j] ;
| x[ℓ] := (b[ℓ] − temp)/a[ℓ, ℓ]
end
```

Exercise 3

Explain why one can, without danger, incorporate the first line into the principal loop by letting ℓ vary from n to 1:

```
for ℓ := n downto 1 do begin
| temp := 0 ;
| for j := ℓ + 1 to n do temp := temp + a[ℓ, j] * x[j] ;
| x[ℓ] := (b[ℓ] − temp)/a[ℓ, ℓ]
end
```

5.5. Rapid Calculation of Powers

Suppose that we wish to calculate

$$x^{59} = \underbrace{x * x * x * x * \cdots * x * x * x}_{\text{58 multiplications}}.$$

It is possible to do better because $x^{59} = x * (x^2)^{29}$ which shows that 30 multiplications will suffice at the cost of storing $y = x^2$:

$$y := x * x; \quad x^{59} = x * \underbrace{y * y * y \cdots y * y * y}_{\text{28 multiplications}}.$$

But, we can do the same thing again and introduce $z = y^2$. Then $x^{59} = x*y*z^{14}$ shows that 17 multiplications suffice:

$$y := x * x; \quad z := y * y; \quad x^{59} = x * y * \underbrace{z * z * z \cdots z * z * z}_{\text{13 multiplications}}.$$

Systematizing with the help of a formula,

$$u^n * v = \begin{cases} (u^2)^{n/2} * v & \text{if } n \text{ is even,} \\ (u^2)^{(n-1)/2} * (u * v) & \text{otherwise.} \end{cases}$$

(The factor v is indispensible because the exponents are not always even.) We can then calculate x^{59} with only 10 multiplications:

$$\begin{array}{lll}
n_0 = 59 & u_0 = x & v_0 = 1 \\
n_1 = 29 & u_1 = u_0^2 & v_1 = u_0 v_0 \\
n_2 = 14 & u_2 = u_1^2 & v_2 = u_1 \\
n_3 = 7 & u_3 = u_2^2 & v_3 = u_2 v_2 \\
n_4 = 3 & u_4 = u_5^2 & v_4 = u_3 v_3 \\
n_5 = 1 & u_5 = u_4^2 & v_5 = u_4 v_4 \\
\text{result} = u_5 v_5 & &
\end{array}$$

The transformation into an algorihm is immediate:

```
u := x ; v := 1 ;
repeat
| if n mod 2 = 1 then v := u * v ;
| u := u * u ; n := n div 2
until n = 1 ;
result := u * v
```

Exercise 4 *(Solution at end of chapter)*

Justify this algorithm and show that the number of multiplications is $O(\log_2 n)$. To do this, one shows that $u^n * v$ is an invariant of the loop and notes that n diminishes by half after each passage through the loop.

5.6. Calculation of the Fibonacci Numbers

How can one calculate the N-th Fibonacci number? We have learned how to transform a recurrence relation of order 1 into an algorithm. Annoyingly, however, the Fibonnaci series is defined by a second order recurrence relation.

Mathematicians have known for a long time that one can *arbitrarily decrease the order of a recurrence relation or a differential equation by working in a larger dimensional space.* In our case, we lower the order by putting:

$$X_n = \begin{pmatrix} F_n \\ F_{n-1} \end{pmatrix}.$$

Then, by replacing F_n by $F_{n-1} + F_{n-2}$, we can write

$$X_n = \begin{pmatrix} F_n \\ F_{n-1} \end{pmatrix} = \begin{pmatrix} F_{n-1} + F_{n-2} \\ F_{n-1} \end{pmatrix} = \begin{pmatrix} 1 & 1 \\ 1 & 0 \end{pmatrix} \begin{pmatrix} F_{n-1} \\ F_{n-2} \end{pmatrix},$$

which is a first order recurrence relation:

$$X_n = AX_{n-1}, \quad A = \begin{pmatrix} 1 & 1 \\ 1 & 0 \end{pmatrix}, \quad X_1 = \begin{pmatrix} F_1 \\ F_0 \end{pmatrix} = \begin{pmatrix} 1 \\ 0 \end{pmatrix}.$$

The transformation to an algorithm is a *piece of cake*[1] once one has solved and understood Exercise 1. If the coordinates of the vector X are x and y, the N-th Fibonacci number is calculated as follows:

```
x := 1 ; y := 0 ;
for i := 1 to N do begin
| temp := x + y ; y := x ; x := temp
end ;
F_N := x
```

Exercise 5 *(Solution at end of the chapter)*

Write a program to rapidly calculate the N-th Fibonacci number by raising the matrix $A = \begin{pmatrix} 1 & 1 \\ 1 & 0 \end{pmatrix}$ to the N-th power.

5.7. The Notion of a Stack

Computer scientists love *stacks*. The picture you should have in mind is that of a stack of plates. You can do the following to a stack.

- You can *put* a new plate on top of the stack – this is often call the *push* operation.
- You can *remove* the top plate from the stack – this is often called the *pop* operation.

A stack is manipulated with the aid of a number of very simple primitives:

- *push*(*stack*, x) puts the object x on top of the stack;
- x = *pop*(*stack*) removes the top element from the stack and stores its value in the variable x;
- *empty*(*stack*) prepares an empty stack: that is, one without plates;
- *is_full*(*stack*) is a boolean which informs us that the stack is saturated,
- *is_empty*(*stack*) is a boolean which tells us that the stack is empty.

The Pascal procedures that follow and that *implement* a stack will probably surprise you and appear unnecessarily ponderous. However, professional programmers abide these gymnastics because they know from experience that this is the price one pays for maximum security (that is, a code that is easy to adjust, that functions the first time and that is independent of the rest of the program).

In this book, we will implement our stacks using arrays:

[1] This phrase means that there is no difficulty: it suffices to let one's hand do the writing.

```
const ht_max_stack = 100 ;
type stack_integers = record
  | top : integer ; plate : array[1..ht_max_stack] of integer
end ;
var stack : stack_integers ;
```

The variable *top* is the *height* of the stack.

```
procedure push(var stack : stack_integers ; x : integer) ;
begin
  | if is_full(stack) then writeln('full stack')
  | else with stack do begin
  |   | top := top + 1 ; plate[top] := x
  | end
end ;

function pop(var stack : stack_integers) : integer ;
begin
  | if is_empty(stack) then writeln('empty stack')
  | else with stack do begin
  |   | pop := plate[top] ; top := top − 1
  | end
end ;

procedure empty(var stack : stack_integers) ;
begin
  | stack.top := 0
end ;

function is_full(var stack : stack_integers) ;
begin
  | if stack.top = ht_max_stack
  | then is_full := true
  | else is_full := false
end ;

function is_empty(var stack : stack_integers) ;
begin
  | if stack.top = 0
  | then is_empty := true
  | else is_empty := false
end ;
```

This manner of proceeding is very efficient: one lays out a set of statements (with tests attached) which one can then forget.[2] It suffices to use the primitives: no error is possible, and there is no need to remember how the stack is implemented.

[2] A good program is made up of independent modules. Remember: one cannot climb stairs while chewing gum.

Another advantage is the independence of this module: if we need to modify the implementation of our stack, it suffies to redefine the type *stack* and to rewrite the primitives. We do not need to touch the rest of the program.

For beginners

One should not think that using an array is the only way of implementing a stack. For example:

- If we wanted to stack characters, we could consider stacks consisting of strings of characters.
- If we want to stack integers between 0 and $b-1$, it may be useful to consider the stack as an *integer* n and the plates as numerals in base b. To push (that is, pile on) the number c we replace n by $b*n+c$; to pop (that is, remove) the number, we divide n by the base b.
- Very frequently, one uses *cells* and *pointers*.

5.8. Linear Traversal of a Finite Set

Some problems require that one linearly traverse a finite set E; mathematically speaking, this amounts to giving E a total order. The traverse is realized by repeated calls to the successor function.

```
x := first_element(E) ;          x := first_element(E) ; finish := false ;
repeat                           while not finish do begin
| ...                            | ...
| successor(x, finish)           | successor(x, finish)
until finish                     end
```

If the set E possesses at least one element, the two codes are equivalent. The boolean *finish* permits one to interrupt the loop at the appropriate moment. The procedure *successor* assigns to x the successor of x when this is not the greatest element of E; otherwise, it gives *finish* the value *true*.

For beginners

It is not particularly necessary to put the motor *successor*$(x, finish)$ at the beginning of the loop. Why?

Exercise 6

What is the boutrophedon order good for? Not a whole lot! There is, however, an interesting instance where its use is natural. Let $n = p^\alpha q^\beta$ be the decomposition of n into prime factors (and suppose that we know p, q, α, β explicitly). Imagine that we want to write the $(\alpha+1)(\beta+1)$ possible divisors of n. Writing the divisors d of n in single file amounts to using a total order. But which total order should we use?

• *First solution:* we can use the natural order on $\mathbb{N}$ and run through the integers in the interval $[1, n]$.

```
for d := 1 to n do
    if n mod d = 0 then writeln(d)
```

• *Second solution:* to be given a divisor d of n is to be given a pair (i, j) satisfying $d = p^i q^j$. We can order the pairs lexicographically, which leads to two nested "for" loops

```
for i := 0 to α do
    for j := 0 to β do
        write(power(p, i) * power(q, j))
```

• *Third solution:* we can order the pairs (i, j) with the boustrephedon order, which gives the list

$$1, p, p^2, \ldots, p^\alpha, p^\alpha q, p^{\alpha-1}q, \ldots, q, q^2, pq^2, \ldots, p^\alpha q^2, p^{\alpha-1}q^2, \ldots$$

where we pass from one divisor to the following with a single operation (multiplication or division).

What to conclude?

• The first solution is the simplest (and, therefore, the most robust). This is what we should think of first.

• The second solution is a Penelope code because when we pass from $d = p^3q^2$ to $d' = p^4q^2$, we forget that we have already calculated p^3 and q^2; it also generalizes poorly when n has more than two prime divisors.

• The third solution is the most delicate to implement. Nevertheless, it is the most attractive because it generalizes to the case of any integer n and because it is the most rapid.

5.9. The Lexicographic Order

5.9.1. Words of fixed length

Consider the set $\mathcal{M}_n$ of words with n letters (or n numerals). We have known since we were children how to arrange these words using the lexicographic order.

Two embedded "for" loops realise this order on pairs of integers and allow us to linearly traverse the set $\mathcal{M}_2$. If we want to linearly traverse the set $\mathcal{M}_3$, we use three nested "for" loops. But would we find it acceptable to nest one hundred "for" loops to traverse $\mathcal{M}_{100}$? Would our compiler support it? Even more seriously, if we don't know the value of n when we are entering the code, we are helpless.

Let us return to our problem and ask ourseles how we find the successor of a word w given in a dictionary. To simplify, suppose that our "words" have five letters.

• The successor of the word "*amies*" is the word "*amiet*", which is obtained by *augmenting* the last letter of the first word; that is, by replacing the letter "*s*" by the following letter "*t*".

• If the word is "*rasez*", we cannot augment the last letter; we augment the second to last and the word that follows is "*rasfa*".

• Similarly, the successor of "*buzzz*" is "*bvaaa*".

Thus, the algorithm that we use is the following:

• The word "*zzzzz*" does not have a successor.

• Otherwise, the word w contains a letter different from "*z*", and we seek the largest index k such that

$$w = w_1 \cdots w_k\, zz \cdots z, \qquad w_k < z.$$

If w'_k is the letter following w_k, the desired successor is the word:

$$w_1 \cdots w_{k-1} w'_k\, aa \cdots a.$$

More precisely, we can calculate the successor of the word w using the following code:

```
finish := false ; k := −1 ;
for i := 1 to n do if w[i] < 'z' then k := i ;
if k = −1 then finish := true
else begin
    w[k] := next_letter(w[k]) ;
    for i := k + 1 to n do w[i] := 'a'
end
```

Generalization

The generalization to other types of words is immediate. Consider, for example, the set of n-tuples of integers:

$$\mathcal{M}_n = [\![1, max_1]\!] \times [\![1, max_2]\!] \times \cdots \times [\![1, max_n]\!] \subset \mathbb{N}^n.$$

Since each interval $[1, max_i]$ is totally ordered, we can endow the product set $\mathcal{M}_n$ with the associated lexicographic order.

This amounts to considering an element $x = (x_1, \ldots, x_n)$ of $\mathcal{M}_n$ as a word whose letters are $x_1, \ldots, x_n$. The smallest word is $(1, \ldots, 1)$, the greatest is

$(max_1, \ldots, max_n)$, and the passage from x to its successor x' is effected as follows:

```
finish := false ; k := −1 ;
for i := 1 to n do if x[i] < max[i] then k := i ;
if k = −1 then finish := true
else begin
    x[k] := x[k] + 1 ;
    for i := k + 1 to n do x[i] := 1
end
```

5.9.2. Words of variable length

Consider now the set $\mathcal{N}_n$ of words containing at most n letters, and let us explore how one computes the successor of a given word:

- the successor of the word w = "*raz*" is the word w' = "*raza*",
- the successor of the word w = "*mise*" is the word w' = "*misea*",
- the successor of the word w = "*misez*" is the word w' = "*misf*",
- the successor of the word w = "*buzzz*" is the word w' = "*bv*",

The process is analogous to the preceding algorithm:

- If the word contains less than n letters, we follow it by the letter "*a*".
- If the word contains n letters, we try to augment the last letter. If this is not possible, we remove all the "*z*"'s at the end; if the word that remains is not empty (that is, $w \neq$ "$zz \cdots z$"), we augment its last letter.

A linear traverse of the set $\mathcal{N}_n$ is accomplished using a stack of characters.

```
finish := false ; empty(stack) ;
push(stack, 'a') ; {because the smallest word is w = 'a'}
repeat
    if not is_full(stack)
    then push(stack, 'a')
    else begin
        while not is_empty(stack) and (topval(stack) = 'z') do
        garbage := pop(stack) ;
        if is_empty(stack)
        then finish := true
        else begin
            character := pop(stack) ;
            push(stack, next_letter(character))
        end
    end
until finish ;
```

Here, we need a supplementary primitive, the function *topval* which returns the value of the plate at the top of the pile without *modifying* it. (If this primitive is not available, one can simulate it by popping the top plate, then pushing it back.) Note also, the use of the auxiliary variable *garbage*, which allows us to get rid of useless plates.

Exercices 7

1) For every integer $n \geq 1$, one has at least one equality of the form

$$n = \pm 1^2 \pm 2^2 \pm \cdots \pm k^2, \tag{5.5}$$

where the integer k depends on n.

Proof. The assertion is true for $n = 1, 2, 3, 4$ since

$$1 = 1^2, \qquad 2 = -1^2 - 2^2 - 3^2 + 4^2,$$
$$3 = -1^2 + 2^2, \; 4 = -1^2 - 2^2 + 3^2.$$

We continue using strong induction. Let $n > 4$ and suppose that the property is true for all integers $< n$. By induction, we can write

$$n - 4 = \pm 1^2 \pm 2^2 \pm \cdots \pm k^2.$$

Moreover, we always have

$$4 = (k+1)^2 - (k+2)^2 - (k+3)^2 + (k+4)^2,$$

which immediately gives

$$n = \pm 1^2 \pm 2^2 \pm \cdots \pm k^2 + (k+1)^2 - (k+2)^2 - (k+3)^2 + (k+4)^2.$$

Write a program which writes *all* decompositions (5.5) of an integer n with k as small as possible.

2) Find all ways of placing eight queens on a chessboard so that no one can take any other.

5.10. Solutions to the Exercises

Exercise 1

Let $A = \begin{pmatrix} a & b \\ c & d \end{pmatrix}$. The desired translation is then:

$$\begin{aligned} &temp := a * x + b * y \; ; \\ &y := c * x + d * y \; ; \\ &x := temp \end{aligned}$$

Exercise 4

Let n_0 be the initial value of n and $n_1, n_2, \ldots, n_\omega$ the values following n. Since $n_{k+1} \le \frac{1}{2}n_k$, the function $f(n) = n$ strictly decreases each time through the loop. It follows that this algorithm never loops. From the inequalites, one deduces that $n_\omega \le 2^{-\omega}n_0$. Since $n_\omega = 1$, the number of passages through the loop satisfies $\omega \le \log_2 n_0$.

Let n', u' and v' be the values taken by n, u and v after a passage through the loop. If $n = 2m$ is even, we have $n' = m$, $u' = u^2$ and $v' = v$; if $n = 2m+1$ is odd, we have $n' = m$, $u' = u^2$ and $v' = uv$. Since $u'^{n'}v' = u^n v$, it follows that $u^n v$ is an invariant of the loop. Thus, $x^{n_0} * 1 = u^{n_\omega} * v = u * v$ since $n_\omega = 1$.

Exercise 5

We have:

$$\begin{pmatrix} F_N \\ F_{N-1} \end{pmatrix} = X_N = AX_{N-1} = \cdots = A^{N-1}X_1 = A^{N-1}\begin{pmatrix} 1 \\ 0 \end{pmatrix}.$$

The desired Fibonacci number is the (1,1) element of the matrix A^{N-1}. We can rapidly calculate A^{N-1} using the algorithm:

```
n := N − 1 ; U := A ; V := I₂ ;
repeat
| if n mod 2 = 1 then V := V * U ;
| U := U * U ; n := n div 2
until n = 1 ;
A^(N−1) := U * V
```

Put $U = \begin{pmatrix} u_1 & u_2 \\ u_3 & u_4 \end{pmatrix}$ and $V = \begin{pmatrix} v_1 & v_2 \\ v_3 & v_4 \end{pmatrix}$. To translate the assignments

$$V := V * U \quad \text{and} \quad U := U * U$$

use a temporary matrix:

$$\textit{temp} := V * U;\ V := \textit{temp};$$
$$\textit{temp} := U * U;\ U := \textit{temp};$$

Then, this gives the algorithm:

```
n := N − 1 ;
u1 := 1 ; u2 := 1 ; u3 := 1 ; u4 := 0 ; {U := A}
v1 := 1 ; v2 := 0 ; v3 := 0 ; v4 := 1 ; {V := I2}
repeat
| if n mod 2 = 1 then
| begin {preliminary calculation of the matrix temp := V * U} ;
| | temp1 := v1 * u1 + v2 * u3 ; temp2 := v1 * u2 + v2 * u4 ;
```

```
    temp₃ := v₃ * u₁ + v₄ * u₃ ;  temp₄ := v₃ * u₂ + v₄ * u₄ ;
    v₁ := temp₁ ;  v₂ := temp₂ ;  v₃ := temp₃ ;  v₄ := temp₄ ;  {V := temp}
  end ;
  {preliminary calculation of the matrix temp := U * U} ;
  temp₁ := u₁ * u₁ + u₂ * u₃ ;  temp₂ := u₁ * u₂ + u₂ * u₄ ;
  temp₃ := u₃ * u₁ + u₄ * u₃ ;  temp₄ := u₃ * u₂ + u₄ * u₄ ;
  u₁ := temp₁ ;  u₂ := temp₂ ;  u₃ := temp₃ ;  u₄ := temp₄ ;  {U := temp}
  n := n div 2
until n = 1 ;
F_N := u₁ * v₁ + u₂ * v₃
```

Remarks

1) Sequences of the type *temp* := $U * V$; V := *temp* can be avoided using the procedure of matrix multiplication. As we shall see in Chapter 6, if we set

```
procedure mult_mat(var Z : matrix ; X, Y : matrix) ;
{return in Z the product X * Y}
```

we can write *mult_mat*(V, U, V).

2) Modern Pascal languages allow the better solution

```
function mult_mat(X, Y : matrix) : matrix ;
```

now we can write the more natural instruction V := *mult_mat*(U, V).

6. The Pascal Language

The goal of this chapter is not to describe the language Pascal in detail — there are excellent books which do this — rather, the goal is to clarify the functioning of a computer and to introduce several programming devices.

6.1. Storage of the Usual Objects

The main memory of a computer (also called the *random access memory* or RAM can be represented as a very long tape partitioned into equal sized compartments numbered from[1] 0 to a very large number (these days, the number is usually between 10^6 and 10^9). Each memory compartment contains eight minuscule condensers (each about as large as a microbe!) which are either charged or discharged. The information contained in a condenser is called a *bit* (short for *b*inary dig*it*). A memory compartment using eight bits is called, quite naturally, an *octet*. Since these microscopic condensers charge and discharge very rapidly: a computer can recharge them at least 50 times a second. Computer scientists speak of a computer *refreshing* its memory – in reality, it heats up because of Joule effects! A break in electrical current, even if fleeting, has catastrophic repercussions because the contents of the RAM memory are erased, much like a message written on a luminous screen which is extinguished.

If we associated to each bit an integer equal to 0 or 1, we can consider an octet as an element $(b_0, \ldots, b_7)$ of the set $\{0, 1\}^8$. Thus we can store $2^8 = 256$ integers in an octet because the number

$$x = b_0 + 2b_1 + 2^2 b_2 + \cdots + 2^7 b_7$$

takes all values between 0 and $2^8 - 1 = 255$.

We will say in what follows that an octet *contains* an integer between 0 and 255. But we shouldn't fool ourselves – this is an *illusion*, nothing more. More generally, when we specify the type of a variable, we are deciding that the contents of one or more variables can be *interpreted* as a letter, a real number, a boolean, etc. Suppose, for example, that an octet contains the number 65: if

[1] Indexes always start from 0 in a computer.

this octet is supposed to contain an integer, we will say that it contains "65"; if it is supposed to contain a letter, we will say that it contains the letter "A".

It is good to know how octets use the objects under discussion:

- A boolean occupies an octet (a bit would suffice, but could not be separately managed.)
- A integer occupies two consecutive octets if it is of *integer* type and four consecutive integers if it is of *longint* type.
- A real number occupies six consecutive octets.

6.2. Integer Arithmetic in Pascal

This surprises many beginners! Knowing that we can store $2^8 = 256$ integers in an octet, we see that:

- the type *integer* permits us to store $256^2 = 2^{16} = 65,536$ distinct integers;
- the type *longint*, which uses four octets, permits us to store $256^4 = 2^{64} = 4,294,967,296$ distinct integers.

Since we work in $\mathbb{Z}$, these integers are divided evenly about the origin. Consequently,

- one can store the integers in the interval $I = [\![-2^{15}, +2^{15}[\![$ in a variable of type *integer*;
- one can store the integers in the interval $I = [\![-2^{31}, +2^{31}[\![$ in a variable of type *longint*.

In practice, one remembers that the type *integer* allows one to work in Pascal with integers between -32000 and $+32000$ and the type *longint* allows one to work with ten digit integers which do not exceed 2×10^{10} in absolute value.

6.2.1. Storage of integers in Pascal

Let $N > 1$ be any integer, and $b > 1$ an even integer. Put:

$$\Omega = b^N, \quad J = [\![0, \Omega[\![.$$

In Pascal, one has $b = 2^8 = 256$, $N = 2$ for the *integer* type and $N = 4$ for the *longint* type.

To store an integer y in the interval $J = [\![0, \Omega[\![$, we can use its base b representation

$$y = y_{N-1}b^{N-1} + \cdots + y_1 b + y_0, \quad 0 \leq y_i < b, \tag{6.1}$$

by placing each y_i in a compartment in memory capable of storing an integer between 0 and $(b-1)$; in Pascal, where $b = 2^8$, the number y then occupies N octets. But there are two serious criticisms of this scheme:

• What does one do with negative integers? It is inadmissible to use a whole octet, that is 256 bits, to store a sign which should only require a single bit.

• Do we need two algorithms for addition and subtraction?

Computer scientists, who never lack for imagination, have found better ways[2] to store the integers. An elegant and well-known solution is to use a bijection

$$\rho : I = [\![-\tfrac{1}{2}\Omega, +\tfrac{1}{2}\Omega [\![\longrightarrow J = [\![0, \Omega[\![$$

and store the representation (6.1) of $\rho(x)$. If we agree to send the positive integers of I to those of J using the condition $\rho(x) = x$ for $x \geq 0$, we obtain very naturally the bijection:

$$\rho(x) = \begin{cases} x & \text{if } 0 \leq x < \frac{1}{2}\Omega, \\ x + \Omega & \text{if } \frac{1}{2}\Omega \leq x < 0. \end{cases} \tag{6.2}$$

In effect, knowing that the positive integers in I fill up the first half of J, we are forced to send the negative integers x to the second half by translation. From this definition, we note the following.

• The bijection ρ^{-1} is given by the formula:

$$\rho^{-1}(y) = \begin{cases} y & \text{if } 0 \leq y < \frac{1}{2}\Omega, \\ y - \Omega & \text{if } \frac{1}{2}\Omega \leq y < \Omega. \end{cases} \tag{6.3}$$

• The sign of x is easily read off $\rho(x)$: an integer x is positive when $\rho(x)$ is *small* (that is when $\rho(x) < \frac{1}{2}\Omega$) and it is negative when $\rho(x)$ is *large* (that is, when $\rho(x) \geq \frac{1}{2}\Omega$).

• In particular, it suffices to remember the congruence

$$\rho(x) \equiv x \pmod{\Omega}. \tag{6.4}$$

Definition 6.2.1. *The bijection (6.2) is called the representation complementary to the base.*

We can now define an addition, denoted $\oplus$, on the set of integers that we can store. To do this, we demand that the following diagram be commutative:

$$\begin{array}{ccc} [\![-\frac{1}{2}\Omega, +\frac{1}{2}\Omega[\![\times [\![-\frac{1}{2}\Omega, +\frac{1}{2}\Omega[\![& \xrightarrow{\oplus} & [\![-\frac{1}{2}\Omega, +\frac{1}{2}\Omega[\![\\ \Big\downarrow {\scriptstyle \rho\times\rho} & & \Big\downarrow {\scriptstyle \rho} \\ [\![0, \Omega[\![\times [\![0, \Omega[\![& \xrightarrow{+\text{mod } \Omega} & [\![0, \Omega[\![\end{array}$$

[2] If this subject interests you, I highly recommend J.-M. Muller's book *Arithmétique des ordinateurs*, Masson (1989). This book, which is very easy to read and as engaging as a crime novel, explains in detail how one constructs algorithms and circuits which implement the four arithmetic operations.

Equivalently, we put

$$a \oplus b = \rho^{-1}\big((\rho(a) + \rho(b)) \bmod \Omega)\big). \tag{6.5}$$

Since $\rho(a) + \rho(b)$ lies between 0 and 2Ω, it is easy to be more precise:

$$a \oplus b = \begin{cases} \rho^{-1}\big(\rho(a) + \rho(b)\big) & \text{if } \rho(a) + \rho(b) < \Omega, \\ \rho^{-1}\big(\rho(a) + \rho(b) - \Omega\big) & \text{if } \rho(a) + \rho(b) \geq \Omega. \end{cases} \tag{6.6}$$

Definition 6.2.2. *The operation $\oplus$ is called addition complementary to the base.*

We demystify this addition: to calculate $x \oplus y$, first calculate $x + y$, then add the approriate multiple of Ω so that $x + y + k\Omega$ lies between $-\frac{1}{2}\Omega$ and $\frac{1}{2}\Omega$.

Examples

Suppose to begin that we have $b = 10$ and $N = 2$, so that $\Omega = 100$ and $\frac{1}{2}\Omega = 50$.

- Since $17 + 31 = 48$, we have $17 \oplus 31 = 48$.
- Since $-43 + 31 = -12$, we have $-43 \oplus 31 = -12$.
- Since $23 + 31 = 54$, we subtract Ω and get $23 \oplus 31 = -46$.
- Since $-22 + -33 = -55$, we add Ω and find that $-22 \oplus -33 = 45$.

Now suppose that we are working with an *integer* in Pascal where $\Omega = 2^{16} = 65\,536$.

- Since we have $27\,856 + 15\,831 = 43\,687 > \frac{1}{2}\Omega$, we subtract Ω and obtain $27\,856 \oplus 15\,831 = -21\,849$.

Finally, suppose that we are working with the *longint* in Pascal so that $\Omega = 2^{32} = 4\,294\,967\,296$.

- Since $-2101234456 + -199999999 = -2301234455 < -\frac{1}{2}\Omega$, we add Ω and find that $-2\,101\,234\,456 \oplus -1\,999\,999\,999 = 1993732841$.

As we have just established, addition complemetary to the base does not at all coincide with ordinary addition! But this is the price we pay if we want the same algorithm for addition and subtraction. What we have is a particularly elegant implementation of the additive group $\mathbb{Z}_\Omega$ since (6.4) and (6.5) show that

$$a \oplus b \equiv a + b \pmod{\Omega}.$$

The principle is the same for multiplication. In summary,

> *The integers in Pascal are not those of $\mathbb{Z}$, but those of $\mathbb{Z}_{2^{16}}$ or, in the case of long integers, those of $\mathbb{Z}_{2^{32}}$. The results obtained are, therefore, only certain modulo 2^{16} or 2^{32}.*

For beginners

• This peculiarity of Pascal can cause errors that are difficult to detect[3] when the capacity is exceeded in an intermediate calculation. We frequently encounter this problem when we want to work modulo an integer n: we must be very careful not to leave the interval $[-\frac{1}{2}\Omega, \frac{1}{2}\Omega]$ so that the machine does not introduce a congruence modulo Ω which could interfere in a disastrous way with our congruence modulo n (since $x + k\Omega \not\equiv x$ modulo n).

• Since the result of a calculation with integers using addition, subtraction, and multiplication is only valid modulo 2^{16} or 2^{32}, one might hope to accompany a program with a theoretical study which assures one that that the results otained are exact provided that one enters integers within some predetermined good intervals. But this is often just a dream ...

6.3. Arrays in Pascal

Consider the array of booleans $toto[0..100]$. To have room to fill up this array, the program reserves a *segment* (that is, consecutive memory compartments) which is 101 octets long in RAM in which it will successively put the contents of $toto[0]$, then $toto[1]$, and so on until $toto[100]$. The program knows the address, which we will call the *base* of the first element of the array. When the program encounters the statement

$$x := toto[i],$$

it first calculates the *offset* of the memory compartment containing $toto[i]$; that is, the amount of the displacement needed to reach it starting from $toto[0]$

$$address = base + offset = base + i.$$

This done, it effects the assignment. (Exercise: what is the offset for $toto[a..b]$?)

The array of booleans $toto[0..100, 0..100]$ occupies a memory segment consisting of $101 \times 101 = 10,201$ octets. This segment starts with the first line, then the second, and so on (*television scanning*). The offset of the element (i, j) is $101 \times i + j$ and, so, its address is:

$$address = base + offset = base + 101 \times i + j.$$

(Exercise: what is the offset for the array $toto[a..b, c..d]$?)

The situation is more complicated when we are not dealing with booleans. Suppose, for example, that $toto[0..100, 0..100]$ is an array of real numbers. Knowing that a real number occupies 6 octets, the program reserves a segment

[3] However, the V+ option in compiling does permit detection of this type of error (at the expense of speed of execution).

of memory $61206 = 6 \times 10201$ octets in length for the array. The first of the six octets where $toto[i, j]$ is lodged then has the address:

$$address = base + offset = base + 606 \times i + 6 \times j.$$

There is now an extra multiplication.

Observe that accessing elements of an array induces *hidden additions and multiplications* and these slow the execution of a program. A programmer who is unaware of this peculiarity might imagine, for example, that the statement

$$x := toto[i, j] + toto[k, \ell]$$

requires a single addition, whereas in reality the program executes five additions and four multiplications!

For beginners

If you have assimilated the above, the following code will dismay you:

$$X[k] := 0 \ ; \ \textbf{for}\ i := 1\ \textbf{to}\ n\ \textbf{do}\ X[k] := X[k] + a[i]$$

Although it takes a little longer to type, a good code is:

```
temp := 0 ;
for i := 1 to n do temp := temp + a[i] ;
X[k] := temp
```

When a program runs too slowly, it can be helpful to replace some small arrays of fixed size by variables.

Exercise 1 *(Solution at end of chapter)*

Consider the array with $k+1$ indices $toto[min_0..max_0, \ldots, min_k..max_k]$. What is the length of the segment of memory that is used? What is the offset of the element $toto[i_0, \ldots, i_k]$? (Use the expansion in a variable base explained in Chapter 2 to generalize television scanning.)

6.4. Declaration of an Array

One particularly disagreeable feature of Pascal is that an array can never change size during the execution of a program. If we want to work with matrices with real coefficients, we must know in advance the largest dimension and declare it:

```
const dim_max = 10 ;
type matrix = array[1..dim_max, 1..dim_max] of real ;
var A : matrix ;
nb_rows, nb_col : integer ;
```

Thus, in Pascal, a matrix is a triple (A, nb_rows, nb_col). There is, unfortunately, no other way to handle this if one desires the comfort of the declaration *array*.

Pascal is an old language; its modern successors do not have this limitation. The reason for this limitation has much to do with the expressed goal of Pascal: it is above all a language designed to inculcate good programming reflexes. It was never designed for the industrial world.[4]

6.5. Product Sets and Types

A mathematical problem often contains complicated objects. Happily, the majority of such objects belong to product sets. When we wish to store such product sets, we must first ask ourselves whether or not the sets are equal.

6.5.1. Product of equal sets

We need the set $M = E^n$. If E_type is the type of the elements of E, we write:

type *power_of_E* = **array**[1..*n*] **of** *E_type* ;
var *M* : *power_of_E* ;

The element with "coordinates" $(i_1, \ldots, i_n)$ is then written as $M[i_1, \ldots, i_n]$.

Examples

1) If we want to use the vectors U, V, W and matrices A, B, C with integer coordinates, we write:

type *vector* = **array**[1..*n*] **of** *integer* ;
matrix = **array**[1..*n*, 1..*n*] **of** *integer* ;
var *U*, *V*, *W* : *vector* ;
A, *B*, *C* : *matrix* ;

The i-th coordinate of U is $U[i]$ and the element (i, j) of A is $A[i, j]$.

2) If a program manipulates the columns of a $p \times n$ matrix, it can be useful to use the type:

type *column* = **array**[1..*n*] **of** *integer* ;
matrix = **array**[1..*n*] **of** *column* ;
var *A*, *B*, *C* : *matrix* ;

The j-th column of the matrix A is $A[j]$ and the i-th element of the j-th column is $A[j][i]$. The compiler will not take offense if you type $A[j, i]$ (but pay attention to the interchange of indices!).

---

[4] A conference delegate from a multinational corporation told to a friend the following: "When I want to certain of an algorithm, I program it in Pascal because I know that the compiler will let nothing pass. If I want something that runs quickly, I program it in C. Finally, when I have three lines of code to write, I use Basic."

6.5.2. Product of unequal sets

We often need the product of unequal sets, say $M = E^2 \times F^3 \times G$. If *E_type*, *F_type*, *G_type* are the types of the elements of E, F, G, we will write:

```
type product_E2_F3_G = record
    pr1, pr2 : E_type ;
    pr3, pr4, pr5 : F_type ;
    pr6 : G_type ;
end ;
var M : product_E2_F3_G ;
```

To store the element $(e1, e2, f1, f3, f3, g)$ of M, we write indifferently:

```
                                                 with M do begin
M.pr1 := e1 ; M.pr2 := e2 ;                          pr1 := e1 ; pr2 := e2 ;
M.pr3 := f1 ; M.pr4 := f2 ;          <==>            pr3 := f1 ; pr4 := f2 ;
M.pr5 := f3 ; M.pr6 := g                             pr5 := f3 ; pr6 := g
                                                 end
```

6.5.3. Composite types

Beginners are often troubled by composite types. Consider, for example, the following impressive declaration:

```
type toto = record
    whole : integer ;
    exist : boolean ;
    re, im : real
end ;
tata = array[1..n, 1..n] of toto ;
titi = record
    u : toto ;
    v : tata ;
end ;
var X : titi ;
```

How can one use the the impossible object that we are calling X? To see how, imagine that we are a compiler:

- $X.u$ is an object of type *toto*. As a result $X.u.whole$ is an integer, $X.u.exist$ is a boolean, and $X.u.re$, $X.u.im$ are two reals;

- $X.v$ is of type *tata*, whence $X.v[i]$ is of type *toto*. As a result, $X.v[i].whole$ is an integer. so that $X.v[i].exist$ is a boolean and $X.v[i].x$, $X.v[i].y$ are reals.

Remark

Pascal allows assignments between *two objects of the same type*:

$$X := Y; \quad \leftarrow \quad \textit{allowable assignment}$$

But, you should not conclude that tests of equality between two objects of the same type are permissible – they aren't:

```
if X = Y ← illegal equality test
then ...
```

For example, if your program manipulates polynomials, you can use a variable of polynomial type which you might call *poly_zero* and in which you store the zero polynomial. Each time that you want to zero out a polynomial, it suffices to type $P := poly\_zero$.

6.6. The Role of Constants

Suppose that we want to translate the statement $t := r + s$ into Pascal where r and s are two fractions. We must *teach* our program fractions because the language Pascal does not contain this type (the only types available are integers and reals). Knowing that a fraction is a pair (numerator, denominator) we use an array, which leads to the following declaration:

```
type fraction = array[0..1] of integer ;
var r, s, t : fraction ;
procedure add_frac(var t : fraction ; r, s : fraction) ;
begin
| t[0] := r[0] * s[1] + r[1] * s[0] ; {calculation of the numerator of t}
| t[1] := r[1] * s[1] ;               {calculation of the denominator of t}
| simplify(t)
end ;
```

But this is the clumsy programming of a beginner! When we write this code, we must constantly remember that $t[0]$ designates the numerator of the fraction t and $t[1]$ its denominator. Sooner or later, aided by fatigue, we will make a mistake[5] because part of our energy is devoted to remembering these conventions.

This is why decent programming languages allow one to name *constants* so that one can remember to what they refer. The following code

```
const
num = 0 ;   ==>  { t[num] := r[num] * s[den] + r[den] * s[num];
den = 1 ;        { t[den] := r[den] * s[den];
```

[5] One cannot climb stairs while chewing gum.

is better because it makes it unnecessary to recall which index represents the numerator. Comments are not necessary; the self-documentation makes the program more certain and legible.

Remark

We could have used a *record* to code our fractions:

```
type fraction = record
  | num, den : integer
end ;
var r, s, t : fraction ;
```

$$\Longrightarrow \begin{cases} t.num := r.num * s.den + r.den * s.num; \\ t.den := r.den * s.den; \end{cases}$$

This solution has a slight advantage over the proceeding: typing *t.num* requires five characters, whereas *t*[*num*] requires six (there are two brackets). The demand on the memory is the same because a *record* is stored in a similar manner to an *array*.

For beginners

Suppose that we want to work in $\mathbb{Q}[i]$; that is, with complex numbers of the form

$$z = \frac{a_1}{b_1} + i\frac{a_2}{b_2}, \quad a_1, a_2 \in \mathbb{Z}, \quad b_1, b_2 \in \mathbb{N}^*.$$

To find a good type, experience shows that it is best to seek *first* the most convenient notation by trying them out in several lines of code. Once this is decided, the construction of the appropriate type then proceeds easily.

If we want to let *z.re.den* denote the real part of the denominator of *z*, we would use the declaration on the left. But if we were to decide that *z*[*den*]*.re* is preferable, we would use the code on the right.

```
type fraction = record
  | num, den : integer
end ;
type complex = record
  | re, im : fraction
end ;
```

```
const num = 0 ; den = 1 ;
type fraction = record
  | num, den : integer
end ;
complex = array[num..den] of fraction ;
var z : complex ;
```

Exercise 2

Reconstruct the declarations which correspond to *z.re.den*, *z*[*re*][*den*], *z.re*[*den*], *z*[*den*]*.re* and *z*[*den*][*re*]. More generally, what are the declarations that allow to write: *P*[*i*, *k*][*j*], *Q*[*i*, *k*]*.coeff*[*j*], *R*[*i*]*.toto*[3]*.alpha*[5], *R*[*i*]*.toto*[3]*.beta*[5] and *toto.tata.titi.thing*.

6.7. Litter

When you declare a variable x, you say to the program: "Reserve a segment in RAM to store the value of x". The program doesn't do anything else: it does not "clean up" afterwards by, for example, setting all the octets to zero. If you do not initialize the variable x, you risk finding remains of other programs.

Here is an analogy: you buy some land several miles from your house. Your lawyer will ask a surveyor to mark off the boundaries of the field and will draw up a sales contract, but he or she will not clear the field which might be covered with litter, broken bottles, or underbrush. This is left to you.

This phenomenon is very easy to demonstrate. Type the following program:

```
program litter ;
var vector = array[1..1000] of real ;
i : integer ;
begin
| for i := 1 to 1000 do writeln(vector[i])
end .
```

If you run this program right after you have turned on your computer, you will probably only obtain zeroes on the screen. However, if you run the program after having previously run some other program which uses a lot of memory, you will see numbers appear randomly on the screen which are the "litter" the preceding program left behind.

For beginners

Think of this any time that you are tempted to complain "I do not understand: my program worked so well yesterday!". Yesterday, you probably used a machine that had been just turned on, so that all the memory was set to zero and the absence of initialization did not manifest itself. Today, the faulty program was not the first to run on the machine, and the variables that you forgot to initialize were initialized by the litter left by the preceding program resulting in aberrant values of the variables.

6.8. Procedures

One should think of a procedure as a *black box* which information enters and leaves.

A *procedure* is a small program which functions inside the main program. For this reason, the syntax of a procedure resembles that of a program:

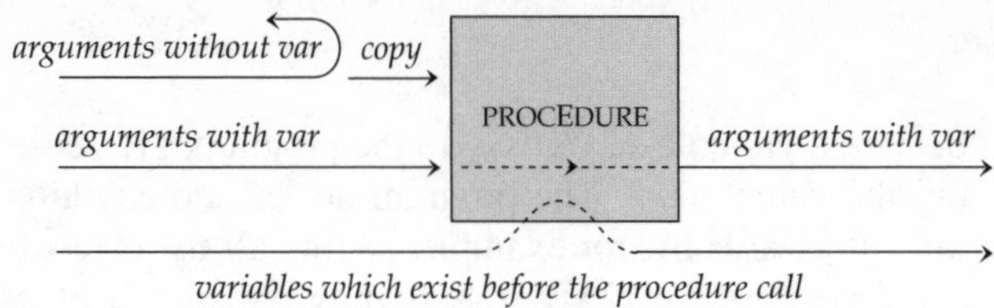

Fig. 6.1.

procedure *toto*(x_1 : $type_1$; x_2, x_3 : $type_2$; **var** x_4 : $type_3$) ;
  declaration of the constants of the procedure ;
  declaration of the types of the procedure ;
  declaration of the variables of the procedure ;
  declaration of procedures and functions ;
    known by the single procedure toto
begin {*procedure*}
  | *body of the procedure*
end ; {*procedure*}

Notice that following two ways in which the syntax of a procedure differs from that of a program:

- if a procedure possesses arguments, the name of the procedure is followed by the list of arguments in parentheses;
- the final "end" of a procedure is followed by a semicolon (since the final "end" of a program is followed by a period).

Before explaining what a program does when it encounters a procedure or function call, we clarify some points of syntax.

6.8.1. The declarative part of a procedure

Here are three declarations of procedures:

procedure *toto* ;
procedure *tata*(x : *real*) ;
procedure *titi*(x, y : *real* ; **var** t : *real* ; n : *integer*) ;

- The procedure *toto* is a procedure without an argument (or without a parameter).
- The precedure *tata* possesses a single argument.
- The procedure *titi* possesses four arguments.

The words *argument* and *parameter* are synonyms. An argument can be preceded by the reserved word "var". The number of arguments of a procedure is always the same.[6] Finally, the arguments of a procedure need not be declared variables. These are "placeholders" in a sense which we will make precise later.

[6] With the exception of certain system procedures such as *write*, *read* and *concat*.

For beginners

Each argument is necessarily followed by its type. Consequently, if you write your declaration as

$$\textbf{procedure}\ toto(x : \textbf{array}[1..10]\ \textbf{of}\ real)\ ;$$

the compiler will complain because "array[1..10] of real" is not a type! From the strict point of view of syntax, the compiler expects to find an identifier after the "x". The presence of the square bracket in "array[" triggers the protest.

6.8.2. Procedure calls

One says that one *calls* the procedure *titi* when one writes

$$\ldots;\ titi(u, v, w, k);\ \ldots$$

A procedure call is a *statement*; the arguments are separated by commas. One can call a procedure anywhere that one can write a statement, so, for example, in the interior of another procedure or a function:

$$x := x + 1;\ \ toto;\ \ y := sin(x);\ \ tata(log(x));\ \ titi(x, 2 * x + 1, x, n);$$

In the chapter on recursion we shall see that one can even call a procedure inside its own code!

There is an essential difference between parameters "with var" and parameters "without var".

- A parameter "without var" can be replaced by any arithmetic expression, in particular by the name of a variable or a constant. Of course, the arithmetic expression in question must only contain variables *known* to the program at the moment of the procedure call.
- In contrast, a parameter "with var" can only be replaced by a *variable* known to the program; any other arithmetic expression is rejected. If we return to the procedure *titi*, we do not have the right to type $titi(x, x, x + y, n)$ or $titi(x, x, 100, n)$ because the third argument is preceded by a "var" in the declaration of *titi*.

One says that a parameter "without var" is passed *by value* and that a parameter "with var" is passed *by address*. We will return to this subject at much greater length in Chapter 13.

For beginners

One meets from time to time, in the *body of the principal program*, horrors such as:

```
begin
| choose_vector(var X : vector ; var n : integer) ;
| procedure toto(var u : integer ; a, b : real) ;
end .
```

These syntactic monstrosities result from a grave confusion between declaring a procedure and calling that procedure. When we *declare* a procedure, we are *educating* our program; when we *call* a procedure, we are demanding that our program *act*, not learn!

6.8.3. Communication of a procedure with the exterior

Remember our black box model (Fig. 6.1).

• The values of the parameters passed by value are "photocopied" into special variables automatically created for this occasion (this is the reason that parameters "without var" need not be declared variables). Since the procedure actually works with copies, the original parameters are therefore *never modified* by the procedure.

• On the contrary, parameters passed by address are really communicated to the procedure which allows the procedure to *actually modify* their value. This is the reason that parameters "with var" can only be names of variables.

In a somewhat more suggestive manner,we say that parameters "with var" *leave* a procedure; we also say that the procedure *returns* its calculations in arguments passed by address.

• The variables *local* to a procedure, which are created at the moment the procedure is called, are destroyed at the end of the call. Thus, their values do not leave the procedure.

Examples

1) We write a procedure to calculate the sum $Z = X + Y$ of two vectors of dimension *dim*.

```
procedure sum_vector(X, Y : vector ; var Z : vector ; dim : integer) ;
var i : integer ;
begin
| for i := 1 to dim do Z[i] := X[i] + Y[i]
end ;
```

We communicate the vectors X, Y to the procedure as well as their common dimension *dim*: these arguments are passed by value. The result Z *must necessarily* be passed by address because it must leave the procedure.

2) Suppose we want to write a procedure to *choose* a vector X.

```
procedure choose_vector(var X : vector ; var dim : integer) ;
var i : integer ;
begin
  repeat
    write('dim = ') ; readln(dim)
  until (1 ≤ dim) and (dim ≤ dim_max) ;
  for i := 1 to dim do begin
    write('X[', i : 1, '] = ') ; readln(X[i])
  end
end ;
```

We do not communicate any information to the procedure. This arrives, via the keyboard, when it is activated, not before. Thus, there is no argument "without var" because the procedure does not need any information to function. On the other hand, the arguments X and *dim*, which are destined to *receive* our messages and leave, are passed by address.

For beginners

1) Here is an error that one often encounters (notice the location of the variable Z):

```
procedure sum_vector(X, Y : vector ; dim : integer) ;
var i : integer ; Z : vector ;
begin
  for i := 1 to dim do Z[i] := X[i] + Y[i]
end ;
```

Syntactically this program is correct, but semantically it is false! As we have already pointed out, a procedure *destroys* its local variables once it finishes its work: the vector Z winds up "in the garbage"...

2) Here is an even larger error (the two vectors Z) committed by individuals who are genuinely indifferent to computer science (these exist) and who refuse to respect the difference between the "var" that one puts before an argument and the "var" which serves to declare the local variables in a procedure.

```
procedure sum_vector(X, Y : vector ; dim : integer ; var Z : vector) ;
var i : integer ; Z : vector ;
begin
  for i := 1 to dim do Z[i] := X[i] + Y[i]
end ;
```

How can one distinguish the two vectors Z? This is not honest! And how will the compiler be able to guess what is going on in the programmer's head?

3) We end with an error that one encounters fairly often. We want to translate the pseudo-statement "$X := X + Y$" into code .

```
procedure add_vector(X, Y : vector ;  var X : vector ;  dim : integer) ;
var i : integer ;
begin
| for i := 1 to dim do X[i] := X[i] + Y[i]
end ;
```

Here the programmer wrongly imagines that what *enters* (the vector X alongside Y) must be distinct from what *leaves* (this is why "*var X* " is present at the end. The compiler is not bothered by the presence of two arguments with the same name.

6.9. Visibility of the Variables in a Procedure

Consider the programs *visibility_1* and *visibility_2*. The program *visibilty_1* writes 1999 three times in succession while *visibility_2* writes 1999, 1515 then 1999.

```
program visibility_1 ;
var x : integer ;
___________________

procedure toto_1 ;
begin
| writeln(x)
end ;

___________________
begin
| x := 1999 ;
| writeln(x) ;
| toto_1 ;
| writeln(x)
end .
```

```
program visibility_2 ;
var x : integer ;
___________________

procedure toto_2 ;
var x : integer ;
begin
| x := 1515 ;  writeln(x)
end ;
___________________
begin
| x := 1999 ;
| writeln(x) ;
| toto_2 ;
| writeln(x)
end .
```

From this observation, we can deduce:

• that the procedure *toto_1* "sees" the global variable x of the program because it is capable of writing its value;

• that the local variable x of the procedure *toto_2* *provisionally* masks the global variable x of the program but this last variable reappears once the procedure ceases functioning.

All variables which exist at the moment of a procedure call are global variables for the procedure. Consequently, the variables of a program are global for all procedures.

A procedure sees all variables which exist at the moment a procedure is called except those whose names are masked by local variables.

Consider now the principal part of a program which chooses vectors X and Y and displays the result:

```
begin                              begin
| choose(X, Y, dim) ;              | choose(X, Y, dim) ;
| sum_vector(X, Y, Z, dim) ;       | sum_vector(X, Y, Z) ;
| display(Z, dim)                  | display(Z)
end .                              end .
```

When we write a program, it is difficult to remember at each instant that *dim* is the true dimension of the vectors. Since all procedures see the variable *dim*, we can suppress the references to *dim* in the procedures *sum_vector* and *display* (as in the program on the right). Of course, you must then modify the procedure *sum_vector*.

```
procedure sum_vector(X, Y : vector ; var Z : vector) ;
var i : integer ;
begin
| for i := 1 to dim do Z[i] := X[i] + Y[i]
end ;
```

Procedures are simplified thereby. There is less "background noise" and the program is easier to follow and functions perfectly. But this freedom also has disadvantages:

- If the program is being written simultaneously by several persons, each programmer must know the list of global variables of the program.
- If we write a procedure that we intend to re-use in another program (as in a library of procedures), we *must absolutely not allow ourselves this freedom* because we do not know in advance what the global variables of the program will be. It is then essential to write *airtight* procedures; that is, procedures which only communicate with the exterior *via* their parameters.

6.10. Context Effects

Consider the program *context_effects* whose main part is the following:

```
begin
| x := 1999 ; writeln(x) ; surreptitious ; writeln(x)
end .
```

Nothing allows us to foresee that the program will write 1999, then 1515! In other words, there is no way we could know that the variable x is modified by the procedure *surreptitious*. This frightening phenomenon, called a *context effect*, is a mechanical consequence of the visibility of global variables in a procedure: because the procedure can see global variables, it can modify them. Context effects are a result of *bad programming* which must be avoided at all costs. A procedure *should only modify parameters which are transmitted to it by address, and should not touch the others.*

```
program context_effect ;
var x : integer ;
procedure surreptitious ;
begin
| x := x - 484
end ;
begin
| x := 1999 ;  writeln(x) ;  surreptitious ;  writeln(x)
end .
```

Nonetheless, context effects are often tolerated. Consider a medium size program that initializes 50 variables when it starts.

```
begin
| message ;
| diverse_initializations ;  ← voluntary context effects
| ...
end .
```

The procedure *diverse_initializations* gives initial values to the 50 variables by context effects: it would be rather painful to declare 50 arguments "with var" (or ten procedures with five arguments). But, be honest: do not forget to document it because you are playing with fire.

For beginners

Context effects are sometimes involuntary (these are the most frightening ones). Consider the main part of a program which chooses a square matrix:

```
begin
| ... ;  choose_matrix(A) ;  ...
end .
```

The programmer, blinded by the matrix, has forgotten that it is a pair (*array*, *dim*). Nevertheless, the program functions correctly because it is written as follows:

```
procedure choose_matrix(var A : matrix) ;
var i, j : integer ;
begin
| write('dim = ') ;  readln(dim) ;  ← involontary context effect
| for i := 1 to dim do
| for j := 1 to dim do begin
| | write('A[', i : 1, ',' j : 1, '] = ') ;  readln(A[i, j])
| end
end ;
```

The right declaration is

procedure *choose_matrix*(**var** *A* : *matrix* ; **var** *dim* : *integer*) ;

Here is another typical example of an involuntary context effect. Suppose that we want to calculate the determinant of a square matrix (the choice of algorithm does not matter). Beginners often propose the following code:

```
choose_matrix(A) ;
determinant(A) ;
```

When one asks the author, he or she, disconcerted (is my identifier not clear enough?), responds that *determinant* calculates the determinant of the matrix. When you remark that the syntax is that of a procedure call, the author replies by promising to introduce a variable, call it *det*, in the interior of the procedure which allows the procedure to store the value of the determinant. But, since this variable is not among the parameters of the procedure, we now have a context effect! What's worse, the program runs! A good solution is to use a function:

```
choose_matrix(A) ;
det := determinant(A) ;
```

Since an arithmetic expression can contain the value of a function, the code on the right is better:

det := *determinant*(*A*) ;	**if** *determinant*(*A*) $\neq 0$
if *det* $\neq 0$ **then** ...	**then** ...

6.10.1. Functions

The syntax is similar to that of a procedure with two differences:

1) a *function* has a type which one must not forget in a declaration;

2) one must not forget to give a value to the function:

```
function sum(x : vector ; n : integer) : real ;
var i : integer ; temp : real ;
begin
| temp := 0 ;
| for i := 1 to n do temp := temp + x[i] ;
| sum := temp
end ;

function is_solution(a, b : integer) : boolean ;
begin
| if a div b then is_solution := true else is_solution := false
end ;
```

The name of a function can occur alone before the assignment sign; in contrast, it must be followed by an open parenthesis and a list of arguments after the assignment sign. The following code is therefore incorrect:

```
function sum(x : vector ; n : integer) : real ;
var i : integer ;
begin
| sum := 0 ; {legal}
| for i := 1 to n do
| sum := sum + x[i] ;
| {sum after ":=" is not allowed alone}
end ;
```

You can redefine the value of a function as many times as you like:

```
function last_place_nonzero(x : vector) : integer ;
var i : integer ;
begin
| last_place_nonzero := −1 ;
| for i := 1 to n do
| if x[i] ≠ + then last_place_nonzero := i
end ;
```

When the name of a function occurs in the code that defines the value of the function, one says that the function is *recursive*. The best known examble is that of the factorial function which we will study in detail in Chapter 12.

6.10.2. Procedure or function?

A function in *old* Pascal can only be of basic type: boolean, integer, real, or a string of characters. It cannot have a more complicated type (that is, a type that one teaches to the compiler). This historical limitation[7] of the Pascal language leads many beginners astray. Suppose for example that I need the product of two matrices. I would like to declare the function

```
function product_matrix(X, Y : matrix) : matrix ;
```

and use it in the following very natural way

```
... ; Z := product_matrix(X, Y) ; ...
```

However, if my Pascal does not accept functions of the type *matrix*, I must use a procedure instead

```
procedure product_matrix(var Z : matrix ; X, Y : matrix) ;
```

[7] This limitation dates to an era when machines were not as powerful as today and when modern languages did not exist.

and type

$$\ldots\ ;\ product\_matrix(Z, X, Y)\ ;\ \ldots$$

in the program. This is reminiscent of a phenomenon which is quite familiar to mathematicians: we can explicitly define a function z of the variables x and y by $z = f(x, y)$ or implicitly by $f(z, x, y) = 0$.

6.11. Procedures: What the Program Seems To Do

The description[8] that follows is not at all realistic (we will see why in Chapter 13). However, it permits us to understand and *predict* the effect of calling a procedure.

```
procedure toto(x, y, z : integer ; var u : real) ;
var i : integer ;
begin
  | i := x ;
  | x := x + 1 ;
  | u := x + y/z
end ;
```

When the program encounters the statement

$$toto(i, 1515, A + B \textbf{ mod } 3, R),$$

what occurs is as if the program were to execute the following sequence of actions:

• creation of the auxiliary variables *x_toto*, *y_toto*, *z_toto* and *i_toto* (this is provoked by the arguments "without var" x, y, z and the local variable i; the variable "with var" u is not involved);

• initialization of the auxiliary variables

$$x\_toto := i;\ \ y\_toto := 1515;\ \ z\_toto := A + B \textbf{ mod } 3;$$

in other words, the procedure "photocopies" the values of the arguments i, 1515 and $A + B$ mod 3 into *x_toto*, *y_toto* and *z_toto* (the local variable *i_toto* is not involved);

• modification of the code of the procedure: the parameter u passed by address is replaced by the argument R and the variables x, y, z, i are replaced by *x_toto*, *y_toto*, *z_toto*, *i_toto* which gives the new code

```
i_toto := x_toto ;
x_toto := x_toto + 1 ;
R := x_toto + y_toto/z_toto
```

[8] This description is the result of a collaboration with Michèle Loday-Richaud.

• execution of the modified code;

• destruction of the auxiliary variables *x_toto*, *y_toto*, *z_toto* and *i_toto* once the new code is executed.

In summary what happens is as if the following actions were carried out:

1) creation of the auxiliary variables *x_toto, y_toto, z_toto, i_toto*;
2) initialization of *x_toto, y_toto* and *z_toto* by the values occupying the locations of the variables "without var" *x, y* and *z*;
3) modification of the code of the procedure:
   - *x, y, z* and *i* are replaced by *x_toto, y_toto*, *z_toto* and *i_toto*;
   - the parameter "with var" *u* is replaced by *R* ;
4) execution of the new code
5) destruction of the new variables *x_toto, y_toto*, *z_toto* and *i_toto.*

Remarks

1) To create a variable means to reserve a free location in memory. Remember that *reserve* does not mean *clean.* It is quite possible that the location chosen by the program is the address of a variable which had been "destroyed" and which contains litter.

2) To destroy a variable simply means to authorize the program to use the address for another procedure call.

This model allows us to understand why:

• the parameters of a procedure need not be declared variables in the program;

• the statement $x := x + 1$ does not modify the variable x: in effect, the procedure works on the copy *x_toto* and not on the original x!

• the program does not confuse the global variable i of the program (if such exists) with the local variable i; the local variable i provisionally masks the global variable i;

• an arithmetic expression can not occupy the place of a parameter transmitted by address and why such a parameter can change value: in our model, the call $toto(x, y, z, R + 1)$ would be transformed into the absurd assignment $R + 1 := x\_toto + y\_toto/z\_toto$ (absurd because $R + 1$ is not an identifier).

Let us test our model with the program:

```
program test ;
var i, A, B : integer ;
procedure toto(x : integer ; var y : integer) ;
var i : integer ;
begin
| writeln('entry into toto') ;
| writeln('x = ', x : 1, ',  y = ', y : 1, ',  i = ', i : 1) ;
```

```
   i := 10 ; x := x + i ; y := x + 2 * y ;
   writeln('x = ', x : 1, ', y = ', y : 1, ', i = ', i : 1) ;
   writeln('exit from toto') ;
end ;
procedure message ;
begin
   writeln('-----') ;
   writeln('main program') ;
   writeln('i = ', i : 1, ', A = ', A : 1, ', B = ', B : 1) ;
   writeln('-----') ;
end ;
begin
   i := 1994 ; A := 3 ; B := 51 ;
   message ; toto(A, B) ;
   message ; toto(A + 9, B) ;
   message ; toto(A, A) ;
   message ; toto(B, A) ;
end .
```

When we let the program run, here is what we obtain:

main program: $i = 1994$, $A = 3$, $B = 51$

---

entry into toto: $x = 3$, $y = 51$, $i = 8196$
exit from toto: $x = 13$, $y = 115$, $i = 10$

---

main program: $i = 1994$, $A = 3$, $B = 115$

---

entry into toto: $x = 12$, $y = 115$, $i = 8196$
exit from toto: $x = 22$, $y = 252$, $i = 10$

---

main program: $i = 1994$, $A = 3$, $B = 252$

---

entry into toto: $x = 3$, $y = 3$, $i = 8196$
exit from toto: $x = 13$, $y = 19$, $i = 10$

---

main program: $i = 1994$, $A = 19$, $B = 252$

---

entry into toto: $x = 252$, $y = 19$, $i = 8196$
exit from toto: $x = 262$, $y = 300$, $i = 10$

---

main program: $i = 1994$, $A = 300$, $B = 252$

We can already see the "litter" phenomenon. Each time the procedure *toto* is called, the procedure creates the local variable i which is called *i_toto* in our model. Since *i_toto* is not yet initialized when we want to see the values of x, y, i (which are called *x_toto*, *y_toto*, *i_toto* in our model), the program displays the unexpected value $i = 8196$ stemming from the residue of earlier activity in the memory allocated to the local variable i.

We also obtain these results *without using our computer* with the aid of our model.

$i := 1994;\ \ A := 3;\ \ B := 51;$

$$\left.\begin{array}{l} \textit{creation of } x\_toto \textit{ and } i\_toto \\ x\_toto := A;\ \ i\_toto := 10; \\ x\_toto := x\_toto + i\_toto; \\ B := x\_toto + 2 * B; \\ \textit{destruction of } x\_toto \textit{ and } i\_toto \end{array}\right\} \text{call } toto(A, B)$$

$$\left.\begin{array}{l} \textit{creation of } x\_toto \textit{ and } i\_toto \\ x\_toto := A + 9;\ \ i\_toto := 10; \\ x\_toto := x\_toto + i\_toto; \\ B := x\_toto + 2 * B; \\ \textit{destruction of } x\_toto \textit{ and } i\_toto \end{array}\right\} \text{call } toto(A + 9, B)$$

$$\left.\begin{array}{l} \textit{creation of } x\_toto \textit{ and } i\_toto \\ x\_toto := A;\ \ i\_toto := 10; \\ x\_toto := x\_toto + i\_toto; \\ A := x\_toto + 2 * A; \\ \textit{destruction of } x\_toto \textit{ and } i\_toto \end{array}\right\} \text{call } toto(A, A)$$

$$\left.\begin{array}{l} \textit{creation of } x\_toto \textit{ and } i\_toto \\ x\_toto := B;\ \ i\_toto := 10; \\ x\_toto := x\_toto + i\_toto; \\ A := x\_toto + 2 * A; \\ \textit{destruction of } x\_toto \textit{ and } i\_toto \end{array}\right\} \text{call } toto(B, A)$$

Executing these statements by hand allows us to recover the values displayed by the machine.

Exercise 3

In the preceding program, replace the global variables A, B by x, y in the declarative part and the main body of the program. Explain, using the model, why the displayed values are the same as before.

6.11.1. Using the model

Let us return to the procedure to choose a vector,

```
procedure choose_vector(var X : vector ;  var dim : integer) ;
var i : integer ;
begin
| write('dim = ') ;  readln(dim) ;
| for i := 1 to dim do begin
| | write('X[', i : 1, '] = ') ;  readln(X[i])
| end
end ;
```

When the variables X and *dim* are transmitted to the procedure, their values are "random" (they are litter: one also says that they are *indefinite*). Our model shows that the procedure *choose* gives the variables X and *dim* the values provided by the keyboard. Without touching the body of the procedure, let us now ask what would happen if we were to modify the declarative part:

- One argument is passed by address and the other by value:

```
procedure choose_vector(var X : vector ;  dim : integer) ;
```

Here, the procedure replaces the random contents of the variable X and of the variable *dim_choose_vector* from our model by information entered from the keyboard. The vector X is then corectly initialized. By contrast, the dimension (which was sent to the auxiliary variable *dim_choose_vector*) is lost and the variable *dim* retains the random value that it had earlier.

- The two arguments are passed by value:

```
procedure choose_vector(X : vector ;  dim : integer) ;
```

Now, the procedure modifies *X_choose_vector* and *dim_choose_vector*: the variables X and *dim* retain their indefinite values.

For beginners

1) When starting, one should be very conscientious about passing parameters and not allow oneself any fantasies. Consider, for example, a procedure which returns as Z the product of the matrices X and Y.

```
procedure product_matrix(var Z : matrix ;  X, Y : matrix) ;
```

This declaration is the *only* possible. The following declaration (which attempts to economize on memory) is incorrect, but in a subtle way,

procedure *product_matrix*(**var** Z, X, Y : *matrix*) ;

In effect, the result is correct each time that $Z \neq X$ and $Z \neq Y$. On the contrary, it is grossly false as soon as we want to calculate $X := XY$ or $X := X^2$. To understand why, suppose that we want to calculate $X := X^2$ when X is of dimension 2. The procedure begins, for example, by calculating

$$X[1, 1] := X[1, 1] * X[1, 1] + X[1, 2] * X[2, 1].$$

The initial value of $X[1, 1]$ having been destroyed, the next $X[i, j]$ will be incorrect ... We draw from this a proverb:

One should never modify the data of a program.

2) If, after a procedure is called, you obtain aberrant values (for example, integers that are too large or negative when you are working in the interval [1, 10], or else real numbers that are "infinitely large" or "infinitely small"), this is because you have forgotten the "var". As you see, these values are litter; the true results have been volatized ...

6.12. Solutions of the Exercises

Exercise 1

The index $i_t \in [min_t, max_t]$ takes $b_t = max_t - min_t + 1$ possible values, so that the array contains $b_0 \cdots b_k$ elements. To set up a bijection between the elements of the array and the interval $[\![0, b_0 \cdots b_k[\![$, we use the expansion in a variable base (Chap. 2) by considering $i_t - min_t$ as a number in the base b_t since it satisfies the conditions $0 \leq i_t - min_t < b_t$. The desired bijection is:

$$(i_1, \ldots, i_k) \longmapsto (i_0 - min_0) + (i_1 - min_1)b_0$$
$$+(i_2 - min_2)b_0 b_1 + \cdots + (i_k - min_k)b_0 b_1 \cdots b_{k-1}.$$

7. How to Write a Program

It seems that the work of the engineers, physicists, and draughstmen is, in appearance, only to polish surfaces and refine away angles, ease this joint or stabilize that wing, render these parts invisible, so that in the end there is no longer a wing hooked to a framework but a form flawless in its perfection, completely disengaged from its matrix, a sort of spontaneous whole, its parts mysteriously fused together and resembling in their unity a poem. It seems that perfection is attained when there is nothing more that can be cut out.[1] *At the height of its evolution the machine dissembles its own existence.*

Antoine de Saint Exupéry, *Terre des hommes*

7.1. Inverse of an Order 4 Matrix

Let A be an $n \times n$ matrix with coefficients in a ring. The *adjoint* of A is the matrix of cofactors of A; that is, the matrix

$$\mathrm{Adj}_{i,j} = (-1)^{i+j}\,\mathrm{minor}_{n-1}(i, j), \tag{7.1}$$

where $\mathrm{minor}_{n-1}(i, j)$ denotes the (i, j)-th minor; that is, the determinant of the $(n-1) \times (n-1)$ submatrix obtained by deleting the i-th row and j-th column of A.

Theorem 7.1.1. *With the notation above,*

$${}^t\mathrm{Adj}(A) \cdot A = A \cdot {}^t\mathrm{Adj}(A) = \det(A) I. \tag{7.2}$$

Proof. When we multiply the i-th row of ${}^t\,\mathrm{Adj}(A)$ by the i-th column of A, we obtain the Laplace expansion of the determinant of A along the i-th column which explains why the diagonal entries are $\det(A)$.

When we multiply the i-th row of ${}^t\,\mathrm{Adj}(A)$ by the j-th column of A, with $i \neq j$, we obtain the Laplace expansion of a determinant whose i-th and j-th columns are equal, which implies that the off-diagonal entries are zero.

---

[1] The emphasis is mine.

Corollary 7.1.1. *If the determinant of A is invertible in the ring in which one is working, then the matrix A is invertible with inverse*

$$A^{-1} = \frac{1}{\det(A)}\,{}^t\mathrm{Adj}(A). \tag{7.3}$$

7.1.1. The problem

We are going to compute the inverse of a 4×4 matrix with real coefficients in a somewhat bizarre manner. The constraints[2] are the following:

- the calculation of A^{-1} must use formula (7.3);
- determinants will always be expanded along the first column (for minors of order 2, this gives the traditional formula $ad - bc$);
- minors of order 2 and 3 must only make reference to the single matrix A; one is not allowed to employ an auxiliary matrix to calculate the minors.

7.1.2. Theoretical study

Before lauching into programming proper, we focus on our algorithms. The determinant of A must be calculated using the formula:

$$\begin{aligned}\det(A) = a_{1,1}\begin{vmatrix} a_{2,2} & a_{2,3} & a_{2,4} \\ a_{3,2} & a_{3,2} & a_{3,4} \\ a_{4,2} & a_{4,2} & a_{4,4}\end{vmatrix} - a_{2,1}\begin{vmatrix} a_{1,2} & a_{1,3} & a_{1,4} \\ a_{3,2} & a_{3,2} & a_{3,4} \\ a_{4,2} & a_{4,2} & a_{4,4}\end{vmatrix} \\ + a_{3,1}\begin{vmatrix} a_{1,2} & a_{1,3} & a_{1,4} \\ a_{2,2} & a_{2,2} & a_{2,4} \\ a_{4,2} & a_{4,2} & a_{4,4}\end{vmatrix} - a_{4,1}\begin{vmatrix} a_{1,2} & a_{1,3} & a_{1,4} \\ a_{2,2} & a_{2,2} & a_{2,4} \\ a_{3,2} & a_{3,2} & a_{3,4}\end{vmatrix}.\end{aligned}$$

Since there is no question of typing this formula in our future program, we use the technique of *naming* the difficult objects; that is, of introducing the function minor_3:

$$\begin{aligned}\det(A) = a_{1,1}\ \mathrm{minor}_3(1,1) - a_{2,1}\ \mathrm{minor}_3(2,1) \\ + a_{3,1}\ \mathrm{minor}_3(3,1) - a_{4,1}\ \mathrm{minor}_3(4,1).\end{aligned} \tag{7.4}$$

The adjoint requires knowing the function $\mathrm{minor}_3(i, j)$ for all values of i and j.

This raises a new problem: how can we calculate minor_3 without using an intermediate matrix of dimension 3?

---

[2] This is *not at all* the way in one would calculate the determinant with a computer! Gaussian elimination is *infinitely more efficient.*

At this stage, we encounter an interesting phenomenon: while it is very easy to explicitly write $\text{minor}_3(1,1)$, $\text{minor}_3(1,2)$, $\text{minor}_3(3,4)$, ..., we have trouble generalising because when we try to write $\text{minor}_3(i, j)$, we obtain:

$$\text{minor}_3(i, j) = a_{?,?}\,\text{minor}_2(?,?) - \cdots .$$

We don't know how to write the appropriate indices! In order to advance, we once again *name* that which causes the problem and postpone a finer study of the stubborn objects. If $\text{minor}_2(i, \alpha, j, \kappa)$ denotes the minor of order 2 obtained by suppressing the rows i, α and columns j, κ of A, we have

$$\begin{aligned}\text{minor}_3(i, j) = {} & a_{\alpha,\kappa}\,\text{minor}_2(i, \alpha, j, \kappa)\\ & -a_{\beta,\kappa}\,\text{minor}_2(i, \beta, j, \kappa)\\ & +a_{\gamma,\kappa}\,\text{minor}_2(i, \gamma, j, \kappa).\end{aligned} \tag{7.5}$$

The indices α, β, γ denote the first, second, and third lines of the matrix obtained from A by suppressing the row i: as a result, these are functions of the single varaible i (in other words, j has no connection with these three indexes). Similarly, the index $\kappa = \kappa(j)$ denotes the first column of A when one removes the column j.

Since there are so many unknowns, we might ask if we can be more efficient with fewer functions.[3] To do this, we are going to use the old trick which consists of replacing the three functions $\alpha(i)$, $\beta(i)$, $\gamma(i)$ by the single function $\lambda(i, k)$ where k indicates the function that one must choose: α when $k = 1$, β when $k = 2$ and γ when $k = 3$.

Economy demands that we extract what is common in the row and column indices. To do this, consider four lottery balls arranged in the following order

(1) (2) (3) (4)

and put

$\lambda(i, k) =$ *number of the k-th ball when one removes the i-th ball.*

If we let $\text{minor}_2(i_1, i_2, j_1, j_2)$ denote the determinant of the 2×2 matrix obtained by deleting the rows i_1, i_2 and the columns j_1, j_2 of A, we have

$$\begin{aligned}\text{minor}_3(i, j) = {} & a_{\lambda(i,1),\lambda(j,1)}\,\text{minor}_2\big(i, \lambda(i,1), j, \lambda(j,1)\big)\\ & -a_{\lambda(i,2),\lambda(j,1)}\,\text{minor}_2\big(i, \lambda(i,2), j, \lambda(j,1)\big)\\ & +a_{\lambda(i,3),\lambda(j,1)}\,\text{minor}_2\big(i, \lambda(i,3), j, \lambda(j,1)\big).\end{aligned} \tag{7.6}$$

---

[3] This common sense principle bears the suggestive name of *Occam's razor* in honor of the medieval English philosopher Occam (1285–1349?) who phrased it as follows "*Entia non sunt multiplicanda praeter necessitatem*" (Entities should not be multiplied needlessly).

Now that it is well-defined, the function λ does not resist our efforts long:

$$\lambda(i,k) = \begin{cases} k & \text{if } k < i, \\ k+1 & \text{if } k \geq i. \end{cases} \tag{7.7}$$

You can see the power of this method.[4] Had we tried to solve the problem and the sub-problem *together*, we would have written a horrible formula combining (7.6) and (7.7).

All that remains is to find the explicit value of the function minor_2. Once again, instead of trying to go too fast, we content ourselves with introducing the function:

$\mu(i_1, i_2, k) =$ *number of the k-th ball when one removes the i_1-th and i_2-th balls.*

Inspired by (7.6), we can now write

$$\begin{aligned} \text{minor}_2(i_1, i_2, j_1, j_2) = {} & a_{\mu(i_1,i_2,1),\mu(j_1,j_2,1)}\, a_{\mu(i_1,i_2,2),\mu(j_1,j_2,2)} \\ & - a_{\mu(i_1,i_2,1),\mu(j_1,j_2,2)}\, a_{\mu(i_1,i_2,2),\mu(j_1,j_2,1)}. \end{aligned} \tag{7.8}$$

The value of the function μ is a little more complicated than that of λ. Supposing that $i_1 < i_2$, we find that

$$\mu(i_1, i_2, k) = \begin{cases} k & \text{if } k < i_1, \\ k+1 & \text{if } i_1 \leq k \text{ and } k+1 < i_2, \\ k+2 & \text{otherwise.} \end{cases} \tag{7.9}$$

The only pitfall would be to forget the test $k+1 < i_2$ when $k \in [\![i_1, i_2]\!]$, because this would give $\mu(i_1, i_2, k) = i_2$ when $i_2 = i_1 + 1$ and $k = i_1$.

Remark 7.1.1. It is possible to find a more compact formulation with $i_1 < i_2$:

$$\mu(i_1, i_2, k) = \left\{ \begin{matrix} k & \text{if } k < i_1 \\ \lambda(i_2, k+1) & \text{otherwise} \end{matrix} \right\} \Longrightarrow \mu(i_1, i_2, k) = \lambda\big(i_2, \lambda(i_1, k)\big).$$

Mathematicians appreciate the latter formulation. On the other hand, computer scientists see many potential booby traps here. Since the code is opaque, all kinds of errors are possible (including typing errors); how could one correct something this obscure?

7.1.3. Writing the program

We now translate our algorithms into code. We begin by sketching the body of the main program.

[4] Tomorrow, things will be better; the day after tomorrow, they will be better still.

```
begin
| message ; choose(A) ;
| matrix_inverse(A, inv_A) ;
| display(inv_A)
end .
```

This code contains a serious flaw: it will crash if the matrix A is not invertible.[5] Hence we must be careful.

```
begin
| message ; choose(A) ;
| if abs(det(A)) < ε
| then writeln('matrix not invertible')
| else begin
| | matrix_inverse(A, inv_A) ;
| | display(inv_A)
| end
end .
```

Remember that the test $u = v$ between two real numbers *will not* give the desired result because of numerical errors. Thus, we must replace the test $\det(A) = 0$ by the test $|\det(A)| < \varepsilon$ where ε is chosen in a realistic manner.

But, we cannot relax too soon! We can (and must) improve on this second attempt, because we can never rely blindly on results displayed by a machine. They could be wrong (but the probability is tiny) or, what is more likely, we could have made an error in coding. So, we will only accept a result after verifying it: we require our program to multiply the matrices A and inv_A and display the result. If we obtain a matrix very near the identity, we know that the probablility of simultaneous errors which cancel one another out is very, very small. So a good main program is the following.

```
begin
| message ; choose(A) ;
| if abs(det(A)) < ε
| then writeln('matrix not invertible')
| else begin
| | matrix_inverse(A, inv_A) ;
| | display(inv_A) ;
| | verification(A, inv_A)
| end
end .
```

As you can see, several lines of code can require lots of time. Do not be in too much of a hurry. Re-read and criticize ...

[5] Even the first time, it is necessary to protect yourself ...

Declarations

We can now specify the declarations used in our program. These use the number ε and the matrices A and inv_A. Since the matrices are arguments in procedures, we must define their type:

```
program matrix_inverse ;
const ε = 0.00000001 ;
type matrix = array[1..4, 1..4] of real ;
var A, inv_A : matrix ;
```

The procedure message

This is a sequence of "writeln('. . . ')" statements that explain what your program is going to do.

The procedure choose

We need two indices i and j. It is essential to declare these variables as local variables of the procedure because they are the control variables of a "for" loop. Note also the declaration "var" which allows the procedure to modify (via the keyboard) the variable A in the program.

```
procedure choose(var A : matrix) ;
var i, j : integer ;
begin
| for i := 1 to 4 do
| for j := 1 to 4 do begin
| | write('A[', i : 1, ',', j : 1, '] = ') ; readln(A[i, j])
| end
end ;
```

7.1.4. The function det

The main program uses the function *det*. Unlike *abs*, this function is not known to Pascal. Hence, we teach our program how to calculate it by copying (7.4).

```
function det(A : matrix) : real ;
begin
| det :=
|   A[1, 1] * minor_3(A, 1, 1)
| −A[2, 1] * minor_3(A, 2, 1)
| +A[3, 1] * minor_3(A, 3, 1)
| −A[4, 1] * minor_3(A, 4, 1)
end ;
```

Note the placement on the page, which simplifies verification.

The function minor_3

Since *det* uses the function $minor_3$, we immediately use (7.5) and write out the code for this function. The placement on the page is very important!

```
function minor_3(A : matrix ; i, j : integer) : real ;
begin
| minor_3 :=
|   A[λ(i, 1), λ(j, 1)] * minor_2(A, i, λ(i, 1), j, λ(j, 1))
| −A[λ(i, 2), λ(j, 1)] * minor_2(A, i, λ(i, 2), j, λ(j, 1))
| +A[λ(i, 3), λ(j, 1)] * minor_2(A, i, λ(i, 3), j, λ(j, 1))
end ;
```

The function lambda

Since *minor_3* uses λ, we must also describe how to calculate λ for our program.

```
function λ(i, k : integer) : integer ;
begin
| if k < i then λ := k else λ := k + 1
end ;
```

The function minor_2

Since the *minor_3* uses the function *minor_2*, we also code (7.8).

```
function minor_2(A : matrix ; i1, i2, j1, j2 : integer) : real ;
begin
| minor_2 :=
|   A[μ(i1, i2, 1), μ(j1, j2, 1)] * A[μ(i1, i2, 2), μ(j1, j2, 2)]
| −A[μ(i1, i2, 1), μ(j1, j2, 2)] * A[μ(i1, i2, 2), μ(j1, j2, 1)]
end ;
```

The function mu

Finally, since the function *minor_2* uses the function μ, we implement μ using (7.9). The two internal "begin … end" statements (and the vertical lines that accompany them) are not needed from the point of view of syntax. We retain them because they greatly facilitate comprehension.

```
function μ(i1, i2, k : integer) : integer ;
begin
| if i1 < i2 then begin
| | if k < i1 then μ := k else
| | if k + 1 < i2 then μ := k + 1 else μ := k + 2
| end
| else begin {case i2 < i1 since i1 ≠ i2}
| | if k < i2 then μ := k else
```

```
    if k + 1 < i₁ then μ := k + 1 else μ := k + 2
  end
end ;
```

The procedure matrix_inverse

Now that we have finished with the calculation of the determinant function and its auxiliaries, we turn to the next action of the main program that has not already been specified: the calculation of the inverse of A. For this, we implement (7.3) and (7.2).

```
procedure matrix_inverse(A : matrix ; var inv_A : matrix) ;
var i, j : integer ; Δ : real ;
begin
  Δ := det(A) ;
  for i := 1 to 4 do
  for j := 1 to 4 do begin
    if (i + j) mod 2 = 0
    then inv_A[i, j] := minor_3(A, j, i)/Δ
    else inv_A[i, j] := −minor_3(A, j, i)/Δ
  end
end ;
```

The code has been polished in a number of places.

- Transposition is accomplished by tinkering with the indices: (i, j) before the assignment sign, (j, i) after;
- The calculation of $(-1)^{i+j}$ is based on the parity of $i + j$
- The determinant is handled so that it does not have to be computed sixteen times.

The procedure display

Notice how the statements "write" and "writeln" alternate and how the display "*write*(*inv_A*[i, j]: 8 : 4)" is formatted for real numbers.

```
procedure display(inv_A : matrix) ;
var i, j : integer ;
begin
  for i := 1 to 4 do begin
    for j := 1 to 4 do write(inv_A[i, j] : 8 : 4) ;
    writeln
  end
end ;
```

The procedure verification

We multiply the matrices A and inv_A, and display the result. If it is not sufficiently close to the identity matrix, we should fear the worst.

```
procedure verification(A, inv_A : matrix) ;
var i, j : integer ;  unit_mat : matrix ;
begin
| mult_matrix(A, inv_A, unit_mat) ;
| display(unit_mat)
end ;
```

The procedure mult_matrix

As the name suggests, this procedure returns the product $C = AB$. The code is a classical calculation of $4 \times 4 = 16$ sums. Turn to Chapter 6 if you do not understand the use of the local variable *temp*.

```
procedure mult_matrix(A, B : matrix ;  var C : matrix) ;
var i, j, k : integer ;  temp : real ;
begin
| for i := 1 to 4 do
| for j := 1 to 4 do begin
| | temp := 0 ;
| | for k := 1 to 4 do temp := temp + A[i, k] * B[k, j] ;
| | C[i, j] := temp
| end
end ;
```

7.1.5. How to type a program

The order in which we have written the procedures is not at all the order required by the compiler; instead, the compiler requires them in the reverse order! Thus, when we are typing a program, we need to run over our notes in reverse.

The order which the compiler requires is very easy to understand: at each instant, it must know the procedures and functions *called* by the code it is currently translating.

Since *det* uses the function *minor_3*, the code for *minor_3* must come before that of *det*. For the same reason, the code for λ and *minor_2* must precede that of *minor_3*, and so on. The principal body of the program is typed after all the procedures and functions. Hence, one possible order is:

```
program matrix_inverse ;
{declarations of constants, types and variables (in this order)}
procedure message ;
procedure choose(var A : matrix) ;
```

```
function λ(i, j : integer) : integer ;
function μ(i1, i2, k : integer) : integer ;
function minor2(A : matrix ; i1, i2, j1, j2 : integer) : real ;
function minor3(A : matrix ; i, j : integer) : real ;
function det(A : matrix) : real ;
procedure matrix_inverse(A : matrix ;
var inv_A : matrix) ;
procedure display(inv_A : matrix) ;
procedure mult_matrix(A, B : matrix ; var C : matrix) ;
procedure verification(A, inv_A : matrix) ;
begin
|main body of the program
end .
```

Each procedure or function is followed by its code.

7.2. Characteristic Polynomial of a Matrix

The following definition allows us to avoid errors which are difficult to detect. There are two ways of denoting polynomials:

$$\begin{aligned} A(X) &= a_n X^n + a_{n-1} X^{n-1} + \cdots + a_1 X + a_0, \\ &= b_0 X^n + b_1 X^{n-1} + \cdots + b_{n-1} X + b_n \end{aligned}$$

The first notation, where the index equals the exponent, is the one that we encounter most often today. The second, where the sum of the index and the exponent is equal to the degree, is encountered more often in older works. For this reason, we call the first notation the *modern notation* and the second the *old notation*.

Let A be a matrix with real coefficients and suppose that we want to calculate its characteristic polynomial. Then (note the sign and the old notation),

$$P(\lambda) = (-1)^n \det(A - \lambda I) = \lambda^n + p_1 \lambda^{n-1} + \cdots + p_n.$$

Unlike Maple or Mathematica, Pascal does not allow *symbolic* calculations carried out with *indeterminates*. Thus we cannot calculate this determinant because it involves arithmetic expressions containing λ. This accounts for the difficulty (but also the charm) of this problem.

Certain programmable calculators proceed as follows: they first determine the *numerical* values of the determinants $P(0), \ldots, P(n)$ and then recover P by Lagrange interpolation.[6]

[6] The reader is strongly encouraged to write the corresponding program

We are going to explain, and then program, a very elegant algorithm due to the mathematician-astronomer Leverrier.[7] Let $\lambda_1, \dots, \lambda_n$ be the eigenvalues of the matrix A so that

$$P(\lambda) = (\lambda - \lambda_1) \cdots (\lambda - \lambda_n).$$

Introduce the Newton sums

$$S_k = \lambda_1^k + \cdots + \lambda_n^k,$$

for $1 \leq k \leq n$. We can deduce the $p_1, \dots, p_n$ from the S_k thanks to the Newton-Girard formulas which we shall establish in Chapter 10.

$$\begin{cases} p_1 + S_1 = 0, \\ 2p_2 + p_1 S_1 + S_2 = 0, \\ \dots \\ np_n + p_{n-1} S_1 + \cdots + p_1 S_{n-1} + S_n = 0. \end{cases}$$

Contrary to what one might think, one can compute the Newton sums without first determining the eigenvalues.

Lemma 7.2.1. *For every integer* $k \geq 0$, $S_k = \mathrm{Trace}(A^k)$.

Proof. We put A in upper triangular form using a matrix Q (with, perhaps, complex coefficients).

$$Q^{-1} A Q = \begin{pmatrix} \lambda_1 & * & \cdots & * \\ & \lambda_2 & \cdots & * \\ & & \ddots & \vdots \\ & & & \lambda_n \end{pmatrix}.$$

Then

$$Q^{-1} A^k Q = \begin{pmatrix} \lambda_1^k & * & \cdots & * \\ & \lambda_2^k & \cdots & * \\ & & \ddots & \vdots \\ & & & \lambda_n^k \end{pmatrix},$$

which immediately gives

$$S_k = \mathrm{Trace}\,(Q^{-1} A^k Q) = \mathrm{Trace}(A^k).$$

[7] Urbain Leverrier (1811–1877) became famous for his discovery in 1846 of the planet Uranus *by calculation alone* from its perturbations of the orbit of the planet Neptune.

The algorithm that Leverrier proposed consists of the following two steps:

- calculate $A, \ldots, A^n$ to get the Newton sums $S_1, \ldots, S_n$;
- solve the traingular system of Newton-Girard equations.

Before encoding this as a program, we pause to ask how we might verify the result. Clearly, we can run a number of preliminary tests with, for example, the triangular matrices. But tests, no matter how sophisticated, *cannot* prove that the program is correct; at best, they can detect a programming error by exhibiting incorrect results.

Theorem 7.2.1 (Hamilton-Cayley). *Let A be a matrix with coefficients in a commutative ring with unit. If* $P(\lambda) = \det(\lambda I - A) = \lambda^n + p_1\lambda^{n-1} + \cdots + p_n$ *is the characteristic polynomial of the matrix A, then the following matrix equation holds:*

$$A^n + p_1 A^{n-1} + \cdots + p_{n-1}A + p_0 I = 0.$$

Proof. Although this result is well-known, its proof is less so. Here is a simple proof which makes no use of vector spaces and which, therefore, holds for matrices with coefficients in any commutative ring (with unit) whatsoever. If we replace A by the matrix $(A - \lambda I)$ in (7.2) and if we put

$$B(\lambda) = (-1)^{n\,t}\,\mathrm{Adj}(A - \lambda I),$$

we obtain the matrix equation (with $p_0 = 1$):

$$(A - \lambda I)B(\lambda) = P(\lambda)I = (p_0\lambda^n + p_1\lambda^{n-1} + \cdots + p_{n-1}\lambda + p_n)I. \quad (7.10)$$

It is clear that the matrix $B(\lambda)$ is a polynomial in λ of degree at most $n - 1$, so that we can write (in modern notation)

$$B(\lambda) = B_{n-1}\lambda^{n-1} + \cdots + B_1\lambda + B_0, \quad (7.11)$$

where the B_i are matrices which we want to specify. Substituting into (7.10) and equating coefficients of the same degree, we obtain

$$\begin{array}{c|rl}
I & p_n I &= AB_0, \\
A & p_{n-1} I &= AB_1 - B_0, \\
A^2 & p_{n-2} I &= AB_2 - B_1, \\
\cdots & & \cdots \\
A^{n-1} & p_1 I &= AB_{n-1} - B_{n-2}, \\
A^n & p_0 I &= -B_{n-1}.
\end{array}$$

It remains to multiply these equations by $I, A, A^2, \ldots, A^n$ respectively and add them term by term to obtain the desired result.

We are going to use this celebrated theorem to verify our program: if $P(A)$ is the zero matrix, we can be reasonably certain that P is the characteristic polynomial of A. In fact, since the verification is totally independent of the calculation of p, the probablility of simultaneous errors that cancel one another out is miniscule.

7.2.1. The program Leverrier

The main body of the program

We enounter again the classical trichotomy:

- introduction of data (preceeded by a message explaining what the program does and what data is required by the computer)
- treatment of the data; that is, calculation of Newton sums and solution of the linear system of Newton-Girard equations;
- display of results after verification.

```
begin
| message ;
| choose(A, dim) ;
| store_traces(A, Newton_Sum, dim) ;
| solve_system(Newton_Sum, char_poly, dim) ;
| Hamilton_Cayley(char_poly, A, dim) ;
| display(char_poly, dim)
end .
```

Declarations

Our program makes use of the matrix A, its dimension *dim*, and the vectors *Newton_Sum* and *char_poly*.

- We want to calculate the characteristic polynomial of a matrix A of any dimension. Since Pascal only allows arrays whose dimension is fixed at the moment of declaration, we reserve a large space in memory even though we usually only use a small part of it. This is the reason for the appearance of the constant *dim_max*.
- The variables *Newton_Sum* and *char_poly* are vectors of the same dimension. For convenience, the indices start at zero.

```
program Leverrier ;
const dim_max = 10 ;
type matrix = array[1..dim_max, 1..dim_max] of integer ;
vector = array[0..dim_max] of integer ;
var A : matrix ;
Newton_sum, char_poly : vector ;
dim : integer ;
```

We have chosen to work with integer coefficients (which simplify some tests); but there is no reason that you cannot modify the program to work with rational, real or complex coefficients.

For beginners

One should avoid delusions of grandeur such as entering without thinking $dim\_max = 100$. Would you be willing to type the $100^2 = 10000$ coefficients in such a matrix?

The procedure choose

This procedure starts by asking for the dimension of the matrix, verifying that it is correct, and prompting for entry of the coefficients.

• Note the "repeat until" designed to check the validity of the dimension of A.

• We increase ease of use by displaying the name of the coefficient that is to be entered. We also adhere to the usual typographic conventions and place a space on each side of the the equals sign "=".

```
procedure choose(var A : matrix ; var dim : integer) ;
var i, j : integer ;
begin
  repeat
    write('dimension = ') ; readln(dim)
  until (2 ≤ dim) and (dim ≤ dim_max) ;
  for i := 1 to dim do
  for j := 1 to dim do begin
    write('A[', i : 1, ',' , j : 1, '] = ') ; readln(A[i, j])
  end ;
end ;
```

Calculating traces

The sequence $M_k = A^k$ satisfies the first order recurrence relation:

$$M_0 = I, \quad M_k = AM_{k-1} \quad \text{if} \quad k \geq 1.$$

We can then calculate the traces using the algorithm:

```
M := I_n ;
for k := 1 to n do begin
  M := AM ; S_k := trace(M)
end
```

To transform this algorithm into a procedure, we observe that we need to have the following at our disposal:

- a procedure *unit_mat*(X, n) which implements the assignment $M := I_n$;
- a procedure *matrix_product*(Z, X, Y, n) which returns $Z = XY$;
- a function *trace*(X, n) which returns the trace of a matrix X.

```
procedure store_traces(A : matrix ;
                       var Newton_Sum : vector ; dim : integer) ;
var k : integer ; M : matrix ;
begin
| unit_mat(M, dim) ; {M = I_dim, whence M = A^0}
| for k := 1 to dim do begin
| | product_matrix(M, A, M, dim) ; {M = AM, whence M = A^k}
| | Newton_Sum[k] := trace(M, dim)
| end
end ;
```

The procedure unit_mat

This procedure returns the unit matrix (do not forget the "var").

```
procedure unit_mat(var M : matrix ; dim : integer) ;
var i, j : integer ;
begin
| for i := 1 to dim do
| for j := 1 to dim do
| if i = j then M[i, j] := 1 else M[i, j] := 0
end ;
```

The procedure product_matrix

Look over the part of Chapter 6 devoted to arrays if you do not understand the role of the variable *temp*.

```
procedure product_matrix(var Z : matrix ; X, Y : matrix ; dim : integer) ;
var i, j, k, temp : integer ;
begin
| for i := 1 to dim do
| for j := 1 to dim do begin
| | temp := 0 ;
| | for k := 1 to dim do temp := temp + X[i, k] * Y[k, j] ;
| | Z[i, j] := temp
| end
end ;
```

The function trace

We use the local variable *temp* because we cannot use the name of a function to accumulate a sum

```
function trace(X : matrix ; dim : integer) : integer ;
var i, temp : integer ;
begin
| temp := 0 ;
| for i := 1 to dim do temp := temp + X[i, i] ;
| trace := temp
end ;
```

The procedure solve_system

This is a classical exercise that we have already encountered.

```
procedure solve_system(Newton_Sum : vector ;
                              var coeff : vector ; dim : integer) ;
var i, k, temp : integer ;
begin
| for k := 1 to dim do begin
| | temp := Newton_Sum[k] ;
| | for i := 1 to k - 1 do
| |      temp := temp + coeff[i] * Newton_Sum[k - i] ;
| | coeff[k] := -temp div k   {"div" because one is working}
| |                            {over the integers}
| end
end ;
```

The procedure display

We observe the following conventions.

- we do not write $0X^k$;
- we write X^k instead of $1X^k$;
- we write $-3X^k$ instead of $+-3X^k$;
- we write $-X^k$ instead of $-1X^k$.

```
procedure display(coeff : vector ; dim : integer) ;
var i, j : integer ;
begin
| write('X^', dim : 1) ;
| for i := 1 to dim do
| if coeff[i] > 1 then write(' + ', coeff[i] : 1, 'X^', dim - i : 1)
| else if coeff[i] = 1 then write(' + X^', dim - i : 1)
| else if coeff[i] = -1 then write(' - X^', dim - i : 1)
| else if coeff[i] < -1 then write(' - ', -coeff[i] : 1, 'X^', dim - i : 1)
end ;
```

Since the procedure "write" does not display the sign of a positive number, we must supply it. Why is it imperative to preceded the last "else" with a the test "if *coeff*[i] < 0"?

You can further improve the display by including the following usages:

- write $6X$ instead of $6X^1$;
- write 5 or -7 instead of $5X^0$ or $-7X^0$.

Exercise 1

Let $A = X^n + a_{n-1}X^{n-1} + \cdots + a_1 X + a_0$ be a monic polynomials with roots $\alpha_1, \ldots, \alpha_n$. Let P be a second monic polynomial. We want to find a monic polynomial B whose roots are the numbers $P(\alpha_1), \ldots, P(\alpha_n)$. In other words, knowing that

$$B(X) = \big(X - P(\alpha_1)\big) \cdots \big(X - P(\alpha_n)\big),$$

can we calculate the coefficients of B using those of A? To solve this classical problem, associate to the polynomial

$$A = X^n + a_{n-1}X^{n-1} + \cdots + a_1 X + a_0$$

its *companion matrix*

$$\tilde{A} = \begin{pmatrix} 0 & 1 & & & \\ & 0 & 1 & & \\ & & 0 & \ddots & \\ & & & \ddots & 1 \\ -a_0 & -a_1 & \cdots & \cdots & -a_{n-1} \end{pmatrix}.$$

Here, the coefficients that are not written are zero.

1) By adding to the first column of $(\lambda I - \tilde{A})$ the successive columns multiplied by $\lambda, \lambda^2, \ldots, \lambda^{n-1}$, show that the characteristic polynomial of $\tilde{A}$ is equal to $\pm A$.

2) Show (by putting $\tilde{A}$ in triangular form) that the eigenvalues of the matrix $P(\tilde{A})$ are precisely $P(\alpha_1), \ldots, P(\alpha_n)$. It follows that the polynomial $B = \det(P(\tilde{A}) - \lambda I)$ is, up to sign, the desired solution and that the coefficients of B are polynomials in the coefficients of A and of P. Write a program that calculates, then displays, the characteristic polynomial of $P(\tilde{A})$.

Exercise 2

Let A, B be two rectangular matrices of arbitrary dimension. Put

$$A \otimes B = \begin{pmatrix} a_{1,1}B & a_{1,2}B & \ldots & a_{1,p}B \\ a_{2,1}B & a_{2,2}B & \ldots & a_{2,p}B \\ \ldots & \ldots & \ldots & \ldots \\ a_{n,1}B & a_{n,2}B & \ldots & a_{n,p}B \end{pmatrix}.$$

Theorem 7.2.2. *Let A and B be square matrices of dimensions n and m with eigenvalues $\alpha_1, \ldots, \alpha_n$ and $\beta_1, \ldots, \beta_m$, respectively. Then the eigenvalues of $A \otimes B$ are the $\alpha_i \times \beta_j$; and the eigenvalues of $A \otimes I + I \otimes B$ are $\alpha_i + \beta_j$.*

Proof. We have the identities

$$(AA') \otimes (BB') = (A \otimes B)(A' \otimes B').$$

If we bring A and B to triangular form $P^{-1}AP = T_A$ and $Q^{-1}BQ = T_B$, and use the fact that $(P \otimes Q)^{-1} = P^{-1} \otimes Q^{-1}$, then it follows that

$$(P \otimes Q)^{-1}(A \otimes B)(P \otimes Q) = (P^{-1} \otimes A \otimes P) \otimes (Q^{-1} \otimes B \otimes Q) = T_A \otimes T_B\,.$$

We finish by noting that $T_A \otimes T_B$ is a triangular dmatrix with coefficients $\alpha_i\, \beta_j$ along the diagonal. For $A \otimes I + I \otimes B$, use the identity:

$$(P \otimes Q)^{-1}(A \otimes I + I \otimes B)(P \otimes Q) = T_A \otimes I + I \otimes T_B\,.$$

□

We say that a complex number z is an *algebraic integer* if it is the root of a monic polynomial with rational coefficients:

$$z^n + a_{n-1}z^{n-1} + \cdots + a_0 = 0, \quad a_i \in \mathbb{Q}.$$

By replacing "rational" by "integral" in the preceding proof, we find that

Theorem 7.2.3. *The algebraic numbers are a subfield of* $\mathbb{C}$.

Exercise 3

Write a program to explicitly calculate a polynomial with having $\alpha + \beta$ or $\alpha\beta$ as roots given polynomials which vanish on α and β. (Use companion matrices and tensor products.)

7.3. How to Write a Program

The advice that you are going to receive is not original. Re-read the proverbs at the beginning as well as Descartes' *Discourse on Method* (1637).

Define the problem

This is not at all simple! There are frequently many implicit hypotheses that need to be made made precise and that are not perceived immediately. A good technique is to imagine that a program is running and to continually pose the questions: "What is the program doing? What am I expecting from it? What information must I communicate to it?"

Adjust the algorithms

This is the stage of first order recurrences and mathematical reasoning. At this point, you have total intellectual freedom because you are not yet programming.

Define the types

Return to Earth ... You need to decide how to represent the objects under consideration in the memory of the machine. Since we are using Pascal without pointers, we can hardly ever use arrays or pointers. On the other hand, the choices are unlimited for a professional. A poorly chosen type can make the writing of a program very painful and heavily penalize performance.

To program is to role-play

A good programmer *structures* his or her program; that is, separates a procedure into statements *on the same level*. It is necessary to learn to separate the incidental from the essential: *one should not climb stairs while chewing gum*. When we write the main body of a program for example, we must not ask ourselves about the contents of the procedures or functions (this is for "tomorrow", or even "the day after tomorrow"); we simply *imagine* the names of procedures, their effect and their interplay.

We are lead to successively play different roles: each level of the program corresponds to a different role.

When I write the main body of the program, I am the CEO. My work consists in imagining, in distributing, and in coordinating the tasks; I do not execute them! (*One does not execute orders which have not been given.*) Once the body of the program is written, I become an engineer when I write the large procedures and give the main orders, a foreman when I write the more detailed procedures, then a janitor, and so on.

A poorly structured program is one that contains orders from different levels. When I am the CEO, I am busy with the future of the business. I do not ask myself if there is toilet paper in the second restroom on the right hand side of the third floor! This will come, but much later, when I am the janitor. Do not be too hurried ...

The beginning

We write our program by beginning with the main body of the program; then we specify the procedures and the functions.

The writing is done by touching up, by *successive refinements*: one surmounts difficulties one after another. In the main body of the program, we do not concern ourselves with useless detail; we deliberately stay at the level of generalities. It is important not to start writing overly detailed code prematurely.

To accomplish this, we *mask*[8] very technical portions of the code by procedures and functions which we only *suppose* exist: *tomorrow, things will be better, and the day after tomorrow, better still.* Have confidence, you will come

[8] One does not execute orders which have not been given.

to the point where you write them; do not give into anxiety and try doing it right away.

Some beginners often have a psychological block against working with a procedure which they have not already written (although these same individuals will apply a theorem they cannot prove). It seems easier to accept a static hypothesis that one *sees* than a dynamic process that one does *not see*. (We shall return to this difficulty when dealing with recursion.)

This way of functioning is familiar to mathematicians: "Let's see whether I can prove my theorem by provisionally supposing that properties A, B and C are true."

The code must remain limpid; one should be able to recognize without effort what algorithm is being used.

The main body of the program

The main body consists of three phases:

- data entry (accompanied in due course by initialization and a message explaining what the user can expect);
- treatment of the data (the program, properly speaking);
- display of results.

Do not be fooled by the apparent simplicity of the main body of the program. As you will discover, some lines require much effort, and often many tries.

The procedures

Multiply procedures and functions so that you will not have to master more than five to ten lines of code at a time. With practice, you will be able to increase the number of lines a little (but do not run over a single screen!)

The names of procedures and functions are very important. For example, avoid calling a procedure "calculation" because all procedures calculate something. A name such as this, which applies to everything, carries no information and will oblige you to comment! Use your imagination; if necessary, spend at least a *minute* to find a good identifier.

Do not hesitate to give a very long (hence very informative) name to a variable or procedure that you will not use too much; reserve short names for objects that you will use repeatedly.

To determine if you have forgotten a procedure or function, re-read what you have done. If you find yourelf making comments to yourself like " Oh, yes! I am calculating the determinant of the matrix", then you have forgotten to define the determinant function.

Be very attentive to the problems of transmitting information from one procedure to another.

Complete procedures or modules separately before assembling them. Build libraries of procedures (manipulation of fractions, of matrices, of polynomials, and so on).

Loops

First put in place external loops without trying to specify their content: choose the type carefully using solid arguments; contemplate the result, mentally verifying their function using a trace and carefully examining the limit cases, because these can cause a program to crash or loop endlessly! *Proceed slowly, take your time.*

When you are satisfied, you can pass to a deeper level using the same approach. The process resembles peeling an onion.

Comments

Never forget your imaginary interlocutor because the best way to understand what you have done is to *explain it to yourself.* This technique is a very powerful device to help you become more conscious of what you are doing and to mentally unblock yourself. And, to be sure you are talking to this imaginary interlocutor, write comments.

Avoid comments which are too long: they disfigure a program and break its unity. The ideal comment is brief and fits on the same line as the statement it clarifies. Nothing is more painful than to be obliged to read three lines of code followed by five lines of comments, then a line of code followed by two lines of comments, etc.

Personally, I prefer to write a large comment *before* the code and place references "see (1)", "see (2)", etc. in the code.

Re-read

Do not be too hurried to run through your program. Re-read each procedure trying to implement mentally what it does: does the code correctly translate your thought? Is information being transmitted correctly? Experience shows that you will avoid losing many hours debugging a program that obstinately refuses to work.

Verify the results

Never believe a program, even your own! Any time that it is possible, seek cross-checks.

7.4. A Poorly Written Procedure

To put the preceeding advice into practice, we are going to dissect a procedure written by a beginner and understand why it is poorly written. This procedure was inserted in a program to solve a linear system $Ax = b$.

```
procedure triangularization(var A : matrice ;
                            var b : vector ; dim : integer) ;
var i, j, k : integer ;
begin
  test := test_singularity(A, 1) ;
  k := 1 ;
  enlarge(A, b, 1) ;
  while (test = false) and (k ≤ dim − 1) do begin
    for i := k + 1 to dim do
    for j := k + 1 to dim do
    A[i, j] := A[i, j] − (A[i, k]/A[k, k]) ∗ A[k, j] ;
    b[i] := b[i] − (A[i, k]/A[k, k]) ∗ b[k] ;
    enlarge(A, b, k + 1) ;
    test := test_singularity(A, k + 1) ;
    k := k + 1
  end
end ;
```

Let us examine the code:

• The variable k denotes the current column. This is certainly not clear at first glance! It would be better to call it *current_column* or *col*.

• The variable *test* is a global variable of the program which is surreptiously modified by the procedure so to inform the program when the matrix is not invertible. Thus, we have a context effect.

• The identifier *test* is poorly chosen: what does it mean? Note that *test_singularity*(A, k) is a boolean which tells us whether or not $(a_{i,j})_{1 \le i,j \le k}$ is invertible. Why not call it *Cramer*?

• The procedure *enlarge* leaves one perplexed: what does it do? Does it augment the matrix by bordering it? Not at all! When I asked this question to the beginner, he replied that it sought the number ℓ of the row with the largest coefficient (for reasons of numerical stability) in the k-th column, then exchanges rows k and ℓ. The choice of this identifier is not judicious. Besides, it conflates in a single procedure two actions of a different nature (finding a pivot, exchanging two rows) and this obscures the algorithm.

• The statements *test* := *test_singularity* and *enlarge* occur outside and inside the "while" loop which indicate a bad choice of loop.

• Finally, lines 8 to 11 (the two embedded "for" loops and the statement that follows) are incomprehensible. One must read them very carefully before realising that they kill the coefficients below the pivot. This suggests *prematurely written* code which should be replaced by a procedure.

These criticisms allow us to improve the procedure.

```
procedure triangularization(var A : matrice ;  var b : vector ;
                            var Cramer : boolean ;  dim : integer) ;
var col, place_pivot, coeff : real ;
begin
  col := 1 ;  Cramer := true ;
  repeat
    seek_largest_pivot(A, col, place_pivot, Cramer, dim) ;
    if Cramer then begin
      exchange_rows(A, b, col, place_pivot, dim) ;
      zero_out_under_pivot(A, b, col, dim)
    end ;
    col := col + 1
  until (col ≥ dim − 1) or not Cramer
end ;
```

Isn't the new version more comprehensible, hence more certain?

8. The Integers

8.1. The Euclidean Algorithm

To calculate the GCD of two numbers, we play "ping pong" with the formulas $\text{GCD}(a, b) = \text{GCD}(a, b - a) = \text{GCD}(a - b, b)$ ending with $\text{GCD}(a, 0) = \text{GCD}(0, a) = |a|$. This gives for example

$$\begin{aligned}\text{GCD}(12, 7) &= \text{GCD}(5, 7) = \text{GCD}(5, 2) = \text{GCD}(3, 2) \\ &= \text{GCD}(1, 2) = \text{GCD}(1, 0) = 1.\end{aligned}$$

Formalizing this, we see that we obtain two sequences of numbers (a_n) and (b_n) such that $\text{GCD}(a, b) = \text{GCD}(a_n, b_n)$ and the first order recurrence:

```
if a_n ≥ b_n
then begin a_{n+1} := a_n − b_n ; b_{n+1} = b_n end
else begin a_{n+1} := a_n ; b_{n+1} = b_n − a_n end
```

The translation into code, called the *additive Euclid algorithm*, is immediate:

```
a, b := integers ≥ 0 ;
while (a ≠ 0) and (b ≠ 0) do
    if a ≥ b then a := a − b else b := b − a ;
if a = 0 then GCD := b else GCD := a
```

Some students suggest replacing the test $a = 0$ by the statement $\text{GCD} := a + b$. This is not really a good idea[1] because a test is much more rapid than an addition. In order to speed things up, we can regroup subtractions by the same number which amounts to introducing Euclidean division. We obtain the *Euclidean algorithm*

```
a, b := integers ≥ 0 ;
while (a ≠ 0) and (b ≠ 0) do
    if a ≥ b then a := a mod b else b := b mod a ;
if a = 0 then GCD := b else GCD := a
```

[1] Above all, no tricks!

This algorithm is usually presented using a single sequence (r_n) obtained from the sequences sequences (a_n) and (b_n) by putting $r_{2n} = a_n$ and $r_{2n+1} = b_n$. The GCD is then the last non-zero remainder.

$$\begin{cases} r_0 = a, \quad r_1 = b, & a \geq b \geq 0, \\ r_0 = r_1 q_1 + r_2, & 0 < r_2 < r_1, \\ r_1 = r_2 q_2 + r_3, & 0 < r_3 < r_2, \\ \dots & \\ r_{n-2} = r_{n-1} q_{n-1} + r_n, & 0 < r_n < r_{n-1}, \\ r_{n-1} = r_n q_n & (r_{n+1} = 0 \text{ and } q_n \geq 2). \end{cases} \tag{8.1}$$

Example

We calculate the GCD of 10,780 and 3,675 as follows:

$$\begin{aligned} 10780 &= 2 \cdot 3675 + 3430, \\ 3675 &= 1 \cdot 3430 + 245, \\ 3430 &= 14 \cdot 245 + 0. \end{aligned}$$

The GCD is the last nonzero remainder, namely $\text{GCD}(10780, 3675) = 245$.

Theorem 8.1.1. *The Euclidean algorithm correctly calculates the* GCD *of two numbers.*

Proof. The algorithm terminates because the sequence (r_i) in (8.1) is strictly decreasing and bounded below by 0. The last nonzero remainder is the GCD because $\text{GCD}(a, b) = \text{GCD}(r_0, r_1) = \cdots = \text{GCD}(r_{n-1}, r_n) = r_n$ (in other words, the function $\text{GCD}(a, b)$ is an invariant of the loop). □

8.1.1. Complexity of the Euclidean algorithm

In the middle of the last century, the French mathematician G. Lamé[2] proved that the Euclidean algorithm was very efficient.

Theorem 8.1.2. *The number of divisions required in the Euclidean algorithm is less than or equal to five times the number of digits of the smallest of the two numbers whose* GCD *is being calculated.*

Proof. The formulas (8.1) contain n divisions. Let $F(n)$ be the Fibonacci sequence. Knowing that $F_2 = 1$ and $F_3 = 2$, we immediately have $r_n \geq F_2$ as well as $r_{n-1} \geq 2r_n \geq 2F_2 \geq F_3$. We deduce that $r_{n-2} \geq r_{n-1} + r_n \geq F_3 + F_2 \geq F_4$ whence $r_1 = b \geq F_{n+1}$ by induction. Let $\gamma = \frac{1}{2}(1 + \sqrt{5})$ be the

[2] Gabriel Lamé (1795–1870), a railroad engineer, was considered by Gauss to be one of best French mathematicians of the era.

golden number; that is the positive root of the equation $X^2 = X + 1$. A simple induction shows that $F_n > \gamma^{n-2}$ for $n \geq 3$. Knowing that $b \geq F_{n+1} > \gamma^{n-1}$ and $\log_{10} \gamma = 0.208\ldots > \frac{1}{5}$, we deduce that

$$\log_{10} b > (n-1) \log_{10} \gamma > \tfrac{1}{5}(n-1).$$

To say that b can be written with k numerals in base 10 means that $\log_{10} b < k$. Consequently, $n - 1 < 5 \log_{10} b < 5k$ shows that $n \leq 5k$. □

Remark

The precise result is not important. It suffices to remember that the number of required divisions is bounded by $C \log b$ where C is a constant and b is the smallest of the two numbers.

8.2. The Blankinship Algorithm

From the definition $d\mathbb{Z} = a\mathbb{Z} + b\mathbb{Z}$ of the GCD, it follows that there exist $u, v \in \mathbb{Z}$ such that $au + bv = d$. But this does not tell us how to calculate u and v. This is a nice example of static mathematics! One way to find u and v is to reverse the steps of the Euclidean algorithm. Doing this, for example, for the calculation of the GCD of 10780 and 3675 gives:

$$\begin{aligned} 245 &= 3675 - 3430, \\ &= 3675 - (10780 - 2 \cdot 3675) \\ &= 3 \cdot 3675 - 10780. \end{aligned}$$

This algorithm does not interest programmers because it requires storing *all* intermediate results. Happily, it is possible to calculate u and v in the course of the Euclidean algorithm by *surfing* on the edge of the calculations. To do this, it suffices to adapt the method of Gauss pivoting to integers.[3] Suppose that we are calculating the GCD of 252 and 198 (see Table 8.1). To calculate u and v, Blankinship proposed constructing a matrix

$$M_0 = \begin{pmatrix} 252 & 1 & 0 \\ 198 & 0 & 1 \end{pmatrix}$$

and applying the Euclidean algorithm to the first column all the while extending the operations to the rows of M. The algorithm terminates when the first column contains the GCD. At this moment, u and v are found across from the GCD :

$$18 = 4 \cdot 252 - 5 \cdot 198.$$

[3] W.A. Blankinship, *A new version of the Eulidean Algorithm*, American Mathematical Monthly 70 (1963), pp. 742–745.

Old matrix	Pivot	Manipulation	New matrix
$M_0 = \begin{pmatrix} 252 & 1 & 0 \\ 198 & 0 & 1 \end{pmatrix}$	198	$L_1 := L_1 - L_2$	$M_1 = \begin{pmatrix} 54 & 1 & -1 \\ 198 & 0 & 1 \end{pmatrix}$
$M_1 = \begin{pmatrix} 54 & 1 & -1 \\ 198 & 0 & 1 \end{pmatrix}$	54	$L_2 := L_2 - 3L_1$	$M_2 = \begin{pmatrix} 54 & 1 & -1 \\ 36 & -3 & 4 \end{pmatrix}$
$M_2 = \begin{pmatrix} 54 & 1 & -1 \\ 36 & -3 & 4 \end{pmatrix}$	36	$L_1 := L_1 - L_2$	$M_3 = \begin{pmatrix} 18 & 4 & -5 \\ 36 & -3 & 4 \end{pmatrix}$
$M_3 = \begin{pmatrix} 18 & 4 & -5 \\ 36 & -3 & 4 \end{pmatrix}$	18	$L_2 := L_2 - 2L_1$	$M_4 = \begin{pmatrix} 18 & 4 & -5 \\ 0 & -11 & 14 \end{pmatrix}$

Table 8.1. *The Blankinship algorithm: the values of u, v and $d =$ GCD(252, 198) such that $252u + 198v = d$ are on the first row of M_4: $d = 18$, $u = 4$ and $v = -5$*

Proof. Consider the *unimodular matrices*[4]

$$U(\lambda) = \begin{pmatrix} 1 & \lambda \\ 0 & 1 \end{pmatrix}, \quad L(\lambda) = \begin{pmatrix} 1 & 0 \\ \lambda & 1 \end{pmatrix}, \quad T = \begin{pmatrix} 0 & 1 \\ 1 & 0 \end{pmatrix}.$$

Let M be a matrix with two rows and put:

$$M' = U(\lambda)M, \quad M'' = L(\lambda)M, \quad M''' = TM.$$

A straightforward calculation shows that one passes from M to M', M'', M''' by the following elementary row operations:

$$M \mapsto M' \ : L_1 := L_1 + \lambda L_2,$$

$$M \mapsto M'' \ : L_2 := L_2 + \lambda L_1,$$

$$M \mapsto M''' : \textit{exchange rows } L_1 \textit{ and } L_2.$$

Since we pass from M to M_4 by a sequence of row operations, we have an equality of the form

$$M_4 = E_3 E_2 E_1 M_0 = E M_0,$$

where the E_i denote unimodular matrices *which we need not know*. If we write $M_4 = E M_0$ in the form

$$\begin{pmatrix} d & u & v \\ 0 & u' & v' \end{pmatrix} = E\left(\begin{matrix} a \\ b \end{matrix}, I\right) = \left(E\begin{pmatrix} a \\ b \end{pmatrix}, E\right),$$

we see why we bordered the vector ${}^t(a, b)$ by the identity matrix: the product of the unimodular matrices appears automatically! The desired result is simply the (1,1) element of the product $E\binom{a}{b}$:

$$au + bv = d. \qquad \square$$

[4] A unimodular matrix is a matrix with integer coefficients and determinant ± 1. The inverse of such a matrix also has integer coefficients. We shall study unimodular matrices more fully in Chapter 11.

If we let $M = \begin{pmatrix} a & ua & va \\ b & ub & vb \end{pmatrix}$ be the matrix to manipulate, Blankinship's algorithm is then:

```
a, b := prescribed integers > 0 ;
ua := 1 ; va := 0 ;
ub := 0 ; vb := 1 ;
while (a > 0) and (b > 0) do begin
  | if a ≥ b then begin
  |   | q := a div b ;
  |   | a := a − b ∗ q ; ua := ua − q ∗ ub ; va := va − q ∗ vb
  | end
  | else begin
  |   | q := b div a ;
  |   | b := b − a ∗ q ; ub := ub − q ∗ ua ; vb := vb − q ∗ va
  | end
end ;
if a > 0
then begin gcd := a ; u := ua ; v := va end
else begin gcd := b ; u := ub ; v := vb end ;
```

For beginners

• Students sometimes want to replace the variables $a, b, u_1, \ldots, v_2$ by a matrix $M[i, j]$ with two rows and three columns. This is not a good idea because it slows the execution (the reason is explained in Chapter 6). We are, however, forced to use this solution when we apply Blankinship's algorithm with more than two integers.

• One can make Blankinship's algorithm much more compact (which speeds it up, but makes it opaque) by only retaining the first two columns of the matrix M (that is, the variables a, b, ua, ub). Once the value of u is known, one finds v by division $v = (d - au)/b$, where a and b are the *original* values.

8.3. Perfect Numbers

One says that an integer $n > 1$ is *perfect* if it is equal to the sum of all its proper divisors; that is, if it satisfies the condition $n = \sum_{d \mid n, d<n} d$ or the equivalent condition $2n = \sum_{d \mid n} d$. The smallest perfect number is 6 because $6 = 1 + 2 + 3$. All the even perfect numbers have been know for a long time: they are numbers of the form

$$n = 2^{p-1}(2^p - 1)$$

where p a prime number such that $2^p - 1$ is also prime. On the other hand, it is still not known whether there are odd perfect numbers: all that is known is that if such exist, their size must be gigantic. Although perfect numbers are not in themselves of great interest to us, the calculation of sums of divisors will provide us with an opportunity to present some techniques for optimizing code. If we introduce the sequence whose general term is

$$S_d = \begin{cases} S_{d-1} + d & \text{if } d \text{ divides } n, \\ S_{d-1} & \text{otherwise,} \end{cases}$$

then the calculation of the sum of divisors is as follows:

```
S := 0 ;
for d := 1 to n do
    if n mod d = 0 then S := S + d
```

Like Laurel and Hardy, divisors come in pairs: if d divides n, then n/d divides n. This suggests using the recurrence:

$$S_d = \begin{cases} S_{d-1} + d + n/d & \text{if } d \text{ divides } n \text{ and } d^2 < n, \\ S_{d-1} + d & \text{if } d \text{ divides } n \text{ and } d^2 = n, \\ S_{d-1} & \text{otherwise.} \end{cases}$$

If we do not allow ourselves recourse to the real numbers here, we cannot write, for example,

```
for d := 1 to round(sqrt(n)) do ...
```

Since we cannot use the function $\sqrt{n}$, we abandon the "for" loop in favor of a "repeat" loop and write:

```
S := 0 ; d := 1 ;
repeat
  if n mod d = 0 then begin
    S := S + d ;
    if d² < n then S := S + n div d
  end ;
  d := d + 1
until d² > n
```

(8.2)

We calculated the sequence d^2 twice in (8.2). To avoid this, we introduce the variable $square = d^2$ which gives (8.3).

```
S := 0 ; d := 1 ; square := 1 ;
repeat
  if n mod d = 0 then begin
    S := S + d ;
    if square < n then S := S + n div d
  end ;
  d := d + 1 ; square := d²
until square > n
```

(8.3)

The above is an example of a "Penelope code" because we forget that we already know d^2 when we calculate $(d+1)^2$. It is more efficient to calculate the new square using the formula *square* $+2d+1$. If we introduce this modification at the right time (that is, before modifying d) and if we replace the statement $square := square + 2*d + 1$ by the slightly faster $square := square + d + d + 1$ we get

```
S := 0 ; d := 1 ; square := 1 ;
repeat
  if n mod d = 0 then begin
    S := S + d ;
    if square < n then S := S + n div d
  end ;
  square := square + d + d + 1 ;
  d := d + 1 ;
until square > n
```
(8.4)

Exercise 1

Compare the speeds of (8.2), (8.3) and (8.4) experimentally.

8.4. The Lowest Divisor Function

Consider the code that we wrote in Chapter 3.

```
function LD1(n : integer) : integer ;
var d : integer ;
begin
  d := 2 ;
  while n mod d > 0 do d := d + 1 ;
  LD1 := d
end ;
```

Since it is pointless to seek an even divisor of an odd number, we could hope to *double* the speed of the function LD_1 by proceeding by steps of 2 starting with 3 when n is odd.

```
function LD2(n : integer) : integer ;
var d : integer ;
begin
  if n mod 2 = 0 then LD2 := 2
  else begin
    d := 3 ;
    while n mod d > 0 do d := d + 2 ;
    LD2 := d
  end
end ;
```

We can push this idea a little further. When n is divisible by neither 2 nor 3, there is no point seeking divisors of the form $2d$ or $3d$. Thus we can start at $d = 5$ and alternately add 2 or 4 (Chap. 2):

$$5 \xmapsto{+2} 7 \xmapsto{+4} 11 \xmapsto{+2} 13 \xmapsto{+4} 17 \xrightarrow{+2} 19$$
$$\xmapsto{+4} 23 \xmapsto{+2} 25 \xmapsto{+4} 29 \xmapsto{+2} 31 \xmapsto{+4} 35$$
$$\xmapsto{+2} 37 \xmapsto{+4} 41 \xmapsto{+2} 43 \xmapsto{+2} 49 \cdots .$$

The integers that remain are those of the form $d_1 = 6n + 1$ or $d_5 = 6n + 5$.

```
function LD3(n : integer) : integer ;
var d1, d5 : integer ;
begin
  if n mod 2 = 0 then LD3 := 2
  else if n mod 3 = 0 then LD3 := 3
  else begin
    d5 := 1 ;
    repeat
      d1 := d5 + 4 ; d5 := d1 + 2
    until (n mod d1 = 0) or (n mod d5 = 0) ;
    if n mod d1 = 0 then LD3 := d1 else LD3 := d5
  end
end ;
```

When n is a prime number, the algorithm for LD_2 tries to divide n by *all* odd integers $\leq n$ whereas we know the response as soon as d^2 exceeds n. So, we introduce the square of d and the rapid calculation that we developed for the perfect numbers:

```
function LD4(n : integer) : integer ;
var d, dd, square : integer ;
begin
  if n mod 2 = 0 then LD4 := 2
  else begin
    d := 3 ; square := 9 ;
    while (n mod d > 0) and (square <= n) do begin
      dd := d + d ;
      square := square + dd + dd + 1 ;
      d := d + 2
    end ;
    if square > n then LD4 := n else LD4 := d
  end
end ;
```

Have our attempts to increase performance been successful? Here are the times (in seconds) that it took a medium size computer to calculate LD for all *odd* numbers in the intervals $I_1 = [\![2\,000, 5\,000]\!]$, $I_2 = [\![5\,000, 10\,000]\!]$, $I_3 = [\![20\,000, 30\,000]\!]$ and $I_4 = [\![50\,000, 100\,000]\!]$.

	I_1	I_2	I_3	I_4
LD_1	7.2	23.8	137.2	303.7
LD_2	3.6	11.8	68.2	151.4
LD_3	2.3	7.8	46.3	100.1
LD_4	0.3	0.5	1.5	4.9

Since we have only looked at odd numbers, we see, in accord with our expectations, that the function LD_2 is close to twice as fast as LD_1. The function LD_3 takes about 66 % of the time taken by LD_2, which is again what we would expect. In fact, the interval $[\![1, N]\!]$, contains $[N/k]$ multiples of k, so there are $N - [N/2] - [N/3] + [N/6]$ integers which are not divisible by either 2 or 3. If N is very large, then $N \approx [N]$ and

$$\frac{N - N/2 - N/3 + N/6}{N - N/2} = \frac{2}{3} \approx 0.66.$$

Finally, in spite of its complicated and delicate code, the function LD_4 surpasses all the others.

8.5. The Moebius Function

Recall the Moebius function $\mu(n)$ is defined as

$$\mu(n) = \begin{cases} 1 & \text{if } n = 1, \\ 0 & \text{if } n > 1 \text{ is divisible by the square of a prime number,} \\ (-1)^k & \text{if } n > 1 \text{ is the product of } k \text{ distinct prime numbers,} \end{cases}$$

To calculate $\mu(n)$, beginners typically store the prime numbers in an array and then inspect the array to calculate $\mu(n)$. This static conception, with its two separate phases, squanders lots of code, time and memory. Let us try a more dynamic approach by calculating "approximations" to $\mu(n)$ as we find prime factors of n. Suppose that we have already found divisors $p_1 < \cdots < p_i$ of n; if $p_{i+1} \geq p_i$ is the next divisor, it is clear that we have:

$$\mu(p_1 \cdots p_i p_{i+1}) = \begin{cases} -\mu(p_1 \cdots p_i) & \text{if } p_{i+1} > p_i, \\ 0 & \text{if } p_{i+1} = p_i. \end{cases} \tag{8.5}$$

This method is justified by the following result whose proof is immediate.

Lemma 8.5.1. *The sequence*

$$p_1 = \mathrm{LD}(n),\ p_2 = \mathrm{LD}(n/p_1),\ p_3 = \mathrm{LD}(n/p_1 p_2), \ldots$$

of prime divisors of n is increasing (not necessarily strictly).

The algorithm for decomposing n into prime factors gives a classical two-column table. We introduce a third column containing $\mu(p_1 \cdots p_k)$:

n	LD	μ	n	LD	μ
$n_0 = 1050$	$p_1 = 2$	$\mu_1 = -1$	$n_0 = 210$	$p_1 = 2$	$\mu_1 = -1$
$n_1 = 525$	$p_2 = 3$	$\mu_2 = 1$	$n_1 = 105$	$p_2 = 3$	$\mu_2 = 1$
$n_2 = 175$	$p_3 = 5$	$\mu_3 = -1$	$n_2 = 35$	$p_3 = 5$	$\mu_3 = -1$
$n_3 = 35$	$p_4 = 5$	$\mu_4 = 0$	$n_3 = 7$	$p_4 = 7$	$\mu_4 = 1$
		stop	$n_4 = 1$		*stop*

We translate this into the language of recurrent sequences.

$$\begin{aligned}
&n_0 = \text{given number} > 1;\\
&p_1 = \mathrm{LD}(n_0); \quad n_1 = n_0/p_1; \quad \mu_1 = -1;\\
&p_2 = \mathrm{LD}(n_1); \quad n_2 = n_1/p_2; \quad \mu_2 = \begin{cases} -\mu_1 & \text{if } p_1 < p_2,\\ 0 & \text{if not;}\end{cases}\\
&\vdots\\
&p_\ell = \mathrm{LD}(n_{\ell-1}); \ n_\ell = n_{\ell-1}/p_\ell; \ \mu_\ell = \begin{cases} -\mu_{\ell-1} & \text{if } p_{\ell-1} < p_\ell,\\ 0 & \text{otherwise;}\end{cases}\\
&\textit{stop when } n_\ell = 1 \textit{ or } \mu_\ell = 0
\end{aligned}$$

Note that we encounter a second order recurrence because it is necessary to know $p_{\ell-1}$ and p_ℓ to calculate μ_ℓ. We reduce to a first order recurrence by passing to dimension two; that is, by introducing the sequence $old\_\mathrm{LD}_\ell = p_{\ell-1}$. If we put, to simplify,

$$S(t) = \begin{cases} old\_\mathrm{LD}_\ell = p_{\ell-1}; \quad p_\ell = \mathrm{LD}(N_{\ell-1}); \quad n_\ell = n_{\ell-1}/p_\ell;\\ \mu_\ell = \begin{cases} -\mu_\ell \text{ if } old\_\mathrm{LD}_\ell = p_\ell,\\ 0 \quad \text{otherwise}\end{cases}\end{cases}$$

we can describe our algorithm as follows

$$\begin{cases} n_0 = \textit{given number} > 1; \ \mu_0 = 1; \ p_0 = 1;\\ S(1); \ S(2); \ \ldots; \ S(\ell)\\ \textit{stop when } \ n_\ell = 1 \ \textit{ or } \ \mu_\iota = 0.\end{cases} \tag{8.6}$$

We suppress the time index ℓ in (8.6) and use a "repeat" loop:

```
function Moebius(n : integer) : integer ;
var μ, old_LD, new_LD : integer ;
begin
| μ := 1 ; new_LD := 1 ;
| if n > 1 then begin
| | repeat
| | | old_LD := new_LD ; new_LD := LD(n) ;
| | | n := n div new_LD ;
| | | if old_LD < new_LD then μ := −μ else μ := 0
| | until (n = 1) or (μ = 0) ;
| end ;
| Moebius := μ
end ;
```

Exercises 2

- If we store the values of the Mœbius function for $1 \le n \le N$ in an array $\mu[1..N]$, we can be more efficient. In effect, because we already know $\mu[1], \ldots, \mu[n-1]$, when $d = \mathrm{LD}(n)$, formula (8.5) and the lemma assure us that

$$\mu(n) = \begin{cases} -\mu(n/d) & \text{if } d \text{ does not divide } d/n, \\ 0 & \text{otherwise.} \end{cases}$$

- Suppose we want to verify the Mœbius inversion formula (Chap. 2). For this, fill the array $f[1..N]$ arbitrarily, then store the values of the functions $g(n) = \sum_{d|n} f(d)$ and $h(n) = \sum_{d|n} \varphi(d)\mu(n/d)$ in the arrays $g[1..N]$ and $h[1..N]$, respectively. Have your program display the values of f and of h on two different rows.

- Calculate the Euler phi function φ by a similar method.

8.6. The Sieve of Eratosthenes

In order to find all prime numbers between 2 and N, one often uses the age old algorithm known as the *Sieve of Eratosthenes*. Write 2 and all *odd integers* smaller than N (we have taken $N = 149$).

```
      2  3   5   7   9  11  13  15  17  19  21  23  25  27  29
 31   .  33  35  37  39  41  43  45  47  49  51  53  55  57  59
 61   .  63  65  67  69  71  73  75  77  79  81  83  85  87  89
 91   .  93  95  97  99 101 103 105 107 109 111 113 115 117 119
121   . 123 125 127 129 131 133 135 137 139 141 143 145 147 149.
```

Now remove all multiples of 3 greater than 3. Since $6 = 2 \cdot 3$, it suffices to begin with $3 \cdot 3 = 9$:

```
      2 3  5   7  .  11  13  .  17  19  .  23  25  .  29
31  .  .  35  37  .  41  43  .  47  49  .  53  55  .  59
61  .  .  65  67  .  71  73  .  77  79  .  83  85  .  89
91  .  .  95  97  . 101 103  . 107 109  . 113 115  . 119
121 .  . 125 127  . 131 133  . 137 139  . 143 145  . 149.
```

Since the prime number that follows 3 is 5, we remove all multiples of 5 greater than 5. Knowing that $2 \cdot 5$, $3 \cdot 5$ and $4 \cdot 5$ have already disappeared because they are multiples of 2 or 3, we begin with $5 \cdot 5 = 25$:

```
      2 3 5  7  .  11  13  .  17  19  .  23  . .  29
31   . . .  37  .  41  43  .  47  49  .  53  . .  59
61   . . .  67  .  71  73  .  77  79  .  83  . .  89
91   . . .  97  . 101 103  . 107 109  . 113  . . 119
121  . . . 127  . 131 133  . 137 139  . 143  . . 149.
```

The first number that follows 5 is 7, so we remove all multiples of 7 greater than 7. But since the numbers of the form $7m$ with $m \leq 6$ have disappeared in the course of the preceding operations, we begin with $7 \cdot 7$:

```
      2 3 5  7  .  11  13  .  17  19  .  23  . .  29
31   . . .  37  .  41  43  .  47   .  .  53  . .  59
61   . . .  67  .  71  73  .   .  79  .  83  . .  89
.    . . .  97  . 101 103  . 107 109  . 113  . .   .
121  . . . 127  . 131   .  . 137 139  . 143  . . 149.
```

The prime number following 7 being 11, we remove numbers of the form $11m$ with $m \geq 11$, which only removes $11 \cdot 11 = 121$.

```
      2 3 5  7  .  11  13  .  17  19  .  23  . .  29
31   . . .  37  .  41  43  .  47   .  .  53  . .  59
61   . . .  67  .  71  73  .   .  79  .  83  . .  89
.    . . .  97  . 101 103  . 107 109  . 113  . .   .
.    . . . 127  . 131   .  . 137 139  .   .  . . 149.
```

The first number following 11 is 13, so we must suppress all numbers of the form $13m$ with $m \geq 13$. Since $13 \cdot 13 = 169 > N$, the array above contains no numbers of this type, and the survivors are all the prime numbers less than or equal to 149.

8.6.1. Formulation of the algorithm

We translate the preceding operations using sequences.

- A first sequence is clearly formed by the successive arrays.

$$T_t = \text{set of integers remaining at the instant } t.$$

The first element of the sequence is:

$$T_1 = \{2\} \cup \{\ \textit{odd integers} \leq N\}$$

- This sequence is not sufficient because to deduce T_{t+1} from T_t we require supplementary information: what is the first element not excluded from T_t? So, we introduce the sequence (p_t) of first non-excluded elements

$$p_{t+1} = \min\{n \in T_t \mid n > p_t\}. \tag{8.7}$$

The first few terms are $p_1 = 2$, $p_2 = 3$, $p_3 = 5$, $p_4 = 7$, $p_5 = 11$. If we let $big\_mults(p)$ designate the set of all multiples of p which are larger than p^2 (that is, of the form mp with $m \geq p$), we can write:

$$T_{t+1} = T_t - big\_mults(p_{t+1}).$$

We can now formalize the sieve algorithm.

$$\begin{array}{l}
T_1 := \{2\} \cup \{\textit{odd integers} \leq N\}\ ; \\
t_1 := 1\ ;\ p_1 := 2\ ; \\
\textbf{while } p_t^2 \leq N \textbf{ do begin} \\
\quad p_{t+1} := min\{T_t \cap \,]\!]\, p_t, N\,]\!]\}\ ; \\
\quad T_{t+1} := T_t - big_mults(p_{t+1})\ ; \\
\quad t := t+1 \\
\textbf{end}
\end{array} \tag{8.8}$$

Theorem 8.6.1. *Algorithm* (8.8) *is correct, which means that it terminates, does not crash, and that the last set T_t is exactly the set of prime numbers less than or equal to N.*

Proof. Consider the induction hypothesis:

$$(\mathcal{H}_t) \begin{cases} \text{(i) the prime numbers} \leq N \text{ all belong to } T_t; \\ \text{(ii) the } t \text{ prime numbers are } p_1, \ldots, p_t; \\ \text{(iii) } T_i \text{ contains no numbers } mp_1, \ldots, mp_t \text{ with } m > 1. \end{cases}$$

We are going to show that $(\mathcal{H}_t)$ is an invariant of the loop, which means that it is true *each time we enter the loop*. It is clear that $(\mathcal{H}_1)$ holds. Suppose that $(\mathcal{H}_t)$ is true on entry into the loop and that $p_t^2 \leq N$, which allows us to re-enter the loop. Let q be the first prime number that is encountered after p_t. The corollary of Bertrand's postulate assures us that $p_t < q < p_t^2 \leq N$. Condition (i) then shows that q belongs to T_t. Consequently, $T_t \cap \,]\!]\, p_t, N]\!]$ is not the empty set, whence p_{t+1} exists and the algorithm does not crash. Since the sequence of numbers p_t is strictly increasing and bounded by $\sqrt{N}$, we conclude that the algorithm must terminate. Now, we establish that $(\mathcal{H}_{t+1})$ is true upon entering the loop again.

• The set T_{t+1} is obtained by removing the strict multiples of p_{t+1} from T_t, and hence conditions (i) and (iii) are satisfied.

• Suppose that (ii) were false; that is, that p_{t+1} is not the smallest prime number greater than p_t. With the notation above, we have $p_t < q < p_{t+1}$ which contradicts the definition of p_{t+1}. A single formality remains: we must show that when the algorithm terminates, the set T_t contains all prime numbers $\leq N$ and nothing else. Condition (i) already shows that T_t contains all primes $\leq N$. Suppose that T_t were to contain a composite integer $n = qn'$ with q prime, $q \leq n'$ and $q^2 \leq n \leq N$. Conditions (ii) and (iii) imply that $q > p_{t-1}$. Since we are at the exit of the loop, we have $p_t^2 > N$, whence $q^2 > p_t > N$ which contradicts $q^2 \leq n \leq N$. □

8.6.2. *Transforming the algorithm to a program*

We get rid of the time t in (8.8) to obtain a "true" algorithm.

```
T = {2} ∪ {odd integers ≤ N} ; p := 2 ;
while p² ≤ N do begin
  | p := min{T ∩ ]]p, N]]} ;                          (8.9)
  | T := T − big_mults(p)
end
```

The discussion so far has not dealt with storage of the sets T. To do this, it is natural to choose an array of booleans called *is_removed*

```
const max = 2000 ;
type vector = array[2..max] of boolean ;
var is_removed : vector ; p : integer ;
```

The meaning and use of the variable *is_removed* is as its name suggests:

$$is\_removed[n] = \begin{cases} true & \text{if } n \text{ is removed,} \\ false & \text{if not.} \end{cases}$$

We are going to work "on site"; that is, with the single vector *is_removed*, where we consider the set T_t to be the state of the vector *is_removed* at the instant t. This said, the main body of our program is:

```
begin
| initialize(is_removed) ;
| p := 1 ;
| while p * p ≤ max do begin
| | p := first_non_removed(p, is_removed) ;
| | remove_large_mults(p, is_removed)
```

```
  end ;
  display(is_removed)
end .
```

The procedure initialize

To remove the even integers > 2, we do not use the statement

```
for m := 2 to max div 2 do is_removed[2 * m] := true
```

because it is much faster to repeatedly add 2:

```
procedure initialize(var is_removed : vector) ;
var m : integer ;
begin
  is_removed[2] := false ;
  m := 3 ;
  while m ≤ max do begin
    is_removed[m] := false ;  m := m + 2
  end ;
  m := 4 ;
  while m ≤ max do begin
    is_removed[m] := true ;  m := m + 2
  end
end ;
```

The procedure remove_large_mults

The first number to remove is p^2. We again replace multiplications by additions in order to speed things up.

```
procedure remove_large_mults(p : integer ;  var is_removed : vector) ;
var m : integer ;
begin
  m := p * p ;
  while m ≤ max do begin
    is_removed[m] := true ;  m := m + p
  end
end ;
```

The function first_non_removed

We use here a programming trick to speed up the execution: the two parameters are passed by address to the procedure (the declarations “var”). This avoids unnecessary recopying. It is also without danger because we cannot consult these parameters without modifying them.

```
function first_non_removed(var p : integer ;
                           var is_removed : vector) : integer ;
var q : integer ;
begin
  if p = 2 then first_non_removed := 3
  else begin
    q := p + 2 ; while is_removed[q] do q := q + 2 ;
    first_non_removed := q
  end
end ;
```

Advice on finishing the program

• Take care with the presentation! Display your prime numbers in $10 \cdot 10$ packets.

```
num_displayed := 0 ; num_rows := 0 ;
for q := 2 to max do
if not is_removed[q] then begin
  write(q : 6) ;
  num_displayed := num_displayed + 1 ;
  if num_displayed = 10 then begin
    writeln ; num_rows := num_rows + 1 ; num_displayed := 0
  end ;
  if num_rows = 10 then begin writeln ; num_rows := 0 end
end ;
```

• Check that 1789 (which is prime) appears on your screen. Also, compare your result with the values of $\pi(x)$ listed in Chapter 2.

Exercise 3

One can speed up the sieve by treating the case $p = 2$ separately. But one can gain more speed and, in particular, *economize on memory* by dealing only with *odd numbers*. We begin with a compact version.

```
for i := 1 to (max - 1) div 2 do is_removed[2 * i + 1] := false ;
P := 3 ;
while P * P <= max do begin
  X := P * P ;
  while X <= max do begin
    is_removed[X] := true ;
    X := X + 2 * P
  end ;
  repeat P := P + 2 until not is_removed[P]
end ;
```

Note the increment "$X := X + 2 * P$": as we are only sieving odd integers, it is not necessary to try to remove the even integer $X + P$. Introduce now $P = 2p + 1$ and $X = 2x + 1$. In the algorithm that follows, *is_removed*$[x]$ indicates whether or not the integer $X = 2x + 1$ is removed.

```
for i := 1 to (max − 1) div 2 do is_removed[i] := false ;
p := 1 ; M := (max − 1) div 4 ;
while p * (p + 1) ≤ M do begin
  x := 2 * p * (p + 1) ;
  while x ≤ max do begin
    is_removed[x] := true ;
    x := x + 2 * p + 1
  end ;
  repeat p := p + 1 until not is_removed[p]
end ;
```

On a medium powered computer, deliberately slowed, the classical sieve with $N = 30{,}000$ took 1.02 seconds compared with 0.40 seconds for the new algorithm.

8.7. The Function pi(x)

Let $(p_i)_{i \geq 1}$ be the strictly increasing sequence of prime numbers ($p_1 = 2$, $p_2 = 3$, etc.). Recall that the function $\pi(x)$ is defined as follows *for any real number* x:

$$\begin{aligned} \pi(x) &= \textit{number of primes} \leq x \\ &= \textit{the largest index } i \textit{ such that } p_i \leq x \end{aligned}$$

If x is not too large, a table of primes suffices to calculate $\pi(x)$. But what happens otherwise? The response to this question is useful because, if we can calculate $\pi(x)$ without knowing in advance all the prime numbers $\leq x$, we know the size of a table of prime numbers $\leq x$.

8.7.1. Legendre's formula

Let $x \geq 2$ be a real number. We can partition the $[x]$ integers between 1 and x into three classes.

- the integer 1,
- prime numbers $\leq x$,
- composite integers $\leq x$.

Taking cardinalities of these classes, we get

$$[x] = 1 + \pi(x) + \mathrm{Card}(\textit{composite integers} \leq x).$$

A composite integer has least divisor less than or equal to $\sqrt{x}$, so that it can be written in the form

$$n = mp, \text{ with } m \geq 2 \text{ and } p \text{ prime} \leq \sqrt{x}.$$

To simplify, put

$$\alpha = \pi\left(\sqrt{x}\right).$$

Since there are $[x/p]$ multiples of p which are $\leq x$, there must be $[x/p]-1$ composite integers $\leq n$ which are divisible by p. It is very tempting to assert that there are

$$\sum_{1 \leq i \leq \alpha} \left(\left[\frac{x}{p_i}\right] - 1\right) = \left(\sum_{1 \leq i \leq \alpha} \left[\frac{x}{p_i}\right]\right) - \alpha$$

composite integers $n \leq x$. This enumeration is incorrect beacuse it *double-counts* multiples of the numbers $p_i p_j$ because they are simultaneously multiples of p_i and p_j. It is necessary, to subtract the number of integers of the form $[x/p_i p_j]$ and introduce a new correction for multiples of $p_i p_j p_k$, and so on. If we put

$$\text{Legendre}(x, \alpha) = \sum_{1 \leq i \leq \alpha} \left[\frac{x}{p_i}\right] - \sum_{1 \leq i < j \leq \alpha} \left[\frac{x}{p_i p_j}\right] + \sum_{1 \leq i < j < k \leq \alpha} \left[\frac{x}{p_i p_j p_k}\right] + \cdots,$$

the correct formula is:

$$\text{Card}(\textit{composite integers} \leq x) = \text{Legendre}(x, \alpha) - \alpha.$$

We obtain the celebrated *Legendre formula*

$$\pi(x) = [x] - 1 + \alpha - \text{Legendre}(x, \alpha) \quad \text{with} \quad \alpha = \pi\left(\sqrt{x}\right). \tag{8.10}$$

The correction term and the words to describe it

Take $x = 50$ so that $\alpha = \pi(\sqrt{50}) = \pi(7) = 4$. The correction term Legendre(x, α) is equal to:

$$\begin{aligned} &\left[\frac{n}{p_1}\right] + \left[\frac{n}{p_2}\right] + \left[\frac{n}{p_3}\right] + \left[\frac{n}{p_4}\right] \\ -&\left[\frac{n}{p_1p_2}\right] - \left[\frac{n}{p_1p_3}\right] - \left[\frac{n}{p_1p_4}\right] - \left[\frac{n}{p_2p_3}\right] - \left[\frac{n}{p_2p_4}\right] - \left[\frac{n}{p_3p_4}\right] \\ +&\left[\frac{n}{p_1p_2p_3}\right] + \left[\frac{n}{p_1p_2p_4}\right] + \left[\frac{n}{p_1p_3p_4}\right] + \left[\frac{n}{p_2p_3p_4}\right] \\ -&\left[\frac{n}{p_1p_2p_3p_4}\right]. \end{aligned}$$

In this formula, the essential role is played by the indices

1, 2, 3, 4, 12, 13, 14, 23, 24, 34, 123, 124, 134, 234, 1234.

because it is possible to reconstitute $[n/p_i p_j p_k]$ from i, j, k. Observe also the appearance of *words* whose *letters* 1, 2, 3, 4 form an increasing sequence. A program that calculates these words in fact defines a total order on the words because it calculates them in order one after the other. Since we are talking about words, this immediately brings to mind the lexicographic ordering:

1, 12, 123, 1234, 124, 13, 134, 14, 2, 23, 234, 24, 3, 34, 4.

Now, trying to find an algorithm to create these words and display them vertically on the screen leads naturally to the structure of a stack.

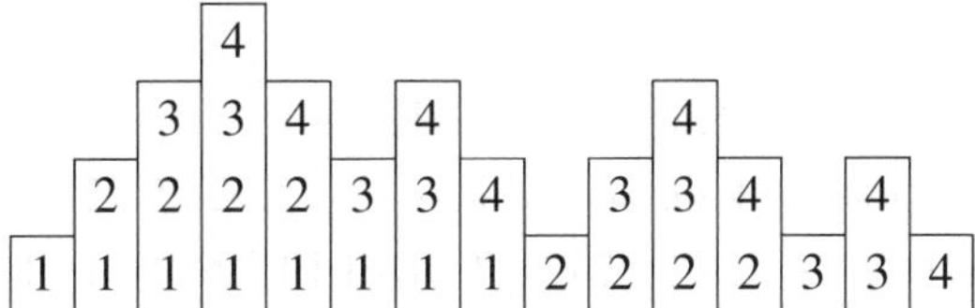

Note that we move from one word to the next using one of the following two operations:

• if the top s of the stack is < 4, we push (insert) $s + 1$ on top of s;

• if the top of the stack is 4, we pop (withdraw) it; if this does not empty the stack, we increment the top of the new stack.

We realize the stack as a pair (*array*, *height*); the variable s contains the next integer to push. The following algorithm produces the words $w = a_1 \cdots a_k$ in increasing order. These words are composed of the letters $1, 2, \ldots, \alpha$ and satisfy the condition $1 \leq a_1 < a_2 < \cdots < a_k \leq \alpha$.

```
s := 1 ; h := 0 ;        {the stack is empty}
repeat
| if s ≤ α
| then push(stack, h, s)
| else pop_increment(stack, h) ;
| if h > 0 then s := stack[h] + 1
until h = 0              {the stack is empty again}
```

The procedures *push* and *pop_increment* are immediate.

```
procedure push(var stack : vector ; var h, s : integer) ;
begin
| h := h + 1 ;
| stack[h] := s
end ;
procedure pop_increment(var stack : vector ; var h : integer) ;
begin
| h := h − 1 ;
| if h > 0 then stack[h] := stack[h] + 1
end ;
```

Exercise 4

We could use another strategy to produce the words w. In effect, to be given indices $1 \leq a_1 < \cdots < a_k \leq \alpha$ amounts to being given a non-empty subset of $[\![1, \alpha]\!]$. To this subset, we associate its characteristc function

$$b_1 + 2b_2 + 2^2 b_3 + \cdots + b_\alpha 2^{\alpha-1}, \tag{8.11}$$

the b_i indicating whether or not the index i occurs in the word w. Thus, it suffices to write in base 2 all integers between 1 and $1+2+\cdots+2^{\alpha-1} = 2^\alpha - 1$ to obtain each word w once (and only once).

8.7.2. Implementation of Legendre's formula

With the integers in Pascal, we can in theory calculate[5] $\pi(x)$ for $x < 2^{15}$. This requires us to store in a vector $p[1..42]$ the list of prime numbers smaller that 181 (since $181^2 < 2^{15} < 182^2$):

2	3	5	7	11	13	17	19	23	29	31
37	41	43	47	53	59	61	67	71	73	79
83	89	97	101	103	107	109	113	127	131	137
139	149	151	157	163	167	173	179	181		

Let $a_1 < a_2 < \cdots < a_h$ be the indices contained in the stack at a given moment and *term* the number associated to the word $w = a_1 \cdots a_h$:

$$term = \left[\frac{x}{p_{a_1} \cdots p_{a_h}}\right].$$

We must be careful because we are entering the delicate world of numerical calculation. The naive approach

"First, I calculate $P = p_{a_1} \cdots p_{a_h}$, then I divide x by P"

comes to a screeching halt, because when we use integers in Pascal, we must never exceed $2^{15} = 32{,}768$ in the course of a calculation. This bound is quickly exceeded because $2 \cdot 3 \cdot 5 \cdot 7 \cdot 11 \cdot 13 \cdot 17 = 510{,}510$. Instead of dividing x by the product of the the p_i, we are going to divide x by p_1, then divide the result by p_2, and so on. This slows the execution of the program, but it *never* results in an overflow of the capacity because the integers produced are decreasing. This strategy is justified by the following, easily established result.

Proposition 8.7.1. *For every integer x,*

$$\left[\frac{x}{p_1 p_2 \cdots p_h}\right] = \left[\left[\cdots\left[\left[\frac{x}{p_1}\right] \times \frac{1}{p_2}\right] \cdots\right] \times \frac{1}{p_h}\right].$$

[5] In theory only, as we will prove in the next paragraph

The code for the function *Legendre* is now simple to write: each time that we modify the stack, we calculate the associated term that we add to or subtract from *Sum*.

```
function Legendre(x, α : integer ; var p : vector) : integer ;
var h, s, Sum : integer ; stack : vector ;
begin
  s := 1 ; h := 0 ; Sum := 0 ;
  repeat
    if s ≤ α then push(stack, h, s) else pop_increment(stack, h) ;
    if h > 0 then begin
      s := stack[h] + 1 ;
      if h mod 2 = 0
      then Sum := Sum + term(p, stack, h, x)
      else Sum := Sum − term(p, stack, h, x)
    end
  until h = 0 ;
  Legendre := Sum
end ;
```

In order to avoid unnecessary recopying and to gain time, we communicate the addresses of the vectors p and *stack*: there is no danger here because we only need to consult the values of these variables.

```
function term(var p, stack : vector ; h, x : integer) : integer ;
var i : integer ;
begin
  for i := 1 to h do x := x div p[stack[i]] ;
  term := x
end ;
```

8.7.3. Meissel's formula

The time needed to calculate the correction term Legendre(x, α) grows rapidly with x. As we have already remarked, the words formed by the indices of the prime numbers are in bijective correspondence with non-empty subsets of $\{1, \ldots, \alpha\}$ which gives a pallette of $2^\alpha - 1$ words to use. If we try to calculate $\pi(10^4)$, since $\alpha = \pi(10^2) = 25$, there would be

$$2^{25} - 1 = 33\,554\,431$$

integer parts to calculate: the response would be a very long time coming ... In 1885, using an improvement of Legendre's theorem, Meissel announced that $\pi(10^9) = 50{,}847{,}534$. Considering the primitive tools for calculation available in this era, this result, although slightly erroneous,[6] was a real *tour de force*.

[6] In 1958, D.H. Lehmer, using a computer, found that $\pi(10^9) = 50{,}847{,}478$.

Let $x \geq 4$ be a real number, and α and β be indices such that

$$p_\alpha^3 \leq x < p_{\alpha+1}^3, \quad p_\beta^2 \leq x < p_{\beta+1}^2.$$

This means that

$$\alpha = \pi\big(\sqrt[3]{x}\big), \quad \beta = \pi\big(\sqrt{x}\big).$$

Having made these choices, we re-partition the integers $\leq x$ into four classes.

- the number 1;
- the prime numbers $\leq x$;
- the composite integers $\leq x$ whose LD is $\leq p_\alpha$;
- the composite integers $\leq x$ whose LD is $> p_\alpha$.

The cardinalities of the the first two sets are obviously 1 and $\pi(x)$. Reasoning as we did for Legendre's formula, it is easy to see that the cardinality of the third class is Legendre$(x, \alpha) - \alpha$. Let n be an integer in the fourth class and decompose it into prime factors $n = p_{i_1} \cdots p_{i_k}$ with $\alpha < i_1 \leq \cdots \leq i_k$ and $k \geq 2$. The bound $p_{\alpha+1}^k \leq p_{i_1} \cdots p_{i_k} \leq x < p_{\alpha+1}^3$ shows that $k = 2$. Finally the condition $n \leq x$, which can be re-written as $p_{i_2} \leq x/p_{i_1}$, shows that $i_2 \leq \pi(x/p_{i_1})$. Thus, there are $\pi(x/p_{\alpha+1}) - \alpha$ integers of the form $p_{\alpha+1}p_i$ with $i \geq \alpha + 1$, hence $\pi(x/p_{\alpha+2}) - (\alpha + 1)$ integers of the form $p_{\alpha+2}p_i$, with $i \geq \alpha + 2$, and so forth up to integers of the form $p_\beta p_i$ because $p_{\beta+1}p_i \geq p\beta + 1^2 > x$. Because these integers are distinct, the cardinality of the third class is equal to

$$\begin{aligned}\sum_{\alpha<i\leq\beta} &\big\{\pi(x/p_i) - (i-1)\big\} \\ &= -\big\{\alpha + (\alpha+1) + \cdots + (\beta-1)\big\} + \sum_{\alpha<i\leq\beta} \pi(x/p_i) \\ &= \tfrac{1}{2}\beta(\beta-1) - \tfrac{1}{2}\alpha(\alpha-1) + \sum_{\alpha<i\leq\beta} \pi(x/p_i)\end{aligned}$$

If we put

$$\text{Meissel}(x, \alpha, \beta) = \sum_{\alpha<i\leq\beta} \pi\Big(\frac{x}{p_i}\Big),$$

we obtain Meissel's formula[7]

$$\begin{aligned}\pi(x) = [x] + \tfrac{1}{2}\beta(\beta-1) &- \tfrac{1}{2}(\alpha-1)(\alpha-2) \\ &- \text{Legendre}(x, \alpha) - \text{Meissel}(x, \alpha, \beta),\end{aligned}$$

where we have put $\alpha = \pi(\sqrt[3]{x})$ and $\beta = \pi(\sqrt{x})$.

[7] One can find much more sophisticated algorithms in Hans Riesel's book *Primes Numbers and Computer Methods for Factorization*, Progress in Mathematics, vol. 57, Birkhäuser, Boston-Basel-Stuttgart (1958).

Example

Let us calculate $\pi(200)$. Since $\sqrt[3]{200} = 5.8$, we have $\alpha = 3$ prime numbers smaller than $\sqrt[3]{200}$, which gives:

$$\begin{aligned}\text{Legendre}(200, 3) = &\left[\frac{200}{2}\right]+\left[\frac{200}{3}\right]+\left[\frac{200}{5}\right]\\ &-\left[\frac{200}{6}\right]-\left[\frac{200}{10}\right]-\left[\frac{200}{15}\right]+\left[\frac{200}{30}\right]= 146.\end{aligned}$$

Since $\sqrt{200} = 14.1$, we have $\beta = 6$ primes less than or equal to $\sqrt{200}$, which gives:

$$\text{Meissel}(200, 3, 6) = \pi\left(\frac{200}{7}\right)+\pi\left(\frac{200}{11}\right)+\pi\left(\frac{200}{13}\right)= 22.$$

Note that the Legendre correction term uses *integer parts* while the Meissel correction term uses the *function* π. Thus desired value is:

$$\pi(200) = 200 + 15 - 1 - 146 - 22 = 46.$$

Exercise 5

Let $M = 2^{15} = 32768$ (the upper limit of the integers in Pascal). The sieve of Eratosthenes on the interval $[\![2, 3000]\!]$ allows us to transfer into the vector $p[1\,..\,430]$ the prime numbers less than or equal to $3 \cdot 10^3$ ($p_{430} = 2999$ and $p_{431} = 3001$).

- If $x \leq 3 \cdot 10^3$, we calculate $\pi(x)$ by direct inspection of the vector p.
- If $3 \cdot 10^3 \leq x \leq M$, then $\pi(\sqrt[3]{3000}) \leq \pi(\sqrt[3]{x}) \leq \pi(\sqrt[3]{M})$, which gives $6 \leq \alpha \leq 11$ and $\beta = \pi(\sqrt{x}) \leq \pi(2^{15/2}) = 42$. If we use Meissel's formula to calculate $\pi(x)$:
  - ▹ the time to calculate Legendre(x, α) becomes reasonable because this function contains no more than $2^\alpha \leq 2^{11} = 2048$ integer parts;
  - ▹ the time needed to calculate Meissel(x, α, β) is very short; in effect, this function contains $\beta - \alpha \leq 36$ terms of the form $\pi(x/p_i)$ that we can calculate directly by inspecting the vector p since $p_i \geq p_{\alpha+1} \geq p_7 = 17$ implies that $x/q \leq M/17 \leq 1928$.

8.8. Egyptian Fractions

The ancient Egyptians, it seems, only liked the fraction $2/3$ and the fractions of the form $1/n$ and they wrote their fractions as sums of inverses of whole numbers. To honor this whim, we call the inverse of an integer an *Egyptian fraction*. Is it always possible to write a given rational number as a sum of Egyptian fractions? And is this expression unique?

Theorem 8.8.1. *Every rational number $a/b > 0$ is a sum of Egyptian fractions:*

$$\frac{a}{b} = \frac{1}{x_1} + \frac{1}{x_2} + \cdots \frac{1}{x_k}, \qquad 1 \le x_1 < x_2 < \cdots < x_k\,, \quad x_i \in \mathbb{N}.$$

Proof. We will use an algorithm first written down by Leonardo de Pisa (Fibonacci) and rediscovered and verified by Sylvester.

- Suppose first that $a/b < 1$ and define $n_1 \ge 2$ by the condition:

$$\frac{1}{n_1} \le \frac{a}{b} < \frac{1}{n_1 - 1}.$$

If a/b equals $1/n_1$, we are done. Otherwise, put $a_1 = an_1 - b$ and $b_1 = bn_1$, so that:

$$\frac{a}{b} = \frac{1}{n_1} + \frac{a_1}{b_1}.$$

Since $a(n_1 - 1) < b$, it follows that $a_1 < a$. Beginning again with a_1/b_1, we let $n_2 \ge 2$ be the integer satisfying the condition :

$$\frac{1}{n_2} \le \frac{a_1}{b_1} < \frac{1}{n_2 - 1}.$$

Putting $a_2 = a_1 n_2 - b_1$ et $b_2 = b_1 n_2$, we now have:

$$\frac{a}{b} = \frac{1}{n_1} + \frac{1}{n_2} + \frac{a_2}{b_2}.$$

As above, we have $a_2 < a_1$. On the other hand, we can write:

$$n_2 \ge \frac{b_1}{a_1} = \frac{bn_1}{an_1 - b} = \frac{n_1}{\frac{a}{b}n_1 - 1} > \frac{n_1}{\frac{n_1}{n_1 - 1} - 1} = n_1(n_1 - 1) > n_1.$$

Since the sequence of the a_i is strictly decreasing, the process must stop after at most a steps, and gives fractions with strictly increasing denomiators.

- If $a/b > 1$, put:

$$H_n = \frac{1}{1} + \frac{1}{2} + \cdots + \frac{1}{n}.$$

Since the harmonic series H_n diverges to $+\infty$, we know that there exists an integer $n \ge 1$ such that:

$$H_n \le \frac{a}{b} < H_{n+1}.$$

If a/b equals H_n, we are done. Otherwise:

$$0 < \frac{a}{b} - H_n < \frac{1}{n+1}.$$

This reduces us to the first case with the condition that the first fraction to introduce must have denominator $> n+1$ since

$$\frac{1}{n_1} \leq \frac{a}{b} - H_n < \frac{1}{n+1} \Longrightarrow n_1 > n+1. \qquad \square$$

Remarks

1) This algorithm presents some serious inconveniences. As the bound $n_{k+1} > n_k(n_k - 1)$ indicates, it tends to choose denominators that are factorials and that grow excessively. For example, the algorithm gives

$$\frac{153}{1001} = \frac{1}{7} + \frac{1}{101} + \frac{1}{11234} + \frac{1}{1135768634} + \frac{1}{227153727} + \frac{1}{257994078222798918}$$

while there exist "better" decompositions such as

$$\frac{153}{1001} = \frac{1}{8} + \frac{1}{36} + \frac{1}{14415} + \frac{1}{346305960}.$$

2) A decomposition of a rational number into a sum of Egyptian fractions is *never unique* because one can always replace the last fraction by a sum of two new fractions thanks to the identity

$$\frac{1}{x} = \frac{1}{x+1} + \frac{1}{x(x+1)}.$$

Restricting the number of fractions does not fix anything. We shall see later that the fraction $\frac{3}{4}$ admits a single decomposition as a sum of two Egyptian fractions, six decompositions as a sum of three fractions and eighty decompositions as a sum of three fractions. This method of representing the rationals is not very practical!

8.8.1. The program

We want to find *all* decompositions of a given fraction as a sum of two or thee Egyptian fractions. The main body of the program is:

```
begin
| message ; choose(a, b) ;
| decomposition_into_two_fractions(a, b) ;
| decomposition_into_three_fractions(a, b)
end .
```

The procedure decomposition_into_two_fractions

We are going to use "brute force" to consider all possible couples (x_1, x_2). But instead of varying x_1 and x_2 independently, we let x_1 take all possible values. Once x_1 is chosen, we check that whether the *rational number* x_2 defined by the condition

$$\frac{1}{x_2} = \frac{a}{b} - \frac{1}{x_1} \tag{8.12}$$

is an integer. In other words, our scheme is the following:

```
for x₁ := 1 to ∞ do begin
| «calculate x₂ using (8.12)» ;
| if x₂ ∈ ℕ then writeln(x₁, x₂)
end
```

We need to get rid of the infinite bound. A short reflection shows that x_1 and x_2 cannot be too large since the sum of two infinitely small numbers is infinitely small. Let (x_1, x_2) be a solution of (8.12) satisfying the condition $1 \le x_1 \le x_2$. The double inequality

$$\frac{1}{x_1} < \frac{a}{b} \le \frac{2}{x_1} \tag{8.13}$$

immediately furnishes the bounds

$$\frac{b}{a} < x_1 \le \frac{2b}{a} \iff \left\lfloor \frac{b}{a} \right\rfloor + 1 \le x_1 \le \left\lfloor \frac{2b}{a} \right\rfloor . \tag{8.14}$$

We note in passing that (8.14) implies $x_1 \le x_2$ since

$$\frac{1}{x_2} = \frac{a}{b} - \frac{1}{x_1} \le \frac{1}{x_1} .$$

We now specify our algorithm. Here x_1 runs over the interval defined by (8.14). To avoid manipulating rationals, we put

$$a_1 = ax_1 - b, \quad b_1 = bx,$$

so that $x_2 = b_1/a_1$ is an integer if and only if a_1 divides b_1.

```
procedure decomposition_into_two_fractions(a, b : integer) ;
var a₁, b₁, x₁, x₂ : integer ;
begin
| for x₁ := lower_bound(a, b, 1) to (2 * b) div a do begin
| | new_fraction(a, b, x₁, a₁, b₁) ;
| | if is_integer(a₁, b₁) then begin x₂ := a₁ div b₁ ; write_2(x₁, x₂) end
| end
end ;
```

The code for the functions and procedures is easy to write. The seemingly bizarre introduction of the parameter k in the funtion *lower_bound* will be justified below.

```
function lower_bound(a, b, k : integer) : integer ;
var temp : integer ;
begin
| temp := 1 + b div a ;
| if temp ≥ k then lower_bound := temp else lower_bound := k
end ;

procedure new_fraction(a, b, x : integer ; var a1, b1 : integer) ;
begin
| a1 := x * a − b ; b1 := b * x ; simplify(a1, b1)
end ;

function is_integer(a, b : integer) : boolean ;
begin
| if b mod a = 0 then is_integer := true else is_integer := false
end ;
```

The procedure decomposition_into_three_fractions

We continue to use brute force by assigning x_1 and x_2 all possible values and checking to see whether the equation

$$\frac{a}{b} = \frac{1}{x_1} + \frac{1}{x_2} + \frac{1}{x_3} \tag{8.15}$$

defines an integer value of x_3. As in the preceding case, we must find reasonable intervals in which x_1 and x_2 live. Consider a solution of (8.16) which satisfies $x_1 \le x_2 \le x_3$. The double inequality

$$\frac{1}{x_1} < \frac{a}{b} = \frac{1}{x_1} + \frac{1}{x_2} + \frac{1}{x_3} \le \frac{3}{x_1}$$

immediately implies that

$$\frac{b}{a} < x_1 \le \frac{3b}{a} \iff \left[\frac{b}{a}\right] + 1 \le x_1 \le \left[\frac{3b}{a}\right].$$

Having chosen x_1 consistent with this, and putting $a_2 = ax_1 - b$ and $b_2 = bx_1$, we are reduced to studying the equation

$$\frac{a_1}{b_1} = \frac{1}{x_2} + \frac{1}{x_3}. \tag{8.16}$$

This is something we know how to do. It suffices to choose x_2, to put $a_2 = a_1x_2 - b_1$ and $b_2 = b_1x_2$, and to test whether $x_3 = b_2/a_2$ is an integer. Experience shows that even $x_2 \leq x_3$ is a solution of (8.16), we cannot be certain of obtaining $x_1 \leq x_2$. For example, when $a = b = 1$ and $x_1 = 3$, we obtain $a_1/b_1 = 1 - 1/3 = 2/3$ which defines the interval $x_2 \in [2, 4]$. To be sure that $x_2 \geq x_1$, we must choose x_2 in the interval:

$$\max\left(x_1, \left[\frac{b_1}{a_1}\right] + 1\right) \leq x_1 \leq \left[\frac{2b_1}{a_1}\right].$$

This precaution is the reason we introduced the parameter k in the function *lower_bound*(a, b, k).

```
procedure decomposition_into_three_fractions(a, b : integer) ;
var a1, b1, a2, b2, x1, x2, x3 : integer ;
begin
  for x1 := lower_bound(a, b, 1) to (3 * b) div a do begin
    new_fraction(a, b, x1, a1, b1) ;      {a1/b1 = a/b - 1/x1}
    for x2 := lower_bound(a1, b1, x1) to (2 * b1) div a1 do begin
      new_fraction(a1, b1, x2, a2, b2) ; {a2/b2 = a1/b1 - 1/x2}
      if is_integer(a2, b2) then begin
        x3 := b2 div a2 ; write_3(x1, x2, x3)
      end
    end
  end
end ;
```

The procedures write_2 and write_3

Because the integers in Pascal are limited, it is prudent to double check before accepting them: one of the integers x_2 and x_3 might be too large and become negative! Thus, we are going to check the equality $a_1/b_1 = 1/x_1 + 1/x_2 + 1/x_3$ by redoing the calculation using real numbers.

```
procedure write_3(x1, x2, x3 : integer) ;
begin
  write(' = 1/', x1 : 1, ' + 1/', x2 : 1, ' + 1/', x3 : 1) ;
  writeln(', precision = ', a/b - 1/x1 - 1/x2 - 1/x3)
end ;
```

If the "solution" is correct, the displayed real number must be very small (let us say of order at most 10^{-7}). If this is not the case, beware! The procedure *write_2* is similar.

8.8.2. Numerical results

The numerical results are impressive. As one might predict, the number of decompositions increases with the number of fractions one allows. The size of integers that appear is surprising. One also observes that some rationals are not the sum of two (or three) Egyptian fractions.

• One sees, for example, that there is a single decomposition of 3/4 as a sum of two fractions:

$$\frac{3}{4} = \frac{1}{2} + \frac{1}{4}.$$

• When one allows three fractions, there are six solutions :

$$\begin{array}{ll|l} \frac{3}{4} & = \frac{1}{2} + \frac{1}{5} + \frac{1}{20} & = \frac{1}{3} + \frac{1}{3} + \frac{1}{12} \\ & = \frac{1}{2} + \frac{1}{6} + \frac{1}{12} & = \frac{1}{3} + \frac{1}{4} + \frac{1}{6} \\ & = \frac{1}{2} + \frac{1}{8} + \frac{1}{8} & = \frac{1}{4} + \frac{1}{4} + \frac{1}{4}. \end{array}$$

Exercise 6

If you have access to long integers, write a Pascal program which finds all decompositions of a/b as a sum of four fractions.

8.9. Operations on Large Integers

If we wish to add, subtract or multiply two integers of 30 digits, the integers already defined in Pascal will not suffice; a special program is necessary. We are going to work in base $b > 2$ with integers that have at most n digits:

$$x = x_0 + x_1 b + x_2 b^2 + \cdots + x_n b^n, \quad 0 \le x_i < b. \tag{8.17}$$

As usual, we use the notation $x = \overline{x_n \cdots x_0}$ to denote (8.17).

8.9.1. Addition

Put $z = x + y$ and add the representations (8.17) of x and y:

$$z = (x_0 + y_0) + (x_1 + y_1)b + \cdots + (x_n + y_n)b^n. \tag{8.18}$$

The sum $x_0 + y_0$ is not a digit if it is greater than or equal to b. Thus, we divide by b

$$x_0 + y_0 = \rho_1 b + z_0, \quad 0 \le z_0 < b. \tag{8.19}$$

to obtain the digit z_0. The quotient ρ_1 is called the *first carry*. Combining (8.18) and (8.19) gives

$$z = z_0 + (x_1 + y_1 + \rho_1)b + (x_2 + y_2)b^2 + \cdots + (x_n + y_n)b^n. \tag{8.20}$$

Dividing in turn $x_1 + y_1 + \rho_1$ by b gives us the *second carry*.

$$x_1 + y_1 + \rho_1 = b\rho_2 + z_1, \quad 0 \le z_1 < b. \tag{8.21}$$

Substituting (8.21) into (8.20) gives

$$z = z_0 + z_1 b + (x_2 + y_2 + \rho_2)b^2 + \cdots + (x_n + y_n)b^n. \tag{8.22}$$

Proceeding little by little, we finally arrive at the equality

$$z = z_0 + z_1 b + z_2 b^2 + \cdots + z_n b^n + \rho_{n+1} b^{n+1}, \quad 0 \le z_i < b. \tag{8.23}$$

The bounds $0 \le x_0 + y_0 \le 2(b-1)$ show that $0 \le \rho_1 \le 1$. More generally, all the carries are equal to 0 or 1, because if $0 \le \rho_i \le 1$ then

$$0 \le x_i + y_i + \rho_i \le 2b - 1 \Longrightarrow 0 \le \rho_{i+1} \le 1. \tag{8.24}$$

This bound is important, because it will allow us to avoid exceeding capacity when we are programming. When the last carry ρ_{n+1} is zero, (8.22) shows that the representation of z in base b is $z = \overline{z_n \cdots z_0}$. The transformation into an algorithm is immediate:

```
ρ := 0 ; {because ρ_{i+1} is a function of ρ_i}
for i := 0 to n do begin
  | temp := x_i + y_i + ρ ; z_i := temp mod b ;
  | if temp < b then ρ := 0 else ρ := 1
end ;
if ρ > 0 then overflow   {x + y has more than n digits}
```

Note that the calculation of the new carry ρ does not use division by b which greatly speeds up the algorithm.

8.9.2. Subtraction

Consider the representations (8.17) of x and y and let $z = x - y$:

$$z = (x_0 - y_0) + (x_1 - y_1)b + \cdots + (x_n - y_n)b^n. \tag{8.25}$$

The difference $x_0 - y_0$ is a digit if and only if it is positive or zero. When it is negative, we "borrow", which amounts to defining z_0 as follows:

$$z_0 = x_0 - y_0 + \rho_1 b, \quad \rho_1 = \begin{cases} 0 & \text{if } x_0 - y_0 \ge 0, \\ 1 & \text{if } x_0 - y_0 < 0. \end{cases} \tag{8.26}$$

Combining (8.26) and (8.27) gives

$$z = z_0 + (x_1 - y_1 - \rho_1)b + (x_2 - y_2)b^2 + \cdots + (x_n - y_n)b^n. \tag{8.27}$$

When we have $-b \le x_1 - y_1 - \rho_1 < b$, we must borrow again to obtain the digit z_1:

$$z_1 = x_1 - y_1 - \rho_1 + \rho_2 b, \quad \rho_2 = \begin{cases} 0 & \text{if } x_1 - y_1 - \rho_1 \ge 0, \\ 1 & \text{if } x_1 - y_1 - \rho_1 < 0. \end{cases}$$

Proceding little by little, we finally obtain

$$z = z_0 + z_1 b + \cdots + z_n b^n - \rho_{n+1} b^{n+1}, \quad 0 \le z_i < b. \tag{8.28}$$

One cannot have $\rho_{n+1} = 1$ when $x \ge y$ because it follows from (8.28) that

$$z \le (b-1)(1 + b + \cdots + b^n) - b^{n+1} = b^{n+1} - 1 - b^{n+1} = -1.$$

Consequently, $x \ge y$ implies that $\rho_{n+1} = 0$, which shows that the expression $\overline{z_n \cdots z_0}$ given by (8.28) is the representation of z in base b. The subtraction algorithm is, therefore:

```
ρ := 0 ; {because ρ_{i+1} is a function of ρ_i}
for i := 1 to n do begin
   | z_i := x_i − y_i − ρ ;
   | if z_i < 0 then begin z_i := z_i + b ;  ρ := 1 end
   | else ρ := 0
end ;
if ρ > 0 then underflow   {that is x < y}
```

8.9.3. Multiplication

Put $Z = x * y$. The formula

$$Z = x * y = \sum_{i=0}^{n} (x_i * y) * b^i$$

allows us to reduce multiplication to adding the results of multiplication of a number and a digit. So, suppose that $0 \le c < b$ is a digit and put

$$z = c * x = cx_0 + cx_1 b + \cdots + cx_n b^n. \tag{8.29}$$

The inequalities $0 \le cx_0 \le (b-1)^2$, where the bounds can be obtained, show that cx_0 is not always a digit. Thus, we divide by the base

$$cx_0 = \rho_1 b + z_0, \quad 0 \le z_0 < b, \tag{8.30}$$

which gives the first carry ρ_1. We now have

$$0 \le \rho_1 = \left[\frac{cx_0}{b}\right] \le \left[\frac{b^2 - 2b + 1}{b}\right] = \left[b - 2 + \frac{1}{b}\right] = b - 2.$$

If we substitute (8.30) into (8.29) we get

$$z = z_0 + (cx_1 + \rho_1)b + \cdots + cx_n b^n. \tag{8.31}$$

Dividing $cx_1 + \rho_1$ by the base gives the next carry:

$$cx_1 + \rho_1 = \rho_2 b + z_1, \quad 0 \le z_1 < b.$$

Its size is

$$0 \le \rho_2 = \left[\frac{cx_1 + \rho_1}{b}\right] \le \left[\frac{b^2 - 2b + 1 + b - 2}{b}\right] = \left[b - 1 - \frac{1}{b}\right] = b - 2$$

and we have

$$z = z_0 + z_1 b + (cx_2 + \rho_2)b + \cdots + cx_n b^n.$$

Proceeding in this manner, we find that the carries can equal at most $(b - 2)$ and we finally obtain

$$z = z_0 + z_1 b + \cdots + z_n b^n + \rho_{n+1} b^{n+1}. \tag{8.32}$$

If the last carry is zero, we obtain the representation $\overline{z_n \cdots z_0}$. Otherwise, we exceed the capacity.

```
ρ := 0 ; {because ρ_{i+1} is a function of ρ_i}
for i := 0 to n do begin
  | temp := c * x_i + ρ ;
  | z_i := temp mod b ;  ρ := temp div b
end ;
if ρ > 0 then overflow   {c * x possesses more than n digits}
```

The estimates

$$0 \le cx_i + \rho_i \le (b - 1)^2 + (b - 2) < b^2 \tag{8.33}$$

are very valuable because they protect us from exceeding the capacity when we program.

8.9.4. Declarations

We are going to work in base $B = 100$. In the program "digit" (in quotes) will refer to a digit in base B, that is, an ordinary integer between 0 and 99. The machine knows knows how to add and multiply two "digits". The bounds $B^2 < 2^{15}$, (8.20) and (8.32) guarantee that the intermediate calculations will never produce negative integers when starting with Pascal *integers*. Knowing that a line on a screen contains 80 characters, we will only be interested in (positive) integers with at most 80 digits. Such an integer can be stored with $n = \frac{1}{2}80 = 40$ ordinary Pascal integers:

$$x = x_0 + x_1 b + \cdots + x_{39} B^{39}.$$

Thus, our declarations will be as follows

```
const B = 100 ; ind_max = 39 ;
type bigint = array[0..ind_max] of integer ;
string80 = string[80] ;
```

8.9.5. The program

The main body employs the boolean variables *finish*, *overflow*, *underflow* and the *bigint* x, y, z. This program repeatedly requests the values of x and y and displays the sum, difference and product whenever possible; that is, whenever the capacity is not exceeded. One leaves the loop when $x = y = 0$ which allows us to test the pairs $(x, 0)$ and $(0, y)$.

```
begin
  message ;
  repeat
    choose('x', x, finish) ;
    if not finish then choose('y', y, finish) ;
    if not finish then begin
      big_sum(x, y, z, overflow) ; display('x + y', z, overflow) ;
      big_subtraction(x, y, z, underflow) ; display('x − y', z, underflow) ;
      big_multiplication(x, y, z, overflow) ; display('x * y', z, overflow) ;
    end
  until finish
end .
```

(Division will be treated separately because it is more complicated.)

The procedures choose and display

As always, the procedures for interfacing with the exterior (*entrances and exits*) are the most delicate to write. Here are some traps and problems:

- It is necessary to convert chains of chosen characters into *bigint*.
- The "digits" with weak weight (those which multiply $1, B, B^2, \ldots$) are at the beginning of the arrays. This explains the appearance of "downto" loops.
- We should display "03" and not "3" when the "digit" is equal to 3.

The procedure choose

When we type 1234567890123456, we send the machine a *chain of characters* which it must convert into a sequence of "digits".

- We cut the chain into two digit pieces, which requires that we adjoin a '0' when it contains an odd number of digits;
- We convert the ordinary digits (which are characters) into the corresponding integers using the statement "*ord*(*chain*[i]) − *ord*('0')";
- A final snare awaits us: the indices of the chain range between 1 and 80; in *bigint* x, they range between 0 and 39.

```
procedure choose(letter : char ; var x : bigint ; var finish : boolean) ;
var i, ℓ : integer ; chain : string80 ;
begin
  write(letter,' =') ;
  readln(chain) ;
  if chain =' 0'
  then finish := true
  else begin
    finish := false ;
    if length(chain) mod 2 = 1 then chain := concat('0', chain) ;
    annul(x) ;
    ℓ := length(chain) div 2 ;
    for i := 1 to ℓ do
    x[ℓ − i] := 10 ∗ (ord(chain[2 ∗ i − 1]) − ord('0'))
                    +ord(chain[2 ∗ i]) − ord('0')
  end
end ;
```

In order to beter understand the subtleties of this procedure, you should run a trace when *chain* = '1234567'. (The procedure *annul* is left to the reader.)

The procedure display

We want to display the "digits" properly on the screen. For example, if we have $x[0] = 0$, $x[1] = 9$ and $x[2] = 7$, the other "digits" being zero, we must display 70900 and not 790 or $00 \cdots 0070900$! (The convenience of the user always comes *before* that of the programmer.)

```
procedure display(word : string80 ; x : bigint ; impossible : boolean) ;
var i, start : integer ;
begin
  if impossible then writeln('no result : overflow or underflow')
  else begin
    writeln(word) ;
    for i := 1 to ind_max do if x[i] > 0 then start := i ;
    write(x[start] : 1) ;
    for i := start − 1 downto 0 do
    if x[i] < 10 then write('0', x[i] : 1) else write(x[i] : 1) ;
    writeln
  end
end ;
```

The procedure big_sum

This procedure returns $z = x + y$ when this number has less than 80 digits; exceeding the capacity is stored in the boolean variable *overflow*.

```
procedure big_sum(x, y : bigint ; var z : bigint ; var overflow : boolean) ;
var i, carry, temp : integer ;
begin
  carry := 0 ;
  for i := 0 to ind_max do begin
    temp := x[i] + y[i] + carry ;
    if temp > B
    then begin z[i] := temp − B ;  carry := 1 end
    else begin z[i] := temp ;  carry := 0 end ;
    if carry = 0 then overflow := false else overflow := true
  end
end ;
```

The procedure big_subtraction

This procedure returns $z = x - y$ when $x \geq y$; otherwise the variable *underflow* becomes true.

```
procedure big_subtraction(x, y : bigint ;
                                  var z : bigint ; var underflow : boolean) ;
var i, carry, temp : integer ;
begin
  carry := 0 ;
  for i := 0 to ind_max do begin
    temp := x[i] − y[i] − carry ;
    if temp < 0
    then begin z[i] := temp + B ;  carry := 1 end
    else begin z[i] := temp ;  carry := 0 end
    if carry = 0 then underflow := false else underflow := true
  end
end ;
```

The procedure big_multiplication

This procedure returns $z = x * y$ when $x * y$ has less that 80 digits. We use the formula $z = \sum_{i=0}^{40} x_i * y * B^i$ and the algorithm for multiplying by a "digit". If $x * y$ has more than n digits, the boolean variable *overflow* becomes true.

```
procedure big_multiplication(x, y : bigint ;
var z : bigint ;  var overflow : boolean) ;
var i : integer ;  temp : bigint ;
begin
  i := 0 ;  annul(z) ;  overflow := false ;
  repeat
    {temp := x[i] * y}
    big_multiplication_by_digit(x[i], y, temp, overflow) ;
    if not overflow then begin
```

```
      shift(temp, i, overflow) ;  {temp := temp * B^i}
      if not overflow
      then big_sum(z, temp, z, overflow)   {z := z + temp}
    end ;
    i := i + 1
  until (i = ind_max) or overflow
end ;
```

The procedure *shift* is left to the reader.

The procedure big_mult_by_digit

Suppose that x is a "digit", that is $x \in [\![0, B-1]\!]$.

```
procedure big_multiplication_by_digit(x : integer ;  y : bigint ;
                                var z : bigint ;  var overflow : boolean) ;
var i, carry, temp : integer ;
begin
  carry := 0 ;
  for i := 0 to ind_max do begin
    temp := x * y[i] + carry ;
    z[i] := temp mod B ;  carry := temp div B ;
  end ;
  if carry = 0 then overflow := false else overflow := true
end ;
```

Exercise 7

Adapt the algorithms and the program to a representation complementing base B (Chapter 6).

8.10. Division in Base b

Let $x, y > 0$ and q, r be the quotient and the remainder upon division of x by y:

$$x = qy + r, \quad 0 \le r < y.$$

As school children, we learned an algorithm to determine the numbers q and r from x and y. This algorithm, however, uses guesses, which makes it unprogrammable, so that a careful theroretical study is in order.

8.10.1. Description of the division algorithm

Suppose that we know the addition and multiplication tables in base b. To be precise, suppose the following:

- We know how to add, subtract and multiply numbers written in base b.

		ξ							y			
x	$\rightarrow$	1	5	6	2	6	9	3	2	3	7	
q_1y	$\rightarrow$	1	4	2	2	.	.	.	6	5	9	3
ξ_1	$\rightarrow$		1	4	0	6	.	.	$\uparrow$	$\uparrow$	$\uparrow$	$\uparrow$
q_2y	$\rightarrow$		1	1	8	5	.	.	q_1	q_2	q_3	q_4
ξ_2	$\rightarrow$			2	2	1	9	.				
q_3y	$\rightarrow$			2	1	3	3	.				
ξ_3	$\rightarrow$					8	6	3				
q_4y	$\rightarrow$					7	1	1				
r	$\rightarrow$					1	5	2				

Fig. 8.1. *Euclidean division of* $x = 1{,}562{,}693$ *by* $y = 237$

• On the other hand, we do not know how to divide anything other than a number having no more that two diigts by a digit (which we do by "reading the multiplication table in reverse").

This said, we recall how we divide $x = 1{,}562{,}693$ by $y = 237$ (working in base 10).

• Let $\xi_0 = 1562$ be the smallest integer $\geq y$ formed from the first digits of x.

• To find the integer part of ξ_0/y, we ask the question "How many times does 237 go into 1562?" Since we do not know the multiplication table, we cannot immediately answer this question. For this reason, we replace the question by a simpler one: "How many times does 2 go into 15?" The estimate $q_1 = 7$ that we obtain turns out to be too large since $7 \cdot 237 = 1659 > 1562$, so we reduce it by one. Since $6 \cdot 237 \leq 1562$, we now know that $q_1 = 6$ is the integer part we seek.

• We calculate next that $r_1 = \xi_0 - q_1 y = 140$, then we "bring down" the digit of x which follows ξ_0. This means that we have made $\xi_1 = 10r_1 + 6 = 1406$.

• We estimate the integral part of ξ_1/y with the question "How many times does 2 go into 14?". The answer $q_2 = 7$ is too large because $7 \cdot 237 = 1659 > 1406$. We decrease q_2 by a unit and find that this still does not work because $6 \cdot 237 = 1422 > 1406$. We begin again and determine finally that the integer part is $q_2 = 5$.

• We calculate $r_2 = \xi_1 - q_2 y = 221$, then we bring down the next digit of x to obtain $\xi_2 = 10r_2 + 9 = 2219$.

• We estimate the integral part of $\xi_2/y = 2219/237$ by asking "How many times does 2 go into 22?". Since the response is greater than the greatest digit,

$$
x \to \overbrace{x_n \quad \cdots \quad x_m}^{\xi_0} \quad x_{m-1} \quad \cdots \quad x_0 \;\Big|\; y
$$

$$
\begin{array}{l|l}
\begin{cases} r_1 = \xi_0 - q_1 y \\ \xi_1 = br_1 + x_{m-1} \end{cases} & q_1 \cdots q_{m-1} \\
\begin{cases} r_2 = \xi_1 - q_2 y \\ \xi_2 = br_2 + x_{m-2} \end{cases} & q_i = [s_{i-1}/y] \\
\quad \vdots & \\
\begin{cases} r_m = \xi_{m-1} - q_m y \\ \xi_m = br_m + x_0 \end{cases} & \\
r_{m+1} = \xi_m - q_{m+1} y &
\end{array}
$$

Fig. 8.2. *Sequences which arise when dividing x by y*

namely 9, we try $q_3 = 9$. This turns out to be correct.

• The next partial remainder is $r_3 = \xi_2 - q_3 y = 86$. We find that $\xi_3 = 10r_3 + 3 = 863$, whence $q_4 = 3$.

• The division comes to an end with $r_4 = \xi_3 - q_4 y = 152$ because there is no digit to bring down. The quotient we seek is 6593 and the remainder 152.

8.10.2. Justification of the division algorithm

The division algorithm uses two auxiliary sequences $(r_i)_{i \geq 1}$ and $(\xi_j)_{j \geq 0}$:

$$
\begin{cases} r_i = \xi_{i-1} - q_i y, \\ \xi_i = br_i + x_{m-i}. \end{cases} \tag{8.34}
$$

The sequence (ξ_j) begins with the smallest integer $\xi_0 \geq y$ formed from the first digits of x; let

$$
\xi_0 = \overline{x_n \cdots x_m}. \tag{8.35}
$$

The sequence of digits of the quotient is defined by

$$
q_i = \left[\frac{\xi_{i-1}}{y} \right], \tag{8.36}
$$

where, as always, the bracket indicates the integral part. Thus, to a first approximation, we can write the division algorithm as follows:

```
«to determine ξ₀ = overline{x_n ··· x_m} » ;
for i := 1 to m do begin
| q_i := [ξ_{i-1}/y] ;
| r_i := ξ_{i-1} − q_i y ;
| ξ_i := b r_i + x_{m−i} ;
end ;
r_{m+1} := ξ_m − q_{m+1} y
```

Theorem 8.10.1. *The division algorithm defined by* (8.34), (8.35) *and* (8.36) *correctly determines the digits of the quotient.*

Proof. Multiply each r_i by b^{m+1-i} and add the resulting equations term by term to get

$$\begin{array}{c|rcl} b^m & r_1 &=& \xi_0 - q_1 y, \\ b^{m-1} & r_2 &=& br_1 + x_{m-1} - q_2 y, \\ \vdots & & \vdots & \\ b & r_m &=& br_{m-1} + x_1 - q_m y, \\ 1 & r_{m+1} &=& br_m + x_0 - q_{m+1} y. \end{array}$$

After simplifying, we get:

$$x = (q_m b^m + q_{m-1} b^{m-1} + \cdots + q_0) y + r_{m+1}. \tag{8.37}$$

If the q_i defined by (8.36) are digits (that is, if they satisfy $0 \leq q_i < b$), and if the last remainder satisfies $0 \leq r_{m+1} < y$, and we know that the expression of the quotient in base b is $q = \overline{q_1 \cdots q_{m+1}}$. We begin by showing that $y \leq \xi_0 < by$. Consider the numbers ξ' formed by all digits of ξ_0 except the last. That is, $\xi' = \overline{x_n \ldots x_{m-1}}$, so that $\xi_0 = b\xi' + x_m$ (if $m = n$, take $\xi' = 0$.) The definition of ξ_0 ensures that $\xi' < y$. Returning to ξ_0, this gives $\xi_0 \leq b(y-1) + b - 1 \leq by - 1 < by$. Using the inequality $y \leq \xi_0 < by$, we deduce that $q_1 = [\xi_0/y]$ is a digit different from 0 and that $r_1 = \xi_0 - q_1 y$ satisfies $0 \leq r_1 < y$. The bound

$$\xi_1 = br_1 + x_{m-1} \leq b(y-1) + b - 1 = by - 1$$

now allows us to conclude that q_2 is a digit and $0 \leq r_2 \leq y$. A simple induction establishes the theorem. □

8.10.3. Effective estimates of integer parts

If we examine our provisional division algorithm, we find an action which we cannot execute with the primitives at our disposal, namely *calculating the integer part of* $[u/v]$ *when* u *possesses more than two digits* (remember that we need to have a multiplication table in order to read it in reverse). Thus, we replace $q = [u/v]$ by the estimate $\overline{q} = [\overline{u}/\overline{v}]$ obtained from the multiplication table. Let u and v be two integers satisfying the condition

$$0 < v \le u < bv,$$

which ensures that $q = [u/v]$ is a digit different from zero. Set

$$u = \overline{u_{n+1}u_n \cdots u_0}, \qquad v = \overline{v_n \cdots v_0},$$

with $u_{n+1} = 0$ when u and v have the same number of digits. Put $B = b^n$ and

$$\begin{aligned} \overline{u} &= \overline{u_{n+1}u_n}, & \overline{\overline{u}} &= \overline{u_{n-1} \cdots u_0}, & u &= B\,\overline{u} + \overline{\overline{u}}, & 0 &\le \overline{\overline{u}} < B; \\ \overline{v} &= v_n, & \overline{\overline{v}} &= \overline{v_{n-1} \cdots v_0}, & v &= B\,\overline{v} + \overline{\overline{v}}, & 0 &\le \overline{\overline{v}} < B. \end{aligned}$$

The number $\bar{u}$ is made up of the first or two first digits of u according as the number of digits of u is equal to or one more than v. If we only have a simple multiplication table in base b, we can only calculate the digit

$$\overline{q} = \min\Big\{b - 1, \Big[\frac{\overline{u}}{\overline{v}}\Big]\Big\}. \tag{8.38}$$

The estimate $\overline{q}$ that we obtain is sometimes catastrophic: in base 10, if $u = 99$ and $v = 19$ then $\overline{q} = 9$ whereas $q = 5$. More generally, if β is the digit $\beta = b - 1$, with $u = \overline{100} = b^2$ and $v = \overline{1\beta} = 2b - 1$ we have $\overline{q} \simeq b$ whereas $q \simeq \frac{1}{2}b$ if b is large!

Lemma 8.10.1. *With the same notation as above, $q \le \overline{q}$.*

Proof. Suppose that $\overline{q} < b - 1$ and write:

$$\frac{u}{v} = \frac{B\,\overline{u} + \overline{\overline{u}}}{B\,\overline{v} + \overline{\overline{v}}} \le \frac{B\,\overline{u} + B - 1}{B\,\overline{v}}.$$

Since the integer part is an increasing function, we will be done if we can prove that the integer part of the upper bound of u/v is $\overline{q}$; that is if we can establish the inequalities:

$$\overline{q} \le \frac{B\,\overline{u} + B - 1}{B\,\overline{v}} < \overline{q} + 1.$$

- The left inequality is immediate:

$$\overline{q} \le \frac{\overline{u}}{\overline{v}} = \frac{B\,\overline{u}}{B\,\overline{v}} < \frac{B\,\overline{u} + B - 1}{B\,\overline{v}}.$$

- To establish the right one, multiply $\overline{u}/\overline{v} < \overline{q} + 1$ by $\overline{v}$ to obtain the inequality $\overline{u} \le \overline{v}(\overline{q} + 1) - 1$. We have

$$\begin{aligned} B\,\overline{u} + B - 1 &\le B\big((\overline{q} + 1)\overline{v} - 1\big) + B - 1 \\ &\le B(\overline{q} + 1)\overline{v} - 1 < B(\overline{q} + 1)\,\overline{v}. \end{aligned}$$

□

We have seen that it can happen that $\overline{q} - q$ has the same order as $\frac{1}{2}b$. Happily, this catastrophe only occurs when the first digit $\overline{v}$ of v is small, as the following result shows.

Lemma 8.10.2. *If $\overline{v} \geq \frac{1}{2}b$, then $\overline{q} - q \leq 2$.*

Proof. We are going to establish the contrapositive

$$\overline{q} - q \geq 3 \implies \overline{v} < \tfrac{1}{2}b.$$

By the definition of the integer part, we can write:

$$3 \leq \overline{q} - q < \frac{\overline{u}}{\overline{v}} - \frac{u}{v} + 1.$$

It follows that

$$2 < \frac{\overline{u}}{\overline{v}} - \frac{u}{v} \leq \frac{\overline{u}v - u\overline{v}}{\overline{v}v}.$$

Since $\overline{u}v - u\overline{v} = \overline{u}(B\overline{v} + \overline{\overline{v}}) - (B\overline{u} + \overline{\overline{u}})\overline{v} = \overline{u}\,\overline{\overline{v}} - \overline{\overline{u}}\,\overline{v}$, we have

$$2 < \frac{\overline{u}\,\overline{\overline{v}} - \overline{\overline{u}}\,\overline{v}}{\overline{v}v} \leq \frac{\overline{u}\,\overline{\overline{v}}}{\overline{v}v} < \frac{\overline{u}B}{\overline{v}v} \leq \frac{u}{\overline{v}v} \leq \frac{b}{\overline{v}}$$

since u/v is bounded above by b. □

When the first digit $\overline{v}$ of v is not too large, we can replace u, v by $\delta u, \delta v$, which does not change q but which multiplies the remainder r by δ. If we choose δ well, the first digit of δv will be sufficiently large.

Proposition 8.10.1. *Suppose that b is even and $\overline{v} < \frac{1}{2}b$, and put $\delta = [b/(\overline{v}+1)]$. Then δv has the same number of digits as v and the first digit of δv is greater than $\frac{1}{2}b$.*

Example 8.10.1. In base 10, when $v = 293,578$, we have $\overline{v} = 2$, $\delta = 3$ and $\delta v = 880\,734$; when $v = 4999$, we have $\overline{v} = 4$, $\delta = 2$ and $\delta v = 9998$.

Proof. The condition $\overline{v} < b$ shows that $\delta \geq 1$ and $\delta v \geq v$. From $\overline{\overline{v}} < B$, we deduce that

$$\delta v < \left[\frac{b}{\overline{v}+1}\right](B\overline{v} + B) \leq B\left[\frac{b}{\overline{v}+1}\right](\overline{v}+1) \leq Bb$$

since $[b/(\overline{v}+1)] \leq b/(\overline{v}+1)$. Thus δv has the same number of digits as v.

Knowing this and that $\delta v = (\delta\overline{v})B + (\delta\overline{\overline{v}})$, we see that the first digit c_1 of δv equals $\delta\overline{v}$ plus, possibly, a contribution from $\delta\overline{\overline{v}}$:

$$c_1 = \delta\overline{v} + \left[\frac{\delta\overline{\overline{v}}}{B}\right].$$

We can write

$$c_1 \geq \delta \overline{v} > \left(\frac{b}{\overline{v}+1} - 1\right)\overline{v} \geq \frac{b}{2} - 1 \geq \left[\frac{b}{2}\right] - 1$$

because $\overline{v}\big(b/(\overline{v}+1) - 1\big) - \frac{1}{2}b + 1 = (\frac{1}{2}b - \overline{v} - 1)(\overline{v} - 1)/(\overline{v}+1) \geq 0$.

We are done upon remarking that the strict inequality $c_1 > [\frac{1}{2}b] - 1$ between integers implies that $c_1 \geq [\frac{1}{2}b]$. □

8.10.4. A good division algorithm

When the base is an even number, we can *prepare* x and y, meaning that we can replace x and y by δx and δy so that the first digit of δy is greater than $\frac{1}{2}b$. We can then be certain that our estimates are good because $\overline{q}$ will equal either q, $q+1$ or $q+2$: the internal loop "while $r_i < 0$ do begin ... end" will be traversed at most twice.

```
δ := 1 ; v̄ := digit of highest weight of y ;
if 2v̄ < b then begin
| δ := b div (1 + v̄) ;
| x := δx ; y := δy
end ;
«determine ξ₀ = x_n ··· x_m» ;
for i := 1 to m do begin
| q_i := estimate(ξ_{i-1}, y) ; {q_i = q̄}
| r_i := ξ_{i-1} − q_i * y
| while r_i < 0 do begin
| | r_i := r_i + y ; q_i := q_i − 1
| end ;
| ξ_i := br_i + x_{m-i}
end ;
r := x − q * y
```

Exercise 8

Write a program that divides one *biginteger* by another.

8.11. Sums of Fibonacci Numbers

In what follows, $x \gg y$ means $x \geq y + 2$ and $x \ll y$ means $x \leq y + 2$.

Theorem 8.11.1 (Zeckendorf, 1972). *Every integer $n \geq 1$ is a sum of Fibonacci numbers:*

$$n = F_{i_1} + F_{i_2} + \cdots + F_{i_k}. \tag{8.39}$$

Moreover, if one requires that the indexes satisfy $i_1 \gg i_2 \gg \cdots \gg i_k \gg 0$, then this decomposition is unique.

Here are the Zeckendorf decompositions of the integers $2 \le n \le 25$.

$2 = F_3$	$10 = F_6 + F_3$	$18 = F_7 + F_5$
$3 = F_4$	$11 = F_6 + F_4$	$19 = F_7 + F_5 + F_2$
$4 = F_4 + F_2$	$12 = F_6 + F_4 + F_2$	$20 = F_7 + F_5 + F_3$
$5 = F_5$	$13 = F_7$	$21 = F_8$
$6 = F_5 + F_2$	$14 = F_7 + F_2$	$22 = F_8 + F_2$
$7 = F_5 + F_3$	$15 = F_7 + F_3$	$23 = F_8 + F_3$
$8 = F_6$	$16 = F_7 + F_4$	$24 = F_8 + F_4$
$9 = F_6 + F_2$	$17 = F_7 + F_4 + F_2$	$25 = F_8 + F_4 + F_2$

Remarks

1) One can show that the Fibonacci sequence is the only sequence with this property.

2) Please refer also to the *Hofstadter function* (Chap. 12, §2).

Lemma 8.11.1. *For all $\ell \ge 2$, one has*

$$F_{\ell+1} = 1 + F_\ell + F_{\ell-2} + F_{\ell-4} + \cdots, \tag{8.40}$$

the sum extending over all indices $\ell - 2i \ge 2$ (the last term is F_2 (resp. F_3) if ℓ is even (resp. odd)).

Proof. The identity follows easily by induction upon repeatedly applying the definition

$$F_{\ell+1} = F_\ell + F_{\ell-1} = F_\ell + F_{\ell-2} + F_{\ell-3} = \cdots.$$

When ℓ is even, one stops at $F_2+F_1 = F_2+1$; when ℓ is odd, at $F_3+F_2 = F_3+1$ (the equality $F_2 = F_1 = 1$ is essential). □

Proof of Zeckendorf's theorm

We begin with existence. Set $n_0 = n$. Since F_t tends to infinity, there exists an index $i_1 \ge 2$ (and only one such) such that:

$$F_{i_1} \le n_0 < F_{i_1+1}.$$

Put $n_1 = n_0 - F_{i_1}$. If n_1 is zero, we are done; otherwise, we have the bound:

$$n_1 = n_0 - F_{i_1} < F_{i_1+1} - F_{i_1} = F_{i_1-1}.$$

Since $1 \le n_1 < F_{i_1-1}$, it follows that there exists a (unique) index $2 \le i_2 < i_1 - 1$ such that:

$$F_{i_2} \le n_1 < F_{i_2+1}.$$

Put $n_2 = n_1 - F_{i_1}$. If n_2 is zero, we are done. Otherwise,

$$1 \leq n_2 = n_1 - F_{i_2} < F_{i_2+1} - F_{i_2} = F_{i_2-1},$$

which allows us to begin again. Since the sequence $(n_i)_{i\geq 0}$ defined in this way is strictly decreasing, the algorithm cannot run endlessly. (Once again, the equality $F_2 = F_1 = 1$ is vital, because it ensures a correct stop.) It is clear that one can extract an algorithm from this argument which will supply the decomposition (8.39). Now we establish uniqueness. The conditions $i_1 \gg i_2 \gg \cdots \gg i_k \gg 0$ imply that for all $t = 1, \ldots, k$, one has:

$$2 + 2(k-t) \leq i_t \leq i_1 - 2(t-1).$$

Since the Fibonacci sequence is increasing, identity (8.40) allows us to write

$$\begin{aligned} n &= F_{i_1} + F_{i_2} + \cdots + F_{i_k} \\ &\leq F_{i_1} + F_{i_1-2} + F_{i_1-4} + \cdots + F_{i_1-2(k-1)} = F_{i_1+1} - 1. \end{aligned}$$

Two cases arise:

• $n = F_\ell$ is a Fibonacci number. It follows from the bound $F_\ell < F_{i_1+1}$ that $\ell \leq i_1$ and thus that $k = 1$.

• n is not a Fibonacci number. The bounds $F_{i_1} < n < F_{i_1+1}$ establish the uniqueness of i_1.

The uniqueness of the other indices now follows by induction.

The Zeckendorf decomposition (8.39) of $n+1$ can be obtained very simply from that of n thanks to (8.40). Let $\ell \in [\![1, k]\!]$ be the smallest index such that $i_{\ell+1} \geq i_\ell + 3$:

$$n = F_{i_1} + \cdots + F_{i_{\ell+1}} + (F_{i_\ell} + F_{i_{\ell-2}} + F_{i_{\ell-4}} + F_{i_{\ell-6}} + \cdots + F_{i_k}).$$

In other words, we put in parentheses those F_i that we encounter when we start from the right letting the indices grow by 2:

- $16 = F_7 + (F_4)$,
- $25 = F_8 + (F_4 + F_2)$,
- $20 = (F_7 + F_5 + F_3)$.

Proposition 8.11.1. *With the notation above:*

$$n + 1 = \begin{cases} F_{i_1} + \cdots + F_{i_k} + F_2 & \text{if } i_k \geq 4, \\ F_{i_1} + \cdots + F_{i_{\ell+1}} + F_{i_\ell+1} & \text{if } i_k \leq 3 \text{ and } \ell > 1, \\ F_{i_1+1} & \text{if } i_k \leq 3 \text{ and } \ell = 1. \end{cases}$$

Proof. This is an immediate application of the identity (3.40). □

Example

Let us start, for example, with the decomposition $25 = F_8 + (F_4 + F_2)$. We have $26 = F_8 + (F_4 + F_2 + 1) = F_8 + F_5$ in view of (3.40). Then we get $27 = F_8 + (F_5 + 1) = F_8 + F_5 + F_2$ and then $28 = F_8 + F_5 + (F_2 + 1) = F_8 + F_5 + F_3$. The next decompositions are:

$26 = F_8 + F_5$	$31 = F_8 + F_6 + F_3$	$36 = F_9 + F_3$
$27 = F_8 + F_5 + F_2$	$32 = F_8 + F_6 + F_4$	$37 = F_9 + F_4$
$28 = F_8 + F_5 + F_3$	$33 = F_8 + F_6 + F_4 + F_2$	$38 = F_9 + F_4 + F_2$
$29 = F_8 + F_6$	$34 = F_9$	$39 = F_9 + F_5$
$30 = F_8 + F_6 + F_2$	$35 = F_9 + F_2$	$40 = F_9 + F_5 + F_2$

Exercises 9

- Write a Pascal program to display all Zeckendorf decompositions of integers in the interval $[\![1, 2000]\!]$ knowing only that $1 = F_2$.
- Let $n = F_{i_1} + \cdots + F_{i_k}$ be *any* sum if Fibonacci numbers. Write a Pascal program which computes the Zeckendorf decomposition of n.
- Write a Pascal program which adds two numbers whose Zeckendorf decompositions are given.

8.12. Odd Primes as a Sum of Two Squares

We know from Chapter 2 that a prime number p of the form $4n + 1$ is a sum of two squares. We are going to prove this theorem differently. This time, our proof will be constructive and be based on Euler's proof.

First step: finding a particular solution of $X^2 + 1 \equiv 0$

We work modulo p. Euler's theorem (Chap. 2) tells us that there are as many squares as nonsquares in $\mathbb{Z}_p^*$ and that if $x \neq 0$, then

$$x^{(p-1)/2} = \begin{cases} +1 & \text{if } x \text{ is a square in } \mathbb{Z}_p, \\ -1 & \text{otherwise.} \end{cases}$$

To find a particular solution of the equation $X^2 + 1 \equiv 0 \bmod p$, we choose an element $x \in \mathbb{Z}_p^*$ at random and raise it to the power $\frac{1}{2}(p-1)$ using the algorithm for fast exponentiation. If squares and nonsquares were uniformly distributed in $\mathbb{Z}_p^*$, we would have exactly one chance in two of choosing a nonsquare; practice shows that we very rapidly obtain a nonsquare. When this is the case, $x^{(p-1)/2} = -1$, which shows that $X = x^{(p-1)/4}$ is a solution of $X^2 + 1 \equiv 0 \bmod p$.

In Pascal one chooses an integer $x \in [\![0, p-1]\!]$ at random by typing $x := random(p)$. To obtain a random integer $x \in [\![1, p-1]\!]$, it suffices to write

$$x := 1 + random(p-1).$$

Second step: decomposing p into a sum of two squares

Let X_0 and Y_0 be two integers such that $X_0 \not\equiv 0$ or $Y_0 \not\equiv 0$ and $X_0^2 + Y_0^2 \equiv 0$ modulo p:

$$X_0^2 + Y_0^2 = pN_0. \tag{8.41}$$

• If $N_0 = 1$, chance is on our side and we are done.

• If $N_0 \geq 2$, we divide X_0 and Y_0 by p using division with centered remainders:

$$\begin{aligned} X_0 &= p\xi + x, \ |x| < \tfrac{1}{2}p, \\ Y_0 &= p\eta + y, \ |y| < \tfrac{1}{2}p. \end{aligned} \tag{8.42}$$

(The inequalities are strict because p is odd.) It follows from (3.41) and (3.42) that $x^2 + y^2 \equiv X_0^2 + Y_0^2 \equiv 0 \bmod p$, whence:

$$x^2 + y^2 = np, \quad n \geq 1. \tag{8.43}$$

We certainly have $n > 0$, because $n = 0$ and (8.42) would imply that both X_0 and Y_0 are multiples of p, contrary to hypothesis.

- If $n = 1$, we are done because $p = x^2 + y^2$.
- If $n > 1$, we divide x and y by n, which is reasonable since the inequality $pn \leq x^2 + y^2 < \frac{1}{4}p^2$ shows that $n < \frac{1}{4}p$:

$$\begin{aligned} x &= n\alpha + a, \ |a| \leq \tfrac{1}{2}n, \\ y &= n\beta + b, \ |b| \leq \tfrac{1}{2}n. \end{aligned} \tag{8.44}$$

It follows from (8.43) and (8.44) that $a^2 + b^2 \equiv x^2 + y^2 \equiv 0 \bmod n$, whence

$$a^2 + b^2 = nN_1. \tag{8.45}$$

Putting $pn = x^2 + y^2$ together with (8.44), using (8.45) and dividing by n, we find that

$$p = n(\alpha^2 + \beta^2) + 2(\alpha a + \beta b) + N_1 . \tag{8.46}$$

If we multiply (8.46) by N_1, we obtain[8]

$$\begin{aligned} pN_1 &= nN_1(\alpha^2 + \beta^2) + 2N_1(\alpha a + \beta b) + N_1^2 \\ &= (a^2 + b^2)(\alpha^2 + \beta^2) + 2N_1(\alpha a + \beta b) + N_1^2 \\ &= (N_1 + \alpha a + \beta b)^2 + (a\beta - b\alpha)^2. \end{aligned} \tag{8.47}$$

[8] Euler had a prodigious capacity for calculation.

Putting $X_1 = |N_1 + \alpha a + \beta b|$ and $Y_1 = |a\beta - b\alpha|$, (8.47) becomes:

$$pN_1 = X_1^2 + Y_1^2. \tag{8.48}$$

Let us estimate the orders of magnitude of n and N_1. If ξ is not zero in (8.42) we can write $|X_0| \geq p|\xi| - |x| \geq p - |x| > \frac{1}{2}p > |x|$; if ξ is zero, we only have $|X_0| \geq |x|$. One shows in the same way that $|Y_0| \geq |y|$. These bounds imply that $1 \leq n \leq N_0$. Since we also have $nN_1 = a^2 + b^2 \leq \frac{1}{4}n^2$, we finally get:

$$1 \leq N_1 \leq \tfrac{1}{4}n \leq \tfrac{1}{4}N_0. \tag{8.49}$$

Let us justify the inequality $N_1 \geq 1$. If N_1 were zero, we would have $a = b = 0$ because of (8.45) as well as $x = n\alpha$ and $y = n\beta$, so that $x^2+y^2 = np$ becomes $n(\alpha^2 + \beta^2) = p$. Since p is prime, this means that $n = p$ or $n = 1$. The bound $pn \leq x^2 + y^2 < \frac{1}{2}p^2$ prohibits the case $n = p$, so that we must have $n = 1$, which is contrary to our working hypothesis.

If $N_1 = 1$, we are done because (8.48) is the decomposition of p as a sum of two squares.

If $N_1 > 1$, we begin anew the calculations above replacing X_0, Y_0 by X_1, Y_1. (If $X_1 \equiv 0$ and $Y_1 \equiv 0 \bmod p$, we deduce from (8.48) that $N_1 \equiv 0$ which is impossible since $1 \leq N_1 \leq \frac{1}{4}n < \frac{1}{16}p$.) This process creates a sequence $N_0 > N_1 > \cdots > N_k = 1$ which decreases very rapidly: $N_{k+1} \leq \frac{1}{4}N_k$. Euler's algorithm for writing p as a sum of two squares is therefore:

```
«seek X such that X² + 1 ≡ 0 mod p» ;
Y := 1 ;
while X² + Y² > p do begin
  | centered_division(X, p, ξ, x) ;
  | centered_division(Y, p, η, y) ;
  | n := (x² + y²) div p ;
  | if n = 1 then begin X := x ;  Y := y end
  | else begin
  |   | centered_division(x, n, α, a) ;
  |   | centered_division(y, n, β, b) ;
  |   | N := (a² + b²) div n ;
  |   | X := abs(N + a * α + b * β) ;
  |   | Y := abs(a * β − b * α) ;
  | end
end
```

Here is the trace of this algorithm when $p = 1913$. Note the very rapid decrease in N which accords with what we would expect theoretically.

N	X	Y	n	x	y
754	1201	1	265	712	1
26	223	3	26	223	3

N	X	Y	n	x	y
5	94	27	5	94	27
1	8	43			

Remarks

There are two ways of leaving the loop: either because $n = 1$ (that is, $p = x^2 + y^2$), or because $N = 1$ (that is, $p = X^2 + X^2$). The two cases arise:

- $p = 401$, $X_0 = 381$ and $Y_0 = 1$ give the solution $x = 20$ and $y = 1$.
- $p = 397$, $X_0 = 334$ and $Y_0 = 1$ give $x = -63$, $y = 1$ which in turn give rise to the solution $X_1 = 19$ and $Y_1 = 6$.

Exercise 10

To find all solutions of the equation $x^2 + y^2 = n$ such that $0 \leq y \leq x$, we can "tack" around the track defined by the circle $x^2 + y^2 = n$ leaving the point $([\sqrt{n}\,], 0)$ and moving vertically when we are inside the circle and diagonally to the left when we are on the exterior of the circle (Fig. 8.1).

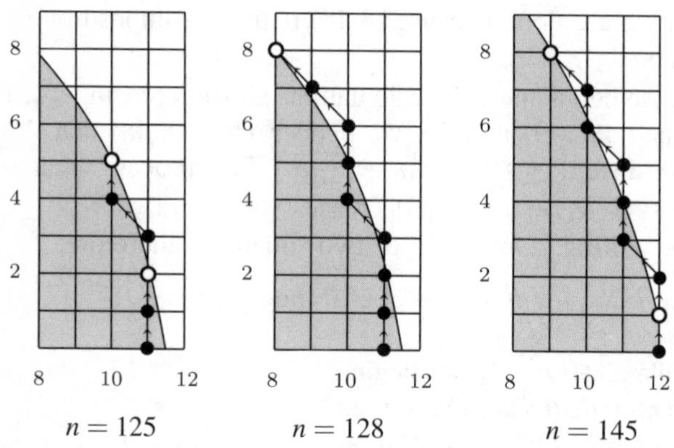

Fig. 8.3. How to find all solutions $0 \leq y \leq x$ *of the equation* $x^2 + y^2 = n$

Translation into Pascal is easy; if we want to speed it up, we can replace the first loop which calculates $\left[\sqrt{n}\right]$ by a dichotomous search.

```
x := 0 ; y := 0 ;
while (x + 1) * (x + 1) ≤ n do x := x + 1 ;
repeat
  | Δ := x * x + y * y − n ;
  | if Δ = 0 then writeln(x, y) ;
  | if Δ > 0 then x := x − 1 ;
  | y := y + 1
until x < y
```

Exercise 11

Show that this algorithm is correct.

8.13. Sums of Four Squares

Theorem 8.13.1 (Lagrange). *Every integer $n \geq 0$ is a sum of four squares.*

Proof. Let U, V be two quaternions. The identity $|UV|^2 = |U|^2 \cdot |V|^2$ can be written:

$$\begin{aligned}(a^2+b^2+c^2+d^2)&(x^2+y^2+z^2+t^2)\\ &= (ax+by+cz+dt)^2+(ay-bx-ct+dz)^2 \\ &\qquad +(az+bt-cx-dy)^2+(at-bz+cy-dx)^2.\end{aligned} \tag{8.50}$$

□

Thus the product of two sums of squares is again a sum of squares and we are reduced to proving the theorem in the case when $n = p$ is a prime. The cases $p = 2$ and $p = 4\ell + 1$ have already been settled, so we can suppose that $p = 4\ell + 3$. We first show (Lemma 8.13.1) that kp is a sum of four squares when k is sufficiently large. We then improve this result by showing (Lemma 8.13.2) that the smallest value of k is 1.

Lemma 8.13.1. *Let $p = 4\ell + 3$ be a prime number. Then the equation $x^2 + y^2 + z^2 \equiv 0$ has a solution $(x, y, z) \neq (0, 0, 0)$ in $\mathbb{Z}_p$. In other words, there exists $k \geq 1$ for which kp is a sum of three squares.*

Proof. We show that the equation $x^2 + y^2 + z^2 = 0$ admits a solution of the form $(x, y, 1)$. Let $d \in [\![2, p-1]\!]$ be the smallest integer which is not a square, so that $d^{(p-1)/2} = d^{2\ell+1} = -1$. Euler's theorem tells us that $-d$ is then a square since $(-d)^{2\ell+1} = 1$. Thus the equations $y^2 = -d$ and $x^2 = d - 1$ each have a solution since $d - 1$ is a square. □

Lemma 8.13.2. *Let $p = 4\ell + 3$ be a prime number and $k \geq 1$ the smallest integer such that kp is a sum of four squares. Then $k = 1$.*

Proof. The first lemma guarantees the existence of an integer k such that $x^2 + y^2 + z^2 + t^2 = kp$. Choose $x, y, z, t \geq 0$ with $k \geq 1$ a minimum. The identity $(p - x)^2 = p^2 - 2px + x^2$ allows us to write

$$[k + p - 2x]p = (p - x)^2 + y^2 + z^2 + t^2.$$

Since k is minimal, we must have $k + p - 2x \geq k$. Thus, $0 \leq x < \frac{1}{2}p$, where the inequality is strict because p is odd. Since this bound also holds for the numbers y, z, t, we have

$$kp < 4\left(\tfrac{1}{2}p\right)^2, \quad i.e. \quad k < p. \tag{8.51}$$

Suppose that $k > 1$:

• If k is even, the possibilities for x, y, z, t are as follows: all are even; two are even, two are odd; all are odd. By relabelling if necessary, we may assume that $x \equiv y \bmod 2$ and $z \equiv t \bmod 2$. We find that

$$\tfrac{1}{2}kp = \left[\tfrac{1}{2}(x+y)\right]^2 + \left[\tfrac{1}{2}(x-y)\right]^2 + \left[\tfrac{1}{2}(z+t)\right]^2 + \left[\tfrac{1}{2}(z-t)\right]^2,$$

which is absurd.

• If k is odd, let $\xi \equiv x \bmod k$ be the remainder after centered division of x by k, which ensures that $|\xi| < \frac{1}{2}k$. Similarly, we associate the remainders η, ζ, τ to the numbers y, z, t. The congruence $\xi^2 + \eta^2 + \zeta^2 + \tau^2 \equiv 0 \pmod{k}$ can be written as

$$k\ell = \xi^2 + \eta^2 + \zeta^2 + \tau^2 \quad \text{with} \quad 0 \le \ell < k. \tag{8.52}$$

If $\ell = 0$, we have $x \equiv y \equiv z \equiv t \equiv 0 \bmod k$ by (8.53). Upon putting $x = kx'$, etc., we obtain $k(x'^2 + y'^2 + z'^2 + t'^2) = p$, which implies that $k = p$ because $k > 1$. But this is forbidden by (8.52). If $\ell > 0$, we use (8.53) and (8.51) to deduce that:

$$\begin{aligned}(kp)(k\ell) = {} & (x\xi + y\eta + z\zeta + t\tau)^2 + (x\eta - y\xi - z\tau + t\zeta)^2 \\ & + (x\zeta + y\tau - z\xi - t\eta)^2 + (x\tau - y\zeta + z\eta - t\xi)^2.\end{aligned}$$

These four squares are divisible by k^2 since, modulo k, we have

$$\begin{aligned} x\xi + y\eta + z\zeta + t\tau &\equiv \xi^2 + \eta^2 + \zeta^2 + \tau^2 &&\equiv 0, \\ x\eta - y\xi - z\tau + t\zeta &\equiv \xi\eta - \eta\xi - \zeta\tau + \tau\zeta &&\equiv 0, \quad \text{etc.} \end{aligned}$$

As a result, we obtain an equation of the form

$$k^2 p\ell = (kX)^2 + (kY)^2 + (kZ)^2 + (kT)^2,$$

which leads to contradiction upon dividing by the term k^2. □

Exercise 12

Transform this proof into an algorithm.

8.14. Highly Composite Numbers

Suppose that we wish to store in a vector $d[1\,..\,dim\_max]$ all divisors of a given number $n \ge 1$. Knowing only, say, that $n \le 200$, what value should we give dim_max ? If $n = p_1^{\alpha_1} \cdots p_r^{\alpha_r}$ is the decomposition of n into prime factors, then our problem amounts to finding the maximum of the function

$$d(n) = \textit{number of divisors of } n = (\alpha_1 + 1) \cdots (\alpha_r + 1) \tag{8.53}$$

as n varies from 2 to 200.

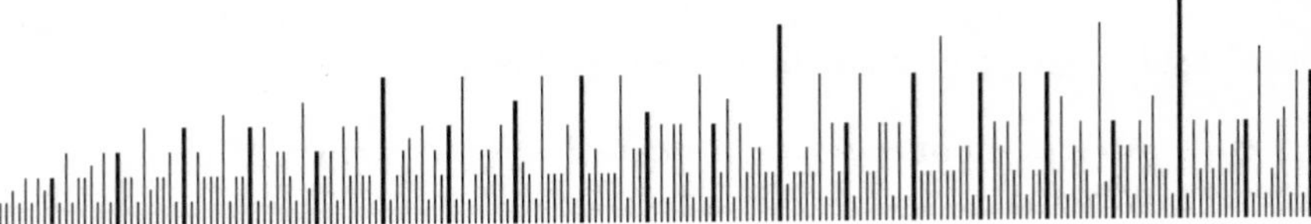

Fig. 8.4. The function $d(n)$ for $2 \leq n \leq 200$.

Imagine that the sticks in our diagram are soldiers which march to the left. Only the soldiers which are taller than all those before them see where they are going. Let us bring these "tall soldiers" to the fore.

```
max := 0 ;
for n := 2 to 200 do begin
  | num_div := 0 ;
  | for d := 1 to n do
  |     if n mod d = 0 then num_div := num_div + 1 ;            (8.54)
  | if num_div > max then begin
  |   | max := num_div ;
  |   | writeln('d(', n : 1, ') = ', num_div : 1)
  | end
end
```

Running this code tells us that there are ten *intermediate viewers* (the "tall soldiers") obtained (Fig. 8.4) when $n = 2, 4, 6, 12, 24, 36, 48, 60, 120, 180$. If follows immediately from this calculation that the number of divisors of an integer $n \leq 200$ is always less than or equal to $d(180) = 18$. When n runs over a very long interval $[\![2, N]\!]$ the code (8.55) is not very fast. In fact, let us estimate the time spent in the two loops.

```
for n := 2 to N do
    for d := 1 to n do
        if n mod d = 0 then num_div := num_div + 1
```

Since the inner loop makes n divisions, the total time, neglecting additions, is proportional to $2 + \cdots + N$, the constant of proportionality depending on the speed with which the divisions are made. If we replace the sum by $\int_2^N x \, \mathrm{d}x$ which does not change the order of magnitude, we conclude that the time required for the calculation is on the order of $\frac{1}{2}N^2$. When $N = 2^{15} = 32768$, we have $\frac{1}{2}N^2 = 2^{30} \approx 5 \cdot 10^8$. If we are using a microcomputer of average power (in 1995) which can do 1000 divisions a second, the calculation will last at least $5 \cdot 10^5$ seconds, that is 139 hours ...

The notion of a highly composite integer, which was introduced by S. Ramanujan[9] leads to an elegant solution of this problem.

Definition 8.14.1. *An integer $n > 1$ is called highly composite if $d(n') < d(n)$ for every integer $n' < n$.*

Proposition 8.14.1. *If H is the largest highly composite integer less than or equal to N, then the maximum number of divisors of an integer less than or equal to N is $d(H)$.*

In fact, for all $n \in [\![H, N]\!]$, we must have $d(n) \leq d(H)$; otherwise the smallest integer $n > H$ such that $d(n) > d(H)$ would be a highly composite integer, contradicting the definition of H.

8.14.1. Several properties of highly composite numbers

Let $(p_i)_{i \geq 1}$ be a strictly increasing sequence of prime numbers. To the decomposition

$$N = p_{i_1}^{\alpha_1} \cdots p_{i_r}^{\alpha_r}, \quad 1 \leq i_1 < \cdots < i_r,$$

into prime factors, we associate the number

$$\widetilde{N} = p_1^{\beta_1} \cdots p_r^{\beta_r}, \quad \beta_1 \geq \beta_2 \geq \cdots \geq \beta_r, \tag{8.55}$$

where the exponents β_i are obtained from the α_i by reordering them into a decreasing sequence. For example:

$$N = 3^5 \cdot 7^2 \cdot 13^3 \cdot 17^2 \cdot 23^9 \Longrightarrow \widetilde{N} = 3^9 \cdot 7^5 \cdot 13^3 \cdot 17^2 \cdot 23^2.$$

Theorem 8.14.1 (Ramanujan). *$d(\widetilde{N}) = d(N)$ and $\widetilde{N} \leq N$.*

Proof. The first equation follows from (8.55). To establish the inequality, note that since $(p^u q^v)/(p^v q^u) = (p/q)^{u-v}$, we have:

$$(u \geq v) \text{ and } (p \leq q) \Longrightarrow p^u q^v \leq p^v q^u.$$

If we permute the α_i to obtain the β_j, we decrease N:

$$p_{i_1}^{\beta_1} \cdots p_{i_r}^{\beta_r} \leq p_{i_1}^{\alpha_1} \cdots p_{i_r}^{\alpha_r} = N.$$

We wind up with the obvious inequality $\widetilde{N} \leq p_{i_1}^{\alpha_1} \cdots p_{i_r}^{\alpha_r}$. □

Corollary 8.14.1. *A highly composite integer necessarily has the form*

$$N = 2^{\alpha_2} 3^{\alpha_3} \cdots p^{\alpha_p}, \qquad \alpha_2 \geq \alpha_3 \geq \cdots \geq \alpha_p \geq 1. \tag{8.56}$$

□

[9] S. Ramanujan, *Highly Composite Numbers*, Proc. London Math. Soc. XIV (1915), pp. 347–409.

Recall that the integers in Pascal do not exceed 2^{15}. Knowing that

$$2 \cdot 3 \cdot 5 \cdot 7 \cdot 11 \cdot 13 < 2^{15} < 2 \cdot 3 \cdot 5 \cdot 7 \cdot 11 \cdot 13 \cdot 17,$$

we are going to try to find all highly composite integers whose greatest prime divisor is less than or equal to 13. We begin by estimating the exponents in the decomposition (8.56).

Theorem 8.14.2 (Ramanujan). *Let $N = 2^{\alpha_2}3^{\alpha_3}\cdots p^{\alpha_p}$ be a highly composite integer and P the prime number that follows p. For every prime $q \leq p$, the following holds:*

$$\left[\frac{\log p}{\log q}\right] \leq \alpha_q \leq 2\left[\frac{\log P}{\log q}\right].$$

Proof. We first bound α_q from below. We suppose here $q < p$; let $x = [\log p/\log q]$ be such that $q^x < p < q^{x+1}$, and consider the integer:

$$N' = \frac{N}{p}q^x < N.$$

We have $d(N') < d(N)$ because N is highly composite and $N' < N$. Eliminating the common terms in the inequality $d(N') < d(N)$, we obtain

$$(\alpha_q + x + 1)\alpha_p < (\alpha_q + 1)(\alpha_p + 1).$$

Upon expanding out and simplifying, we get $x < \alpha_q + 1$.

To bound α_q from above, we put $y = [\log P/\log q]$, which implies that $q^y < P < q^{y+1}$. If $\alpha_q \geq y + 1$, we can consider the integer

$$N' = \frac{N}{q^{y+1}}P < N.$$

As before, we have $d(N') < d(N$, which can be written

$$2(\alpha_q - y) < \alpha_q + 1 \implies \alpha_q < 2y + 1.$$ □

Remarks

1) This theorem shows that there are only a finite number of highly composite integers whose greatest prime divisor is p.

2) For $p = 3, 5, 7, 11, 13$, the theorem above gives the following ranges for the α_i:

p	α_2	α_3	α_5	α_7	α_{11}	α_{13}
3	1..4	1..2				
5	2..4	1..2	1..2			
7	2..6	1..4	1..2	1..2		
11	3..6	2..4	1..2	1..2	1..2	
13	3..8	2..4	1..2	1..2	1..2	1..2

We see from this table that the highly composite integers whose greatest prime is 13 is not accessible using the integers in Pascal because $2^3 \cdot 3 \cdot 5 \cdot 7 \cdot 11 \cdot 13 = 120120 > 2^{15}$. When the greatest divisor is 11, we must eliminate the integers $2^{\alpha_2} \cdots 11^{\alpha_{11}}$ where $\alpha_2 \geq 4$ because $2^4 \cdot 3^2 \cdot 5 \cdot 7 \cdot 11 = 55440 > 2^{15}$.

3) Ramanujan was able to show that the last exponent of (8.56) is

$$\alpha_p = 1 \quad \text{when} \quad N \neq 4 \text{ or } N \neq 36.$$

8.14.2. Practical investigation of highly composite integers

Thanks to (8.57), we know that highly composite integers are hidden among the integers of the form

$$N = 2^{\alpha_2} 3^{\alpha_3} \cdots p^{\alpha_p}, \quad \alpha_2 \geq \alpha_3 \geq \cdots \geq \alpha_p \geq 1.$$

To decide if N is highly composite, we need only verify that $d(M) < d(N)$ for all integers $M < N$. Moreover, Theorem 8.15.1 allows us to restrict our tests to integers of the form:

$$\widetilde{M} = 2^{\beta_2} 3^{\beta_3} \cdots q^{\beta_q}, \quad \beta_2 \geq \beta_3 \geq \cdots \geq \beta_q \geq 1, \quad \widetilde{M} < N. \tag{8.57}$$

To bound β_r from above, we put $\widetilde{M} = r^{\beta_r} W$. Then $rW \geq 2 \cdot 3 \cdot 5 \cdot \cdots \cdot q$ shows that

$$r^{\beta_r} < \frac{\widetilde{N} \cdot r}{2 \cdot 3 \cdot 5 \cdot \cdots \cdot q}. \tag{8.58}$$

Proposition 8.14.2. *The integer* $N = 2^{\alpha_2} 3^{\alpha_3} \cdots p^{\alpha_p}$ *is highly composite if and only if* $d(\widetilde{M}) < d(N)$ *for all integers* $\widetilde{M}$ *satisfying* (8.57) *and* (8.58).

Example

Is the integer $N = 50{,}360 = 2^4 \cdot 3^4 \cdot 5 \cdot 7$ highly composite? Since

$$2 \cdot 3 \cdot 5 \cdot 7 \cdot 11 \cdot 13 < N < 2 \cdot 3 \cdot 5 \cdot 7 \cdot 11 \cdot 13 \cdot 17,$$

we must perform the test $d(\widetilde{M}) < d(N)$ for all integers $\widetilde{M} < N$ of the form

(i) 2^{β_2},

(ii) $2^{\beta_2} 3^{\beta_3}$,

(iii) $2^{\beta_2} 3^{\beta_3} 5^{\beta_5}$,

(iv) $2^{\beta_2} 3^{\beta_3} 5^{\beta_5} 7^{\beta_7}$,

(v) $2^{\beta_2} 3^{\beta_3} 5^{\beta_5} 7^{\beta_7} 11^{\beta_{11}}$,

(vi) $2^{\beta_2} 3^{\beta_3} 5^{\beta_5} 7^{\beta_7} 11^{\beta 11} 13^{\beta_{13}}$,

where the β_i are a decreasing sequence whose size is controlled by (8.59). A computer capable of handling long integers gives the response in a few seconds: $N = 50{,}360$ is highly composite.

We make a note of this result: an integer less than or equal to 50,360 possesses at most

$$d(2^4 \cdot 3^4 \cdot 5 \cdot 7) = 5 \cdot 5 \cdot 2 \cdot 2 = 100 \text{ divisors.}$$

Exercise 13

Find all Pascal integers which are highly composite.

8.15. Permutations: Johnson's' Algorithm

How can one list all $n!$ permutations of the integers $1, 2, \ldots, n$? The following algorithm, due to Johnson in 1963, uses integers decorated with a "weathervane" such as:

$$\overleftarrow{1} \quad \overleftarrow{3} \quad \overleftarrow{5} \quad \overrightarrow{7} \quad \overleftarrow{6} \quad \overrightarrow{4} \quad \overrightarrow{2}\ .$$

One says that a integer with a weathervane is *mobile* it it can "see" a smaller integer or if it "looks outside" (that is, if it sees no one). In our example, the integers $3, 5, 7, 4$ are mobile and $1, 6, 2$ are not.

Johnson's algorithm proceeds as follows:

(i) Start with the permuation $\overleftarrow{1}, \overleftarrow{2}, \ldots, \overleftarrow{n}$.

(ii) Look for the largest mobile integer. If there is not one, the algorithm terminates; otherwise, let m be the largest mobile integer and v the integer seen by m.

(iii) Interchange m and v *without changing their weathervanes.*

(iv) Change the direction of the weathervanes on all integers $k > m$ and return to (ii).

Example

Let us see what this gives when $n = 3$. The first three permutations are in the table on the left: in the first two, the largest mobile integer is 3, which explains why it moves from right to left; no weathervane changes direction when 3 moves because there is no integer greater than 3.

$\overleftarrow{1}$	$\overleftarrow{2}$	$\overleftarrow{3}$	$\overrightarrow{3}$	$\overleftarrow{2}$	$\overleftarrow{1}$
$\overleftarrow{1}$	$\overleftarrow{3}$	$\overleftarrow{2}$	$\overleftarrow{2}$	$\overrightarrow{3}$	$\overleftarrow{1}$
$\overleftarrow{3}$	$\overleftarrow{1}$	$\overleftarrow{2}$	$\overleftarrow{2}$	$\overleftarrow{1}$	$\overrightarrow{3}$

In the third permutation, the largest (and only) mobile integer is $m = 2$. We *first* interchange $\overleftarrow{1}$ and $\overleftarrow{2}$ (without modifying their weathervanes); we *then* change the direction of the weathervane which decorates 3 since $3 > m$. The effect of this operation is to unblock the number 3, allowing it to move to the right. The algorithm stops when no mobile integer remains and we do indeed obtain the $6 = 3!$ permutations of $1, 2, 3$.

Exercise 14

When $n = 4$ (Fig. 8.5), one finds that the number 4 zigzags across four line blocks. When one takes 4 out of the six blocks of four lines, one is left with the permutations made by Johnson's algorithm when $n = 3$.

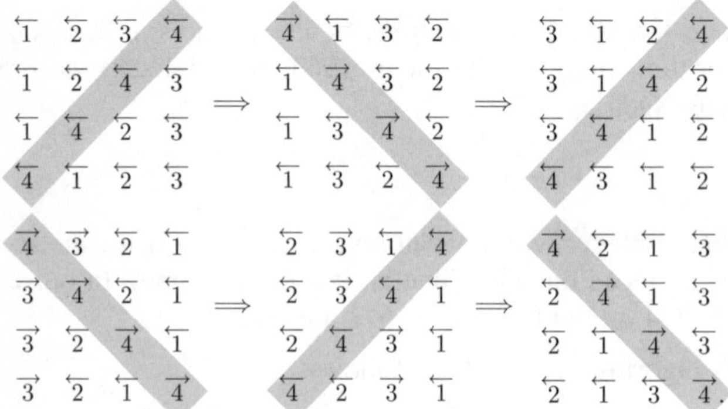

Fig. 8.5. *Johnson's algorithm when $n = 4$*

Show that this behavior is general. In more modern terms, Johnson's algorithm is *fractal*, which is to say that it contains different scales.

Exercise 15

Let $k \in [\![1, n!]\!]$ be the number of the permutation s in Johnson's algorithm. Knowing that the permutations decompose into blocks of n permutations, let B_j be the block containing s and $i \in [\![1, n]\!]$ the row on which s occurs.

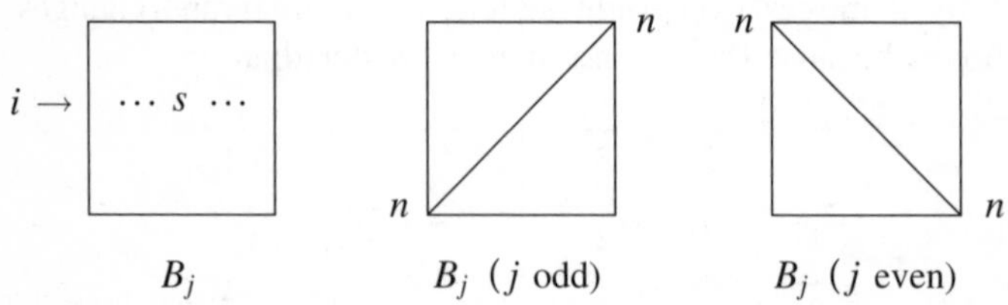

Since $k - 1 = n(j - 1) + i - 1$ and $0 \leq i - 1 < n$, we conclude that

$$j = 1 + (k - 1) \text{ div } n, \quad i = 1 + (k - 1) \bmod n.$$

If j is odd, we know that the place p of n in the permutation is $p = n-i+1$; if j is even, the place of n in s is $p = i$. Call δs the permutation of $1, \ldots, n-1$ obtained from s by suppressing the integer n. We know that δs is the j-th Johnson permutation of $1, \ldots, n-1$, which allows us to put the element $n-1$ into s, *etc.*

$n=7$	$k=1994$	$j=285$	$i=6$	$p=2$	• 7 • • • • •
$n=6$	$k=285$	$j=48$	$i=3$	$p=3$	• 7 • 6 • • •
$n=5$	$k=48$	$j=10$	$i=3$	$p=3$	• 7 • 6 5 • •
$n=4$	$k=10$	$j=3$	$i=2$	$p=3$	• 7 • 6 5 4 •
$n=3$	$k=3$	$j=1$	$i=3$	$p=1$	3 7 • 6 5 4 •
$n=2$	$k=1$	$j=1$	$i=1$	$p=2$	3 7 • 6 5 4 2
$n=1$	$k=1$	$j=1$	$i=1$	$p=1$	3 7 1 6 5 4 2

Fig. 8.6. *Reconstitution of the permutation $s \in \mathfrak{S}_7$ whose number in Johnson's algorithm is $k = 1994$. The number p denotes the place of n in the permutation s.*

Write a Pascal program which lists the $n!$ permutations of $1, \ldots, n$ using this algorithm.

8.15.1. The program Johnson

The types

In order to teach a computer what a weathervane integer is, one thinks immediately of the pair (integer, boolean). But one quickly changes one's mind when trying to specify the integer seen by k. Systematic tries establish the superiority of the pairs $(integer, vane)$ where $vane = \pm 1$ symbolises the weathervane (with the convention $+1$ if it points to the right and -1 if it points to the left). After this delicate choice, another difficulty awaits us: how can we express simply that an integer which looks outside is not mobile? In order not to complicate the programming with distracting tests, we *border*[10] our permutation to the left and right with the integer $n+1$ as, for example:

$$5 \ \overleftarrow{2} \ \overrightarrow{1} \ \overleftarrow{4} \ \overrightarrow{3} \ 5.$$

(We do not need to endow $n+1$ with a weathervane.) Now, an integer which looks outside sees $n+1$ and cannot be mobile. We now use an array whose indices vary from 0 to 11 in order to be able to treat the cases $n \in [\![2, 10]\!]$.

[10] Yes, this is a trick. But it is vital, which justifies it. We remark that mathematicians often say that a *method* is a trick that occurs at least three times. We will use this trick again in Chapter 12

```
type integer_vane = record num, vane : integer end ;
permutation = array[0..11] of integer_vane ;
var s : permutation ;
```

With these declarations, $s[k].num$ sees $s[k+s[k].vane].num$ which is what we want.

The main body of the program

The main body is a simple loop. The auxiliary variable *counter* will be used by the procedure display to vertically separate blocks of n permutations.

```
begin
  message ;
  counter := 0 ; finish := false ;
  initialize(s, n) ;
  repeat
    display(s, counter) ;
    next(s, finish)
  until finish
end .
```

The procedure display

The permutations are displayed in blocks of n, which allows one to inspect the movement of the integer n across successive blocks.

```
procedure display(s : permutation ; var counter : integer) ;
var i : integer ;
begin
  for i := 1 to n do write(s[i].num : 5) ;
  writeln ;
  counter := counter + 1 ;
  if counter mod n = 0 then writeln
end ;
```

The procedure initialize

As mentioned above, we border the permutation by the integer $n+1$ (there is no reason to define weathervanes at $s[0]$ and $s[n+1]$).

```
procedure initialize(var s : permutation ; var n : integer) ;
var i : integer ;
begin
  repeat write('n = ') ; readln(n) until (n ≥ 2) and (n ≤ 10) ;
  s[0].num := n + 1 ; s[n + 1].num := n + 1 ;
  for i := 1 to n do
      with s[i] do begin num := i ; vane := −1 end
end ;
```

The procedure next

This procedure determines the next permutation (when it exists). When there is a next permutation, the boolean variable *finish* remains false and s contains the following permutation; otherwise, *finish* remains true and s does not represent anything.

```
procedure next(var s : permutation ;  var finish : boolean) ;
var place_bm, value_bm, i : integer ;
begin
| biggest_mobile(s, place_bm, value_bm, finish) ;
| if not finish then begin
| | move_biggest_mobile(place_bm, s) ;
| | the_wind_turns(value_bm, s)
| end
end ;
```

Here is a rather subtle error: the fragment of code that follows is false because information in it circulates badly.

```
if not finish then begin
| move_biggest_mobile(place_bm, s) ;
| the_wind_turns(place_bm, s) ← erroneous statement!
end
```

In effect, after *move_biggest_mobile(place_bm, s)*, the biggest mobile is not in *place_bm*!

The procedure biggest_mobile

A sweep allows one to find the placement and value of the biggest mobile integer. The procedure gives the boolean *finish* the value *false* when it does not find a mobile integer.

```
procedure biggest_mobile(s : permutation ;
          var place_bm, value_bm : integer ;  var finish : boolean) ;
var i : integer ;
begin
| finish := true ;  value_bm := 0 ;
| for i := 1 to n do with s[i] do
| if (num > s[i + vane].num) and (num > value_bm) then begin
| | place_bm := i ;  value_bm := num ;  finish := false
| end
end ;
```

The procedure move_biggest_mobile

This procedure exchanges the mobile integer which is *leaving* with the integer that is *arriving* which it sees. It does not change the weathervanes.

```
procedure move_biggest_mobile(leaving : integer ; var s : permutation) ;
var temp : integer_vane ; arriving : integer ;
begin
| arriving := leaving + s[leaving].vane ;
| temp := s[arriving] ;
| s[arriving] := s[leaving] ;
| s[leaving] := temp
end ;
```

Here is a vicious pitfall which causes many programmers to stumble. Can you explain why the following fragment of code is false?

```
temp := departing ;
s[departing] := s[departing] + s[departing].vane ;
s[departing + s[departing].vane] := temp
```

The procedure the_wind_turns

This procedure changes the direction of the weathervanes on the integers which are greater than m. To avoid useless work, we only examine an permutation when we are certain that there is a weathervane that needs to be changed; that is, when $m < n$.

```
procedure the_wind_turns(m : integer ; var s : permutation) ;
var i : integer ;
begin
| if m < n then for i := 1 to n do
|     with s[i] do if num > m then vane := -vane
end ;
```

8.16. The Count is Good

This section is inspired by a popular French TV game. Suppose that we are given five integers $a_1, \dots, a_4 > 0$ and *goal*. We want to obtain the integer *goal* using a succession of operations on the a_i, the operations being chosen from among the four possible ones. The constraints are as follows.

- Each number a_i must be used once and only once.
- The result of a subtraction must be greater than 0; division must be defined and without remainder.

Let us take, as an example, $a_1 = 2$, $a_2 = 5$, $a_3 = 7$ and $a_4 = 10$. A *succession of operations* amounts to being given a parenthesized arithmetic progression, the parentheses specifying the order of the operations. For example:

$$((a_2 + a_3)/a_1) * a_4 = ((5 + 7)/2) * 10 = 60,$$

$$((a_1 + a_3) + a_2) - a_4 = ((2 + 7) + 5) - 10 = 4,$$

$$(a_4/a_1) + (a_3 - a_2) = (10/2) + 7 - 5 = 7.$$

On the other hand, we are not allowed to use the expressions

$$((a_1/a_2) + a_3) * a_4), \quad ((a_1 - a_2) + a_3) + a_4)$$

because a_1/a_2 is not an integer and because $a_1 - a_2$ is negative. To solve the problem, we will use "brute force" and consider all possible expressions and compare the values of those that are legal to *goal*. (We shall see in Chapter 12 that it is possible to proceed in a more intelligent manner.)

8.16.1. Syntactic trees

We can associate to an arithmetic expression a *binary tree* which describes the order in which the calculations are made. Conversely, we can reconstruct a totally parenthesized arithmetic expression from a syntactic tree.

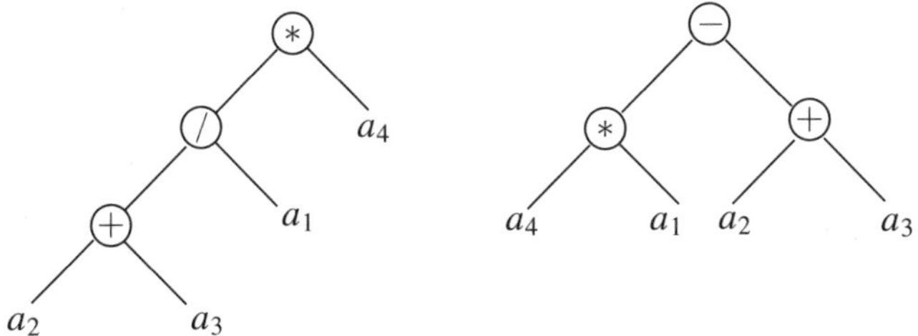

Fig. 8.7. *Syntactic trees associated to* $((a_2 + a_3)/a_1) * a_4$ *and* $(a_4 * a_1) - (a_2 + a_3)$.

Knowing this, we can break the search for all parenthesized arithmetic expressions into two subproblems:

- the search for all binary trees with *four leaves*;
- the search for all *decorations* of a binary tree with four leaves.

The leaves are the a_i; the decoration is formed by the operators. It is easy to sketch five binary trees with four leaves (see Fig. 8.4). But are there others?

A little theory will reassure us.

Definition 8.16.1. *Let* $n \geq 1$. *The* n*-th Catalan number* c_n *is defined to be the number of binary trees with* n *leaves.*

tree 1 tree 2 tree 3 tree 4 tree 5

$$\text{tree}_1 = ((a\ \theta_1\ b)\ \theta_2\ c)\ \theta_3\ d,\ \text{tree}_2 = (a\ \theta_2\ (b\ \theta_1\ c))\ \theta_3\ d,$$

$$\text{tree}_3 = (a\ \theta_1\ b)\ \theta_3\ (c\ \theta_2\ d),\ \text{tree}_4 = a\ \theta_3\ ((b\ \theta_1\ c)\ \theta_2\ d)),$$

$$\text{tree}_5 = a\ \theta_3\ (b\ \theta_2\ (c\ \theta_1\ d)).$$

Fig. 8.8. *The five binary trees with four leaves*

Considering a leaf as a binary tree with one leaf, we have $c_1 = 1$. We also have $c_2 = 1$ and $c_3 = 2$.

Theorem 8.16.1. *For any $n > 1$,*

$$c_n = c_1 c_{n-1} + c_2 c_{n-2} + \cdots + c_{n-1} c_1 = \frac{1}{2n-1}\binom{n}{2n-1}.$$

Proof. Consider a binary tree with $n \geq 2$ leaves. The branch on the left of the root is binary tree with p leaves (reduced eventually to one leaf) and that on the right is a binary tree with $q = n - p$ leaves. To make a binary tree with $n \geq 2$ leaves, it suffices to take a root and to attach two binary trees with p and q leaves, which gives $c_p c_q$ choices. One obtains all binary trees once and only once by letting p vary from 1 to $(n - 1)$. The explicit expression using the binomial coefficient is proved by induction or with the aid of formal series. □

The first few Catalan numbers are now $c_1 = 1$, $c_2 = 1$, $c_3 = 2$ and, most importantly, $c_4 = 5$. We can estimate the total number of arithmetic expressions: there are five binary trees with four leaves; to decorate a tree, we need three operations and four leaves, which gives

$$5 \cdot 4^3 \cdot 4! = 7{,}680 \text{ arithmetic expressions.}$$

The declarations and the main body of the program

We shall store $a_1, \ldots, a_4$ in an array.

```
type data = array[1..4] of integer ;
var a, data, goal : integer ;
begin
| message ;
| choose(a, goal) ;
```

```
  | the_count_is_good(a, goal)
  end .
```

The procedure the_count_is_good

To understand the following it is necessary to realise that a tree is only a representation of a *schema for calculation.* We choose a tree which we "decorate" with the operations op_1, op_2, op_3 and the leaves $a_{s(i)}$ where $s \in \mathfrak{S}_4$ is a permutation. We then communicate these parameters to the procedure *calculate*: if we obtain a legal arithmetic expression which possess a value, we compare this value with *goal* and we display the result if they are equal. Changing permutations is realised by the procedure *next* of Johnson's program with $n = 4$.

```
procedure the_count_is_good(a : data ; goal : integer) ;
const addition = 1 ; subtraction = 2 ;
multiplication = 3 ; division = 4 ;
type {add here the type permutation (see Johnson)} ;
var tree, op1, op2, op3, value : integer ;
s : permutation ; finish, exist : boolean ;
begin
 | for tree := 1 to 5 do
 | for op1 := addition to division do
 | for op2 := addition to division do
 | for op3 := addition to division do
 | for op4 := addition to division do begin
 | | permutation_identity(s) ; finish := false ;
 | | repeat
 | | | calculate(value, exist, tree, op1, op2, op3, a, s) ;
 | | | if exist and (val = goal) then write(value, op1, op2, op3, a, s) ;
 | | | next(s, finish)
 | | until finish
 | end
end ;
```

The procedure calculate

Since a tree is a little program, we have five programs to manage. In order not to try the patience of the reader (and to leave something for him or her to write), we only detail the case of the first tree. When an operation gives an integer less than 0 or if a division has nonzero remainder, the procedure *partial_result* informs us *via* the boolean *exist*: we know that there is no point in pursuing such a case.

```
procedure calculate(var value : integer ; var exist : boolean ;
tree, op1, op2, op3 : integer ; a : data ; s : permutation) ;
var i, temp1, temp - 2 : integer ; leaf : data ;
```

```
begin
  for i := 1 to 4 do leaf[i] := a[s[i].num] ;
  exist := true ;
  case tree of
    1 :
    begin {calculate value = ((f1 op1 f2)op2 f3)op3 f4 and exist}
      partial_result(temp1, exist, leaf[1], op1, leaf[2]) ;
      if exist then partial_result(temp2, exist, temp1, op2, leaf[3]) ;
      if exist then partial_result(value, exist, temp2, op3, leaf[4]) ;
    end ;
    2 :
    begin «calculate value = (f1 op2(f2 op1 f3)) op3 f4 and exist» end ;
    3 :
    begin «calculate value = (f1 op1 f2)op3(f3 op2 f4) and exist» end ;
    4 :
    begin «calculate value = f1 op3 ((f2 op1 f3) op2 f4)) and exist» end ;
    5 :
    begin «calculate value = f1 op3 (f2 op2 (f3 op1 f4)) and exist» end ;
  end ; {case}
end ;
```

The procedure partial_result

We know the operands and the operation. We examine the four cases of possible figures and signal incorrect partial results (subtraction giving a number less than 0 or division that leaves a nonzero remainder) Recall that the boolean *exist* has been initialized to *true* in the procedure *calculate*.

```
procedure partial_result(var temp : integer ;
var exist : boolean ;  a, op, b : integer) ;
var temp : integer ;
begin
  case op of
    addition : temp := a + b ;
    substraction :  if a > b then temp := a − b else exist := false ;
    multiplication : temp := a ∗ b ;
    division :  if a mod b = 0 then temp := a div b else exist := false ;
  end ; {case}
end ;
```

The procedure display

We do not attempt to sketch a tree on our screen; we content ourselves with displaying the corresponding arithmetical expression. For this it suffices to consider each tree.

```
procedure display(value, tree, op1, op2, op3 : integer ;
                                    a : data ;  s : permutation) ;
var i : integer ;  leaf : data ;
begin
  for i := 1 to 4 do leaf[i] := a[s[i].num] ;
  write('value  =  ') ;
  case tree of
    1 : writeln('((', leaf[1] : 1, op(op1), leaf[2] : 1, ')',
                  op(op2), leaf[3] : 1, ')', op(op3, leaf[4] : 1) ;
    2 : ... ;
    3 : ... ;
    4 : ... ;
    5 : ... ;
  end {case}
end ;
```

The function *op* converts the integer op_i into the corresponding character:

```
function op(operation : integer) : char ;
begin
  case operation of
    addition : op := '+' ;
    subtraction : op := '−' ;
    multiplication : op := '*' ;
    division : op := '/' ;
  end ;  {case}
end ;
```

Remark

The interest of this program resides in the ideas that it puts into play. We note in particular the *distance* that separates concepts from their translation; that is, theory from code. We also remark that without theory we would not be able to create and understand the code at all.

9. The Complex Numbers

9.1. The Gaussian Integers

Let $\mathbb{Z}[i]$ denote the set of complex numbers of the form $x + iy$ with $x, y \in \mathbb{Z}$. If we endow this set with addition and multiplication inherited from the complex numbers, we obtain a ring. This ring is a commutative integral domain and is called the ring of Gaussian integers in honor of their creator, Gauss, who introduced them around 1830.

• If $\alpha = x + iy$ belongs to $\mathbb{Z}[i]$, we set

$$N(\alpha) = |\alpha|^2 = x^2 + y^2;$$

and call $N(\alpha)$ the *norm* of α. (This is the norm in the sense that algebraists use the word. It should not be confused with the modulus of a complex number.) It is clear that the norm is multiplicative:

$$N(\alpha\beta) = N(\alpha)\,N(\beta).$$

• The *units* of $\mathbb{Z}[i]$ are the invertible Gaussian integers (that is, those Gaussian integers ε such that there exists ε' satisfying $\varepsilon\varepsilon' = 1$). There are four units 1, -1, i and $-i$. They are characterized by the condition

$$N(\varepsilon) = 1.$$

• We say that α and β are *associates* if there exists a unit ε such that $\alpha = \varepsilon\,\beta$. Thus, the associates of $\alpha = x + iy$ are the four complex numbers

$$\alpha = x + iy, \quad i\alpha = -y + ix, \quad -\alpha = -x - iy, \quad -i\alpha = y - ix$$

obtained by successive rotations of α about 0 through angle $\frac{1}{2}\pi$.

• Finally, one says that $\omega \neq 0$ is *irreducible* if it is not a unit and if it cannot be written as a product of two nontrivial factors. That is, if

$$(\omega = \alpha\beta) \implies (\alpha \text{ or } \beta \text{ is a unit}).$$

9.1.1. Euclidean division

Theorem 9.1.1. *Let α and $\beta \neq 0$ be Gaussian integers. There exists at least one pair (χ, ρ) of Gaussian integers satisfying the conditions:*

$$\alpha = \beta\chi + \rho \quad \textit{and} \quad N(\rho) < N(\beta).$$

Proof. Among the points on the plane with integer coordinates, let χ be as

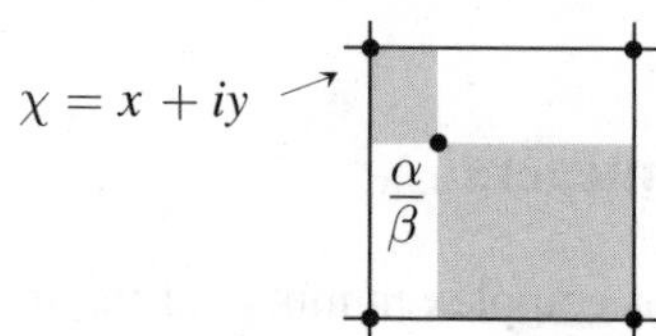

close as possible to the complex number α/β, so that:

$$|\chi - \alpha/\beta| \leq \tfrac{1}{2}\sqrt{2}.$$

Clearing the denominator and removing the square, we obtain

$$N(\alpha - \chi\beta) \leq \tfrac{1}{2}N(\beta).$$

It follows that $\rho = \alpha - \chi\beta$ satisfies the condition $N(\rho) < N(\beta)$. □

Remarks

- The pair (χ, ρ) is not unique. Each Gaussian integer χ sufficiently close to α/β gives a solution. There up to four possible values for χ when the real and imaginary parts of α/β are of the form $n + \frac{1}{2}$.
- The existence of Euclidean division allows us to carry over to $\mathbb{Z}[i]$ the theory of the GCD, Bézout's theorem, the algorithms of Euclid and Blankinship, the existence and uniquenesss of the decomposition into irreducible factors. The same proofs hold with ordinary integers replaced by Gaussian integers. Algebraists express this by saying that the Gaussian integers are a Euclidean ring, hence *factorial*.

9.1.2. Irreducibles

Recall (see Chapters 2 and 8) that every *prime number* of the form $4n+1$, and no *integer* of the form $4n+3$, is a sum of two squares. Moreover, if p is an odd prime, the equation $x^2 + 1 = 0$ has a root in $\mathbb{Z}_p$ if and only if $p \equiv 1 \pmod 4$ and has no root if $p \equiv 3 \pmod 4$.

Theorem 9.1.2. *A Gaussian integer is irreducible if and only if one of its associates belongs to the following list:*

(i) $1+i$;

(ii) $a+bi$, *where* a^2+b^2 *is a prime of the form* $4n+1$;

(iii) p, *where* p *is a prime number of the form* $4n+3$.

Proof. We begin by showing that the three lists in the statement are made up of irreducibles.

• *Gaussian integers of type (i) or (ii) are irreducible.* To see this, it suffices to prove the implication:

$$N(\omega) \text{ a prime number in } \mathbb{Z} \implies \omega \text{ irreducible in } \mathbb{Z}[i].$$

In effect, $\omega = \alpha\beta$ implies that $N(\omega) = N(\alpha)N(\beta)$. If $N(\omega)$ is a prime number, then $N(\alpha) = 1$ or $N(\beta) = 1$; that is, $\alpha = \varepsilon$ or $\beta = \varepsilon$.

• *The Gaussian integers of type (iii) are irreducible.* Let $p = 4n+3$ be a prime number in $\mathbb{Z}$. If p were not irreducible in $\mathbb{Z}[i]$, then $p = \alpha\beta$ with $N(\alpha) > 1$ and $N(\beta) > 1$. Taking norms gives $p^2 = N(\alpha)N(\beta)$. Since p is prime, we must have $p = N(\alpha) = N(\beta)$. But this is impossible since $N(\alpha)$ is a sum of two squares and $p \equiv 3 \pmod 4$.

Now we show that an associate of an irreducible $\omega = x+iy$ belongs to one of the lists (i), (ii) or (iii).

• If $N(\omega)$ is prime, we necessarily have $N(\omega) = 2$ with ω of type (i) or $N(\omega) \equiv 1 \pmod 4$ since $N(\omega)$ is a sum of two squares, which proves that ω is of form (ii).

• If $N(\omega)$ is not a prime number, we write $N(\omega) = p_1 \cdots p_k$ with p_i primes and $k \geq 2$. Since ω is irreducible, ω divides one of the p_i; that is $p_i = \alpha\omega$. Taking norms gives

$$p_i^2 = N(\alpha)N(\omega) \implies N(\omega) = p_i^2 \text{ and } N(\alpha) = 1$$

which proves that ω and p_i are associates. The cases $p_i = 2$ and $p_i \equiv 1 \pmod 4$ are ruled out because p_i is not irreducible. Thus $p_i \equiv 3 \pmod 4$ and ω belongs to the list (iii). □

Remarks

• This theorem shows that it is very important to distinguish between the prime numbers in $\mathbb{Z}$ and the irreducible Gaussian integers since a prime number is not necessarily an irreducible Gaussian integer.

• We insist on the following. By definition, the units $\varepsilon = \pm 1$ and $\varepsilon = \pm i$ are not irreducible!

• Irreducibles of type (ii) are $\pm 2 \pm i$ and $\pm 1 \pm 2i$ since $5 = 2^2 + 1^2$. Another easily remembered example is $5 + 42i$ because $1789 = 5^2 + 42^2$ is a prime number of the form $4n+1$.

• Irreducibles of type (iii) are 3, 7, 11, 19, 23, 31, etc.

• The norm of an irreducible belongs to a very particular class of integers: *it is either a prime number or a square of a prime number*. This remark will be very useful in what follows.

Corollary 9.1.1. *Let p be a prime number of the form $4n+1$ and $(a,b) \in \mathbb{Z}^2$ a particular solution of the equation $p = x^2 + y^2$. Then all solutions of this equation are $(\pm a, \pm b)$ and $(\pm b, \pm a)$.*

Proof. We write $p = (a+ib)(a-ib) = (x+iy)(x-iy)$. Since $a \pm ib$ and $x \pm iy$ are irreducibles of type (ii), the uniqueness of the decomposition of p into irreducible factors implies $x+iy = \varepsilon(a+ib)$ or $x+iy = \varepsilon(a-ib)$. □

Choice of representatives for irreducibles

In the decomposition into irreducible factors, we can replace ω by one of its associates and write, for example,

$$\alpha = \omega_1\omega_2\omega_3 = (-\omega_1)(i\omega_2)(-i\omega_3).$$

Is there a reasonable way to choose a *representative* from among the four associates $\omega, i\omega, -\omega, -i\omega$ of ω to normalize decompositions into irreducible factors?

Since the $i^k\,\omega$ can be obtained from ω by successive rotations through angle $\frac{1}{2}\pi$ centered at the origin, a natural first thought is to choose the irreducibles in the first quadrant as representatives. This is not a good idea, however. To see why, consider the decomposition of 5 into irreducible factors:

$$5 = (2+i)(2-i). \tag{9.1}$$

To get $2-i$ into the first quadrant, we must rotate by $\frac{1}{2}\pi$; that is, multiply by i. The decomposition is then

$$5 = -i(2+i)(1+2i). \tag{9.2}$$

The decomposition (9.2) is much less natural than (9.1) because it is not at all evident at first glance that $-i(2+i)(1+2i)$ is a real number.

For this reason, we adopt the following conventions:

• The irreducibles of type (i) are represented by $1+i$.

• The irreducibles of type (ii) or (iii) are represented by irreducibles $\omega = x+iy$ situated in the part of the half-plane $x \geq 0$ between the two quadrant bisectors; that is, by those that satisfy the condition $0 < |y| < x$.

Let $p = a^2+b^2$ be the unique decomposition of the prime number $p = 4n+1$ satisfying the condition $0 < b < a$. The Gaussian integers $a+ib$ and $a-ib$ *are not* associates because they make an angle $< \frac{1}{2}\pi$ with the origin. It follows from this remark that the eight solutions of the equation $p = x^2 + y^2$ are associates of $a \pm ib$.

Here are some decompositions into irreducible factors:

$$\begin{aligned}
105 &= 3(2+i)(2-i)7, \\
1789 &= (42+5i)(42-5i), \\
1+57i &= i(1+i)(2-i)^3(3+2i), \\
1+58i &= i(2+i)(23-12i), \\
1+59i &= (1+i)(30+29i), \\
1+60i &= i(3-2i)(14+9i), \\
7+15i &= (1+i)(11+4i), \\
15+45i &= (1+i)3(2+i)^2(2-i), \\
31+63i &= (1+i)(2+i)(4+i)(5-2i), \\
101+47i &= (1+i)(2-i)(4-i)(8+3i).
\end{aligned}$$

An algorithm for decomposition into irreducible factors

If we try to recognize a person whom we know only by his or her shadow on a wall, the results will be uncertain. On the contrary, if we know that we have to choose between Laurel and Hardy, it will be easy! By the *shadow*[1] of an irreducible Gaussian integer ω, we mean the unique prime number that divides the norm of ω.

- If the shadow of ω is 2, we know that ω is an assotiate of $1+i$.
- If the shadow of ω is equal to $p \equiv 1 \bmod 4$ and if (a, b) is the unique solution of the equation $p = a^2 + b^2$ such that $a > b > 0$, we know that ω is an associate of $a + bi$ or of $a - bi$.
- If the shadow of ω is equal to $p \equiv 3 \bmod 4$, then ω is an associate of p.

Consider now a Gaussian integer $\alpha = a + bi$ with norm greater than 1. Let $\alpha = \varepsilon\omega_1 \cdots \omega_n$ be its decomposition in irreducible factors, the ω_i being situated between the two quadrant bisectors (this explains the presence of the unit ε). Consider the norm of α:

$$N(\alpha) = N(\omega_1) \cdots N(\omega_n).$$

We know that the shadows of the ω_i are the prime divisors of the norm of α and that each prime divisor of $N(\alpha)$ is a shadow of an ω_i. We can sketch an algorithm as follows:

- decompose $N(\alpha)$ into prime factors;
- reconstruct the irreducible divisors of α from their shadows.

[1] This is not at all classical terminology.

This sketch is not entirely satisfactory because it requires that one store the prime factors of $N(\alpha)$ in advance. We are going to work more dynamically (by "surfing" once more on the wave of calculations) and recycle the algorithm for decomposition into prime factors in Chapter 4.

Example

Decompose $\alpha_0 = 3 + 21i$ into irreducible factors.

• The norm of α_0 is 450. Since this is divisible by 2, we know that α_0 is divisible by $\omega_1 = 1 + i$, an irreducible of type (i) :

$$\alpha_0 = 3 + 21i = (1 + i)\alpha_1.$$

• The norm of $\alpha_1 = \alpha_0/\omega_1 = 12 + 9i$ is 225. Since this is divisible by 3, we know that α_1 is divisible by $\omega_2 = 3$, an irreducible of type (iii) :

$$3 + 21i = (1 + i)(3)\alpha_2.$$

• The norm of $\alpha_2 = \alpha_1/\omega_2 = 4 + 3i$ is 25. Since this is divisible by 5, we know that α_2 is divisible either by $2 + i$ or $2 - i$, the only irreducibles of type (ii) of norm 5 situated between the quadrant bisectors. One try shows that α_2 is divisible by $\omega_3 = 2 - i$:

$$3 + 21i = (1 + i)(3)(2 - i)\alpha_3.$$

• The norm of $\alpha_3 = \alpha_2/\omega_3 = 1 + 2i$ is 5, which shows that α_3 is divisible by $2+i$ or $2-i$. Computation shows that $\alpha_3/(2+i)$ does not belong to $\mathbb{Z}[i]$. We conclude that α_3 is necessary divisible by $\omega_4 = 2 - i$:

$$3 + 21i = (1 + i)(3)(2 - i)^2 i.$$

Here, then, is our algorithm

$$
\begin{aligned}
&\alpha_0 = \alpha \quad \text{Gaussian integer of norm} > 1,\\
&\begin{cases}
p_1 = \text{smallest divisor} > 1 \text{ of } N(\alpha_0),\\
\omega_1 = \text{irreducible divisor of } \alpha_0 \text{ with shadow } p_1,\\
\alpha_1 = \alpha_0/\omega_1,
\end{cases}\\
&\quad\vdots\\
&\begin{cases}
p_n = \text{smallest divisor} > 1 \text{ of } N(\alpha_{n-1}),\\
\omega_n = \text{irreducible divisor of } \alpha_{n-1} \text{ with shadow } p_n,\\
\alpha_n = \alpha_{n-1}/\omega_n,
\end{cases}\\
&\text{stop when } \alpha_n \text{ is unit.}
\end{aligned}
$$

The translation into code is immediate:

```
α := given Gaussian integer of norm > 1 ;
repeat
| p := LD(norm(α)) ;
| ω := irreducible divisor of α with shadow p ;
| α := α/ω ;
until norm(α) = 1
```

9.1.3. The program

We are going to use a *record* to store a Gaussian integer.

```
type Gaussian_integer =  record re, im : integer end ;
var α : Gaussian_integer ;
```

Consequently, $\alpha.re$ and $\alpha.im$ denote the real and imaginary parts of α.

The body of the program

Our program asks repeatedly for Gaussian integers and does not stop until we offer it a Gaussian integer equal to zero or a unit.

```
begin
| message ;
| finish := false ;
| repeat
| | writeln ;  choose(α) ;
| | if norm(α) > 1 then factor(α) else finish := true
| until finish
end .
```

The procedure choose and the function norm

These are the "mindless" parts of the program that one types in directly without preliminary reflection.

```
procedure choose(var α : Gaussian_integer) ;
begin
| write('real part = ') ;  readln(α.re) ;
| write('imaginary part = ') ;  readln(α.im) ;
end ;

function norm(β : Gaussian_integer) : integer ;
begin
| norm := β.re * β.re + β.im * β.im
end ;
```

The procedure factor

This procedure implements the algorithm that we have developed. To check the calculations, we accumulate in the variable *prod* the product of the irreducible factors and units. When the factorization terminates this variable must be equal to the initial value of α.

```
procedure factor(α : Gaussian_integer) ;
var p : integer ; ω, prod : Gaussian_integer ;
begin
| prod.re := 1 ; prod.im := 0 ;                  {prod = 1}
| repeat
| | p := LD(norm(α)) ;                           {p is the shadow of ω}
| | reconstruct_irreducible_divisor(ω, p, α) ;
| | display(ω) ;
| | mult_Gaussian_integer(prod, ω, prod) ;      {prod := ω * prod}
| | divide_Gaussian_integer(α, α, ω) ;          {α := α/ω}
| until norm(α) = 1 ;
| display(α) ;                                   {now α is a unit}
| mult_Gaussian_integer(prod, α, prod) ;        {prod := α * prod}
| write('verification = ') ; display(prod)      {one must recover α}
end ;
```

The procedure reconstruct_irreducible_divisor

This procedure reconstructs the irreducible divisor ω from its shadow p.

- The cases $p = 2$ and $p \equiv 3 \bmod 4$ are trivial.

- When $p = N(\omega) \equiv 1 \bmod 4$, we decompose p as a sum of two squares $p = x^2 + y^2$, with $0 < y < x$ so that $\omega = x + iy$ or $\omega = x - iy$. To know if α is divisible by $\omega = x + iy$, we see if the complex number

$$\frac{\alpha}{\omega} = \frac{(a + bi)(x - iy)}{\omega\,\overline{\omega}} = \frac{ax + by + i(bx - ay)}{p}$$

is a Gaussian integer. If not, we know that $\omega = x - iy$ is the desired divisor. To simplify the tests, we note from the identity

$$y(ax + by) + x(bx - ay) = b(x^2 + y^2) = bp$$

that $ax + by$ are $bx - ay$ are simultaneously divisible or not divisible by p.

```
procedure reconstruct_irreducible_divisor
                (var ω : Gaussian_integer ; p : integer ;
                         α : Gaussian_integer) ;
var x, y : integer ;
begin
| case p mod 4 of
```

```
      1 : begin
        decompose_sum_squares(p, x, y) ;
        ω.re := x ;
        if (α.re * x + α.im * y) mod p = 0 then ω.im := y else ω.im := −y ;
      end ;
      2 : begin ω.re := 1 ; ω.im := 1 end ;
      3 : begin ω.re := p ; ω.im := 0 end ;
    end {case}
  end ;
```

The procedure decompose_sum_squares

Since p is a prime number of the form $4n + 1$, we seek the unique solution to the equation $p = x^2 + y^2$ satisfying the condition $0 < y < x$. We use brute force (two more sophisticated algorithms are written in Chapter 8).

```
x := 1 ;
repeat
  x := x + 1 ; y := 0 ;
  repeat y := y + 1 until (y ≥ x) or (x² + y² = p)
until x² + y² = p
```

But using brute force does not mean that we have to abandon our intelligence: we can usefully amuse ourselves by speeding up the code using the auxiliary variables *square_x*, *square_y* and $\Delta = p - x^2$.

```
procedure decompose_sum_squares(p : integer ; var x, y : integer) ;
var square_x, square_y, Δ : integer ;
begin
  x := 1 ; square_x := 1 ;
  repeat
    square_x := square_x + x + x + 1 ; x := x + 1 ;
    Δ := p − square_x ;
    y := 0 ; square_y := 0 ;
    repeat
      square_y := square_y + y + y + 1 ; y := y + 1
    until (y ≥ x) or (square_y = Δ) ;
  until square_y = Δ
end ;
```

The procedure display

Notice the effort directed at presentation.

```
procedure display(ω : Gaussian_integer) ;
begin
  write('(') ;
```

```
  if ω.re ≠ 0 then write(ω.re : 1, ' ') ;
  if ω.im = 1 then write('+ i') else
  if ω.im = −1 then write('− i') else
  if ω.im > 0 then write('+ ', ω.im : 1, 'i') else
  if ω.im < 0 then write('− ', −ω.im : 1, 'i') ;
  write(')')
end ;
```

The procedures mult_Gaussian_integer and divide_Gaussian_integer

```
procedure mult_Gaussian_integer(var γ : Gaussian_integer ;
                                α, β : Gaussian_integer) ;
begin {returns γ = α · β}
  γ.re := α.re * β.re − α.im * β.im ;
  γ.im := α.im * β.re + α.re * β.im
end ;

procedure divide_Gaussian_integer(var γ : Gaussian_integer ;
                                  α, β : Gaussian_integer) ;
var N : integer ;
begin {returns γ = α/β}
  N := norm(β) ;
  γ.re := (α.re * β.re + α.im * β.im) div N ;
  γ.im := (α.im * β.re − α.re * β.im) div N
end ;
```

Exercise 1

Implement Euclid's and Blankinship's algorithms in $\mathbb{Z}[i]$. (Recall that this is possible because $\mathbb{Z}[i]$, like $\mathbb{Z}$, is a principal domain.)

9.2. Bases of Numeration in the Gaussian Integers

Is it possible to generalize the numeration system for ordinary integers with respect to a given base $b > 1$ to the Gaussian integers? What conditions should $\beta = a + bi$ satisfy in order to define a base of numeration? How should one choose the digits relative to this base?

9.2.1. The modulo beta map

Suppose that $\beta = a + bi$ has norm > 1. Given $\xi \in \mathbb{Z}[i]$, we have seen that there exists a pair (χ, ρ) satisfying the conditions

$$\xi = \beta\chi + \rho, \quad N(\rho) < N(\beta). \tag{9.3}$$

Because (χ, ρ) is not necessarily unique, one cannot speak of a map $\xi \mapsto \xi \bmod \beta$ in $\mathbb{Z}[i]$ without taking precautions. If we reflect a moment, however, we realize that this phenomenon is not new: we know at least two maps "modulo b" in $\mathbb{Z}$, depending on whether we consider remainders between 0 and $b-1$ or centered remainders.

Definition 9.2.1. *We say that two Gaussian integers φ and ψ are congruent modulo β if $\varphi - \psi$ is divisible by β. We say that a set Σ of Gaussian integers in an exact system of representatives modulo β if for each integer ξ, there exists a unique pair $(\chi, \rho) \in \mathbb{Z}[i] \times \Sigma$ such that $\xi = \beta\chi + \rho$.*

Any time that we have such a system, each Gaussian integer is congruent mod β to an element of Σ and only one such. We obtain a map "mod β": $\mathbb{Z}[i] \to \Sigma$ which associates to each $\xi \in \mathbb{Z}[i]$ the unique $\rho \in \Sigma$ to which it is congruent. Therefore, once we have a system Σ, we have a map "mod β" .

The absence of uniqueness actually did us a favor by requiring us to deepen the question. (One often thinks that it is the absence of a unique remainder upon division which forbids speaking of the map "modulo".)

Example

Choose $\beta = 2 + i$. Euclidean division (9.3) shows that ξ is always congruent to a Gaussian integer ρ of norm $N(\rho) < N(\beta) = 5$. The open disk $x^2 + y^2 < 5$ contains the numbers:

$$\pm 2i, \quad \pm 1 \pm i, \quad \pm i, \quad 0, \quad \pm 1, \quad \pm 2.$$

Knowing that $2 + i \equiv 1 - 2i \equiv -1 + 2i \equiv -2 - i \equiv 0 \pmod{2+i}$, we see that among the preceding integers, we only have the congruences:

$$2i \equiv -1 - i \equiv 1, \quad -1 + i \equiv -i \equiv 2,$$
$$-2i \equiv 1 + i \equiv -1, \quad 1 - i \equiv i \equiv -2.$$

As a result, $\Sigma = \{0, 1, 1 \pm i, 2\}$ is an exact system of representatives mod β.

9.2.2. How to find an exact system of representatives

Euclidean division (9.3) already shows that an exact system of representatives Σ is finite. We specify the cardinality of such sets.

Lemma 9.2.1. *Let $\varphi : \mathbb{Z}^2 \to \mathbb{Z}^2$ be a group homomorphism. If φ is injective, then $\mathbb{Z}^2/\varphi(\mathbb{Z}^2)$ is a finite group of cardinality $|\det \varphi|$.*

Proof. Let A be the matrix of φ in the canonical basis. One knows (see the Smith reduction, Exercise 2, Chap. 11) that there exist unimodular matrices E and F such that $EAF = \mathrm{diag}(d_1, d_2)$. In other words, $\mathbb{Z}^2/\varphi(\mathbb{Z}^2)$ is isomorphic to $\mathbb{Z}/d_1\mathbb{Z} \oplus \mathbb{Z}/d_2\mathbb{Z}$, which shows that its cardinality is $d_1 d_2$. Since $\det E = \pm 1$ and $\det F = \pm 1$, we have $\det \varphi = \det A = \pm d_1 d_2$.

Corollary 9.2.1. *An exact system of representatives* Σ *modulo* $\beta = a + ib$ *is a finite set of cardinality* $N(\beta) = a^2 + b^2$.

Proof. It suffices to apply the lemma to the injective linear map $\varphi : \xi \mapsto \beta\xi$ whose matrix in the canonical basis is $A = \left(\begin{smallmatrix} a & -b \\ b & a \end{smallmatrix}\right)$.

Remark

There is no reason to expect that a set Σ of cardinality $N(\beta) = a^2 + b^2$ should be an exact system of representatives, because nothing forbids two elements of Σ from being congruent modulo β. We should be wary of doubtful generalizations: for example, the interval $[\![0, N(\beta) - 1]\!] \subset \mathbb{N}$ *is not* in general an exact system. In effect, if we choose $\beta = 2 + 2i$, then $x + iy \equiv x' + iy'$ mod β implies that $x \equiv x'$ and $y \equiv y'$ mod 2: as a result, $1+i$ is not congruent to any element of Σ.

Proposition 9.2.1. *For an interval* $\Sigma = [\![0, N(\beta) - 1]\!]$ *to be an exact system of representatives of classes modulo* $\beta = a + bi$, *it is necessary and sufficient that* $\text{GCD}(a, b) = 1$.

Proof. The counter-example we gave for $\beta = 2 + 2i$ generalizes and shows that if $\text{GCD}(a, b) > 1$, there exist Gaussian integers which are not congruent to any element of Σ.

Now suppose that a and b are relatively prime: let $\alpha = x + iy$ be arbitrary and try to find $\chi = u + iv$ such that $\alpha - \beta\chi = \rho \in \Sigma$. The condition $\text{Im}(\rho) = y - (av + bu) = 0$ implies that $u = u_0 + ka$ and $v = v_0 - kb$ with $k \in \mathbb{Z}$ arbitrary and (u_0, v_0) a particular solution (which exists since $\text{GCD}(a, b) = 1$). Consequently, $\text{Re}(\rho) = x - (au - bv) = x - (au_0 - bv_0) - k(a^2 + b^2)$, which shows that there exists a ρ, and only one such, in Σ. □

9.2.3. Numeration system in base beta

We return to our initial problem. Given a Gaussian integer $\xi \in \mathbb{Z}[i]$, do there exist "digits" c_i such that one has a unique expression

$$\xi = c_0 + c_1\beta + \cdots + c_n\beta^n \text{ ?} \tag{9.4}$$

What digits should we choose? In what follows we shall see that it is natural to take the digits to be ordinary integers:

$$c_i \in [\![0, N(\beta) - 1]\!]. \tag{9.5}$$

Theorem 9.2.1 (I. Kátai and Szabó, 1975). *With the convention* (9.5), *a Gaussian integer* $\beta = a + bi$ *is a base of numeration if and only if it satisfies the conditions*

$$a < 0 \text{ and } b = \pm 1.$$

Proof. Suppose first that β is a base of numeration and write $1+i$ in this base:

$$1+i = c_0 + c_1\beta + \cdots + c_n\beta^n. \tag{9.6}$$

This expression contains valuable information. In effect, because the imaginary part of $c_1\beta + \cdots + c_n\beta^n$ is a multiple of b, we have $b = \pm 1$.

We cannot have $a = 0$, because $\beta = \pm i$ cannot be a system of numeration (uniqueness in (9.6) would not hold).

We show that we cannot have $a > 0$. We know that the interval $\Sigma = [\![0, N(\beta) - 1]\!]$ is an exact system of residues modulo β, and this allows us to talk of the map " mod β": $\mathbb{Z}[i] \to \Sigma$. We express the Gaussian integer $\xi = 1 - \overline{\beta} = (1-a) + ib$ in the base β:

$$\xi = c_0 + c_1\beta + \cdots + c_n\beta^n, \quad 0 \leq c_i < N(\beta).$$

Multiplying this equality by $1 - \beta$ gives

$$(1-\beta)\xi = c_0 + (c_1 - c_0)\beta + \cdots + (c_n - c_{n-1})\beta^n - c_n\beta^{n+1},$$

and we conclude that $(1-\beta)\xi \equiv c_0$ modulo β.

On the other hand,

$$(1-\beta)\xi = (1-\beta)(1-\overline{\beta}) = N(1-\beta) = (1-a)^2 + b^2.$$

The condition $a > 0$ implies that $0 \leq (1-a)^2 + b^2 < N(\beta)$ from which it follows that $(1-\beta)\xi$ belongs to Σ. Since Σ is an exact system of representatives, the congruence $(1-\beta)\xi \equiv c_0$ is an equality $(1-\beta)\xi = c_0$, which we can again write as:

$$(c_1 - c_0)\beta + \cdots + (c_n - c_{n-1})\beta^n - c_n\beta^{n+1} = 0.$$

Upon dividing by β, we find that $c_0 \equiv c_1 \bmod \beta$. Since c_0 and c_1 are two digits, this implies that $c_0 = c_1$. Beginning again, we obtain $c_0 = c_1 = \cdots = c_n$ and finally $c_n = 0$. That is, $\beta = 1$, which is impossible. □

The converse will be established in the next section.

9.2.4. An algorithm for expression in base beta

Lemma 9.2.2. *Let $\beta = -N + i$, with $N \geq 1$, be such that $\Sigma = [\![0, N^2]\!]$ is an exact system modulo β. Set $\xi = x + iy$. The remainder $c \in \Sigma$ upon division of ξ by β is given by the formula:*

$$c = (x + Ny) \bmod (1 + N^2). \tag{9.7}$$

Proof. Separating real and imaginary parts in $\xi = \xi'\beta + c$, we obtain $y = x' - Ny'$ and $x = -Nx' - y' + c$. Therefore $x + Ny = -(1+N^2)y' + c$. □

We will use this lemma to calculate the digits of ξ with respect to the base β in the usual way:

- to begin, we determine the smallest weight digit c_0 by dividing $\xi_0 = \xi$ by β (so $\xi_0 = \xi_1\beta + c_0$) ;
- we then determine c_1 by dividing ξ_1 by β (let $\xi_1 = \xi_2\beta + c_1$), etc.

Using sequences, our algorithm is:

$$
\begin{array}{l}
\ell := 0\ ;\ x_0 := x\ ;\ y_0 := y\ ;\ \{\xi_0 = x_0 + iy_0\} \\
\textbf{repeat}\ \{\textit{we suppose}\ \xi_0 \neq 0\} \\
\quad c_\ell := (x_\ell + Ny_\ell)\ \textbf{mod}\ (1 + N^2)\ ; \\
\quad \xi_{\ell+1} := (\xi_\ell - c_0)/\beta \quad \{\xi_{\ell+1} = x_{\ell+1} + iy_{\ell+1}\} \\
\quad \ell := \ell + 1 \\
\textbf{until}\ \xi_\ell = 0
\end{array}
$$

Be careful! – here, we use the mathematician's quotient and remainder (that is, $a = bq + r$ with $0 \leq r < b$).

Examples

To save space we write $\xi \xrightarrow{c} \chi$ instead of $\xi = \chi\beta + c$.

1) Choose $\beta = -1 + i$, *i.e.* $N = 1$. If we begin with $\xi = -1$, successive divisions by β give:

$$-1 \xrightarrow{1} 1 + i \xrightarrow{0} -i \xrightarrow{1} i \xrightarrow{1} 1 \xrightarrow{1} 0.$$

Thus $-1 = \overline{11101}$ in the base $\beta = -1 + i$.

2) Choose $\beta = -2 + i$, *i.e.* $N = 2$. Successive divisions of $\xi = 4 + 6i$ by β give:

$$4 + 6i \xrightarrow{1} -3i \xrightarrow{4} 1 + 2i \xrightarrow{0} -i \xrightarrow{3} 1 + i \xrightarrow{3} 1 \xrightarrow{1} 0.$$

Thus $4 + 6i = \overline{133041}$ in base $\beta = -2 + i$.

3) Choose $\beta = -3 + i$, *i.e.* $N = 3$. Divisions of $\xi = -59 + 72i$ by β give:

$$-59 + 72i \xrightarrow{7} 27 - 15i \xrightarrow{2} -9 + 2i \xrightarrow{7} 5 + i \xrightarrow{8} 1 \xrightarrow{1} 0.$$

Thus $-59 + 72i = \overline{18727}$ in base $\beta = -3 + i$.

Proposition 9.2.2. *The algorithm for expression in base β is correct.*

Proof. If the algorithm terminates, it is clear that the c_n are the sequence of digits of ξ in base β. Thus, we only need to show that the algorithm terminates. Consider the sequence of norms $N(\xi_n) = |\xi_n|^2$:

- when $N = 1$ and $\xi = -1$, the sequence is $1, 2, 1, 1, 1$;
- when $N = 1$ and $\xi = 3 - i$, the sequence is $10, 5, 5, 2, 1, 2, 1, 1, 1$;
- when $N = 2$ and $\xi = -3 + i$, the sequence is $10, 10, 2, 2, 1$;
- when $N = 2$ and $\xi = 4 + 6i$, the sequence is $52, 9, 5, 1, 2, 1$;
- when $N = 3$ and $\xi = 3 + 5i$, the sequence is $34, 5, 5, 1$.

This behavior puts us on our way.

- If $N(\xi_r) = |\xi_r|^2$ is strictly decreasing the algorithm terminates
- If $N(\xi_r) = |\xi_r|^2$ is not a strictly decreasing sequence, consider the smallest r such that $N(\xi_{r+1}) \geq N(\xi_r)$; that is, the first index at which the sequence "rebounds". To simplify, put $\xi_r = x + iy$, $\xi_{r+1} = u + iv$ and $c_r = c$. Passing to norms, $\xi_r = \xi_{r+1}\beta + c_r$ gives:

$$\begin{aligned} x^2 + y^2 &= (Nu + v - c)^2 + (Nv - u)^2 \\ &= (Nu + v)^2 + (Nv - u)^2 - 2c(Nu + v) + c^2 \\ &= (N^2 + 1)(u^2 + v^2) - 2c(Nu + v) + c^2. \end{aligned}$$

Since $u^2 + v^2 \geq x^2 + y^2$, we have

$$u^2 + v^2 \geq (1 + N^2)(u^2 + v^2) - 2c(Nu + v) + c^2,$$

which can be written, after simplification and division by N^2 :

$$u^2 + v^2 - 2c\Big(\frac{u}{N} + \frac{v}{N^2}\Big) + \frac{c^2}{N^2} = \Big(u - \frac{c}{N}\Big)^2 + \Big(v - \frac{c}{N^2}\Big)^2 - \Big(\frac{c}{N^2}\Big)^2 \leq 0.$$

In other words, $u + vi$ belongs to the closed disk with radius c/N^2 centered at $(c/N, c/N^2)$. Knowing that $0 \leq c \leq N^2$, we get $0 \leq u \leq c/N + c/N^2$ and $0 \leq v \leq 2c/N^2$, which give:

$$0 \leq u \leq N + 1, \quad 0 \leq v \leq 2. \tag{9.8}$$

We are going to show that the algorithm terminates after a finite number of steps when u and v satisfy satisfy (9.8). Let u', v' be the new integers produced by the algorithm from u and v (that is $u + iv = (u' + iv')\beta + c'$ where c' is the new digit), so that:

$$v' = -\Big\lceil \frac{u + Nv}{1 + N^2} \Big\rceil, \quad u' = v + Nv'.$$

If $u + vN \leq N^2$, we have $v' = 0$ and $u' = v \in [\![0, 2]\!]$, and the algorithm terminates because $u' + iv'$ is a digit if $N \geq 2$. The condition $u + vN \leq N^2$ is always satisfied when $N \geq 4$ because we can write:

$$u + Nv \leq (N + 1) + 2N \leq 4N \leq N^2.$$

If $N = 1, 2, 3$ we have only a finite number of cases and the algorithm terminates for all of them. □

Exercises 2

1) Write a Pascal program which displays a given Gaussian integer in base $\beta = -N + i$.
2) If you have mastered graphical output, display the set of Gaussian integers on the screen that have less than k digits.

9.3. Machin Formulas

For many years[2], one calculated decimal places of π using formulas such as:

$$\begin{aligned}
\frac{1}{4}\pi &= 4\operatorname{Arctan}\left(\frac{1}{5}\right) - \operatorname{Arctan}\left(\frac{1}{239}\right) && \text{(John Machin, 1706)},\\
&= \operatorname{Arctan}\left(\frac{1}{2}\right) + \operatorname{Arctan}\left(\frac{1}{3}\right) && \text{(Hutton, 1776)},\\
&= 2\operatorname{Arctan}\left(\frac{1}{3}\right) + \operatorname{Arctan}\left(\frac{1}{7}\right) && \text{(Clausen, 1847)},\\
&= \operatorname{Arctan}\left(\frac{1}{2}\right) + \operatorname{Arctan}\left(\frac{1}{5}\right) + \operatorname{Arctan}\left(\frac{1}{8}\right) && \text{(Dase, 1884)}.
\end{aligned}$$

In 1974 for example, several million decimal places of π were calculated using the following formula (due to Gauss)

$$\frac{1}{4}\pi = 12\operatorname{Arctan}\left(\frac{1}{18}\right) + 8\operatorname{Arctan}\left(\frac{1}{57}\right) - 5\operatorname{Arctan}\left(\frac{1}{239}\right),$$

and checked using Störmer's formula (1896):

$$\frac{1}{4}\pi = 6\operatorname{Arctan}\left(\frac{1}{8}\right) + 2\operatorname{Arctan}\left(\frac{1}{57}\right) + \operatorname{Arctan}\left(\frac{1}{239}\right).$$

If $x > 0$, then

$$\operatorname{Arctg}\left(\frac{1}{x}\right) = \frac{1}{2}\pi - \operatorname{Arctan}(x) = 2\operatorname{Arctan}(1) - \operatorname{Arctan}(x),$$

which allows us to rewrite the preceding formulas in a more natural way as follows:

$$\begin{aligned}
\operatorname{Arctan}(3) &= 3\operatorname{Arctan}(1) - \operatorname{Arctan}(2) && \text{(Hutton)},\\
\operatorname{Arctan}(7) &= \operatorname{Arctan}(1) + 2\operatorname{Arctan}(2) && \text{(Clausen)},\\
\operatorname{Arctan}(8) &= 5\operatorname{Arctan}(1) - 2\operatorname{Arctan}(2) - \operatorname{Arctan}(5) && \text{(Dase)},\\
\operatorname{Arctan}(239) &= -5\operatorname{Arctan}(1) + 4\operatorname{Arctan}(5) && \text{(Machin)},\\
&= 17\operatorname{Arctan}(1) - 6\operatorname{Arctan}(8) - 2\operatorname{Arctan}(57) && \text{(Störmer)}.
\end{aligned}$$

[2] Nowadays, the search for decimals of π uses another strategy based on the *Brent-Salamin formula* and its offspring which converge vertiginously fast.

Definition 9.3.1. *A Machin formula is an equality of the form*

$$\operatorname{Arctan}(n) = c_1 \operatorname{Arctan}(1) + c_2 \operatorname{Arctan}(2) + \cdots + c_{n-1} \operatorname{Arctan}(n-1) \qquad (9.9)$$

where the c_i are integers. An integer n is said to be decomposble if a formula of type (9.9) *holds for* $\operatorname{Arctan}(n)$.

A Machin formula has an interesting geometric interpretation. Since $\operatorname{Arctan}(n)$ is the argument of $1+in$, formula (9.19) simply says that the complex numbers

$$1+in \quad \text{and} \quad \Pi = (1+i)^{c_1}(1+2i)^{c_2}\cdots(1+(n-1)i)^{c_{n-1}}$$

have the same argument, or what is the same thing, that $(1+in)/\Pi$ is a real number.

Although Gauss had investigated Machin formulas, it wasn't until the middle of the XX-th century[3] that the situation was completely cleared up.

Theorem 9.3.1 (J. Todd, 1949). *An integer n is decomposable (i.e. gives rise to a Machin formula) if and only if it satisfies the following condition:*

$$\text{(T)} \qquad \left\{ \begin{array}{l} \textit{every prime divisor of } 1+n^2 \textit{ is also a divisor of} \\ \textit{an integer of the form } 1+d^2 \textit{ with } 1<d<n. \end{array} \right.$$

Thus, the first decomposable integers are 3, 7, 8, 13, 18, 21, 30, ... and the corresponding Machin formulas are:

$$\begin{aligned}
\operatorname{Arctan}(3) &= 3\operatorname{Arctan}(1) - \operatorname{Arctan}(2),\\
\operatorname{Arctan}(7) &= -\operatorname{Arctan}(1) + 2\operatorname{Arctan}(2),\\
\operatorname{Arctan}(8) &= 5\operatorname{Arctan}(1) - \operatorname{Arctan}(2) - \operatorname{Arctan}(5),\\
\operatorname{Arctan}(13) &= 5\operatorname{Arctan}(1) - \operatorname{Arctan}(2) - \operatorname{Arctan}(4),\\
\operatorname{Arctan}(17) &= \operatorname{Arctan}(1) + 2\operatorname{Arctan}(2) - \operatorname{Arctan}(12),\\
\operatorname{Arctan}(18) &= 3\operatorname{Arctan}(1) - 2\operatorname{Arctan}(2) + \operatorname{Arctan}(5),\\
\operatorname{Arctan}(21) &= 2\operatorname{Arctan}(1) + \operatorname{Arctan}(4) - \operatorname{Arctan}(5),\\
\operatorname{Arctan}(30) &= 7\operatorname{Arctan}(1) - \operatorname{Arctan}(2) - \operatorname{Arctan}(4) - \operatorname{Arctan}(23).
\end{aligned}$$

Since criterion (T) is not very practical, we give an equivalent criterion which is easier to use.

Theorem 9.3.2 (J. Todd, 1949). *An integer n satisfies condition* (T) *if and only if all prime divisors p of $1+n^2$ satisfy $p \le 2n$.*

Proof. Let p be a prime number that divides $1+n^2$ and $1+d^2$. Since n and d are two solutions of the equation $x^2+1=0$ in $\mathbb{Z}_p$, there exists an integer

[3] John Todd, *A Problem on Arctangent Relations*, American Math. Monthly 56 (1949), pp. 517–528.

$k \in \mathbb{Z}$ such that $d = \pm n + kp$. Now suppose that $|d| < n$ and $2n < p$. Then $|kp| = |d \mp n| \leq |d| + |n| < 2n < p$ implies $k = 0$; that is, $d = \pm n$, which is absurd. Conversely, suppose that all odd prime divisors p of $1 + n^2$ are bounded by $2n$. We divide n by p using centered remainders:

$$n \equiv r \pmod{p}, \quad |r| < \tfrac{1}{2}p.$$

Then p divides $1 + r^2$. Since $|r| < \frac{1}{2}p \leq n$, condition (T) is satisfied.

9.3.1. Uniqueness of a Machin formula

Before explaining Todd's work, we ask whether a decomposable integer n can occur in several Machin formulas (9.9) The answer is *yes* as the following example shows:

$$\begin{aligned}\operatorname{Arctan}(342) &= -\operatorname{Arctan}(1) + 2\operatorname{Arctan}(2) - \operatorname{Arctan}(5) \\ &\qquad + \operatorname{Arctan}(44) - \operatorname{Arctan}(129) \\ &= -3\operatorname{Arctan}(1) + 2\operatorname{Arctan}(2) - \operatorname{Arctan}(5) \\ &\qquad + \operatorname{Arctan}(28) + \operatorname{Arctan}(44).\end{aligned}$$

Here the Machin formula $\operatorname{Arctan}(129) = 2\operatorname{Arctan}(1) + \operatorname{Arctan}(23) - \operatorname{Arctan}(28)$ allows one to pass from the first decomposition to the second.

Hence, we must refine our question. If we have a Machin formula for an integer n, we can, as above, replace a decomposable integer $m < n$ in the formula by an expression of the type (9.9). If this substitution gives rise to new decomposable integers, we can do the same thing again. Each time, the decomposable integers that appear get smaller so that after a finite number of steps, we obtain a Machin formula that only contains indecomposable integers.

We now ask whether a decomposable integer can give rise to two different expressions of the form (9.9) which involve only indecomposable integers. This time, the answer is *no.*

Proposition 9.3.1 (E. Kern, 1987). *If $n \geq 2$ is indecomposable, there does not exist a relation of the form*

$$c_n \operatorname{Arctan}(n) + \sum_{i=1}^{n-1} c_i \operatorname{Arctan}(i) = 0, \quad c_1, \ldots, c_n \in \mathbb{Z}, \; c_n \geq 2. \tag{9.10}$$

To understand where this result leads, suppose for a moment that it is true, and that we have two *distinct* Machin formulas

$$\operatorname{Arctan}(n) = \sum_{n_\alpha < n} c_\alpha \operatorname{Arctan}(n_\alpha) = \sum_{m_\beta < n} d_\beta \operatorname{Arctan}(m_\beta),$$

with indecomposable n_α and m_β. Combining the two sums gives

$$\sum_{\ell_\gamma < n} e_\gamma \operatorname{Arctan}(\ell_\gamma) = 0.$$

Let γ_0 be the largest index for which $\varepsilon_{\gamma_0} \neq 0$. Since ℓ_{γ_0} is indecomposable, we cannot have $\varepsilon_{\gamma_0} = \pm 1$. Thus, we would get an equality of the type (9.10) which is impossible by the proposition.

9.3.2. Proof of Proposition 9.3.1

The proof of the following result is easy.

Lemma 9.3.1. *Let $a + ib \neq 0$, $x + iy$ and $u + iv$ be any complex numbers. Then*

$$t = \frac{(u+iv)(x+iy)}{a+ib} \in \mathbb{R}$$
$$\implies u^2(a^2+b^2)(x^2+y^2) = (ax+by)^2(u^2+v^2). \qquad (9.11)$$

Proof. We have $at = ux - vy$ and $bt = uy + vx$. Therefore

$$(ax+by)t = u(x^2+y^2).$$

Taking norms of both sides of $(a+ib)t = (u+iv)(x+iy)$ gives

$$(a^2+b^2)t^2 = (u^2+v^2)(x^2+y^2).$$

A little algebra gives the conclusion. □

Let u and v be the real and imaginary parts of $(1+in)^{c_n}$:

$$(1+in)^{c_n} = u + iv. \qquad (9.12)$$

If $u = 0$, multiply c_n by 2 so that $(1+in)^{2c_n} = \big((1+in)^{c_n}\big)^2 = -v^2 \neq 0$. Now, we may suppose that $u \neq 0$, because if the result is true for c_n, it also holds $2c_n$.

Consider the complex number

$$\frac{a+bi}{x+iy} = (1+i)^{c_1}(1+2i)^{c_2}\cdots\big(1+(n-1)i\big)^{c_{n-1}},$$

where $a + bi$ collects the factors in the product with positive exponents and $x + yi$ those with negative exponents. Thus,

$$(a+bi)(x+yi) = (1+i)^{|c_1|}(1+2i)^{|c_2|}\cdots(1+(n-1)i)^{|c_{n-1}|}.$$

Taking norms gives

$$(a^2+b^2)(x^2+y^2) = (1+1^2)^{|c_1|}(1+2^2)^{|c_2|}\cdots(1+(n-1)^2)^{|c_{n-1}|}. \quad (9.13)$$

As we have already remarked, c_n Arctan(n) is the argument of $(1+in)^{c_n}$ and $\sum_{i=1}^{n-1} c_i$ Arctan(i) is the argument of $(a+ib)/(x+iy)$. It results from (9.10) that the quotient of $u+iv$ by $(a+ib)/(x+iy)$ is a real number. So, formula (9.11) holds. Using (9.12) and (9.13), we can rewrite (9.11) as:

$$u^2(1+1^2)^{|c_1|}(1+2^2)^{|c_2|}\cdots(1+(n-1)^2)^{|c_{n-1}|} = (ax+by)^2(1+n^2)^{c_n}. \quad (9.14)$$

By condition (T), there exists a prime p which divides $1+n^2$ but none of the numbers $1+d^2$ for $d = 1, \ldots n-1$. It also follows from (9.14) that p divides u.

Expanding $(1+in)^{c_n}$ using the binomial formula gives

$$u = 1 - \binom{2}{c_n}n^2 + \binom{4}{c_n}n^4 - \binom{6}{c_n}n^6 + \cdots$$

and, if we use the congruence $n^2 \equiv -1 \bmod p$, we obtain

$$u \equiv 1 + \binom{2}{c_n} + \binom{4}{c_n} + \binom{6}{c_n} + \cdots \equiv 2^{c_n-1} \bmod p.$$

Thus $p = 2$, which contradicts the Todd condition since $p = 2$ divides $1+d^2$ when $d = 1$.

9.3.3. *The Todd condition is necessary*

Consider a Machin formula in which the n_i satisfy $1 \le n_i < n$:

$$\text{Arctan}(n) = c_{n-1}\,\text{Arctan}(n_1) + c_{n-2}\,\text{Arctan}(n_2) + \cdots + c_s\,\text{Arctan}(n_s).$$

Since the complex numbers $1+in$ and $(1+in_1)^{c_1}\cdots(1+in_s)^{c_s}$ have the same argument, there exists a real number $M > 0$ such that:

$$M(1+in) = (1+in_1)^{c_{n-1}}\cdots(1+in_s)^{c_s}.$$

Comparing real parts shows that M is an integer. Passing to norms, we obtain

$$M^2(1+n^2) = (1+n_1^2)^{|c_{n-1}|}(1+n_2^2)^{|c_{n-2}|}\cdots(1+n_s^2)^{|c_s|},$$

which shows that condition (T) holds.

9.3.4. *The Todd condition is sufficient*

We say that a Gaussian integer Φ is *n-adapted* if there exists an integer $M \ge 1$ such that $M\Phi$ is a product of Gaussian integers of the form $1+iw$

with $|w| < n$. We also say that $\Phi^{(1)}, \ldots, \Phi^{(t)}$ is an *adapted factorisation* of $1 + in$ if each $\Phi^{(i)}$ is n-adapted, that is if there exist integers $M_i \geq 1$ and integers $|w_\ell| < n$ such that

$$M_1 \cdots M_t(1 + in) = \varepsilon\big(M_1\Phi^{(1)}\big) \cdots \big(M_t\Phi^{(t)}\big) = \varepsilon(1 + iw_1) \cdots (1 + iw_k).$$

An adapted factorization is a precursor of a Machin formula. This is because $(1 + in)$ and $\varepsilon(1 + iw_1) \cdots (1 + iw_k)$ have the same argument, so

$$\operatorname{Arctan}(n) = k \operatorname{Arctan}(1) + \operatorname{Arctan}(w_1) + \cdots + \operatorname{Arctan}(w_k).$$

A calculator (*i.e.* numeric approximations) allows one to specify the right value of $k \in \mathbb{Z}$ (which collects ε and the factors $1 \pm i$).

9.3.5. Kern's algorithm

To show that condition (T) is sufficient, Todd exhibits an adapted decomposition of $1 + in$. We are going to use the same method, but we will prefer a very fast algorithm due to Éric Kern (1986, unpublished), which rests on two simple ideas.

First, let $\mu = \omega_1 \cdots \omega_r$ be a decomposition into irreducible factors in $\mathbb{Z}[i]$. If the norm $N(\mu) = N(\omega_1) \cdots N(\omega_r)$ is not divisible by any prime number $q \equiv 3 \bmod 4$, we know that the $N(\omega_i)$ are either $p = 2$, or prime numbers $p \equiv 1 \bmod 4$. Hence, if we are given a factorization $N(\mu) = AB$, there exist Gauss integers α, β such that $\mu = \alpha\beta$, $A = N(\alpha)$ and $B = N(\beta)$:

$$N(\mu) = AB \implies \mu = \alpha\beta,\ A = N(\alpha),\ B = N(\beta).$$

This argument holds, in particular, for Gaussian integers of the form $1 + in$. In effect, an odd prime number p which divides $1 + n^2$ is necessarily of the form $p \equiv 1 \bmod 4$ since we know that the equation $x^2 + 1 = 0$ has no roots in $\mathbb{Z}_p$ when $p \equiv 3 \bmod 4$.

The next lemma is the second idea.

Lemma 9.3.2. *Let $n > 1$ an integer such that $1 + n^2$ is not a prime number; let Φ, Φ' Gaussian integers such as*

$$1 + in = \Phi{\cdot}\Phi', \quad 1 + n^2 = N(\Phi){\cdot}N(\Phi'), \quad N(\Phi), N(\Phi') > 1.$$

Now divide n by $N(\Phi)$ using centered remainders:

$$n = N(\Phi)q + w, \quad |w| \leq \tfrac{1}{2}N(\Phi).$$

Then Φ divides $1 + iw$ and there exists $\Phi^{(1)}$ such that

$$1 + iw = \Phi \cdot \Phi^{(1)}, \quad N(\Phi^{(1)}) \leq \tfrac{1}{2}N(\Phi).$$

Proof. Write $1 + n^2 = dd'$ where $d, d' > 1$ and lift this equality to $\mathbb{Z}[i]$:

$$1 + in = \Phi \cdot \Phi', \quad 1 + n^2 = N(\Phi) \cdot N(\Phi'), \quad N(\Phi), N(\Phi') > 1.$$

Now divide n by $N(\Phi)$ using centered remainders; since

$$1 + in = iqN(\Phi) + (1 + iw)$$

and since Φ divides both $1 + in$ and $N(\Phi) = \Phi \cdot \overline{\Phi}$, we know that Φ divides $1 + iw$. Thus, there exists $\Phi^{(1)}$ such that

$$1 + iw = \Phi\Phi^{(1)}.$$

Taking norms gives

$$N(\Phi) \cdot N(\Phi^{(1)}) = 1 + w^2 \leq 1 + \frac{1}{4} N(\Phi)^2.$$

Dividing this inequality by $N(\Phi) \geq 2$ gives $N(\Phi^{(1)}) \leq \frac{1}{2} N(\Phi)$. □

Description of Kern's algorithm

Let $n > 1$ be a decomposable integer and

$$p_1 \cdots p_t = 1 + n^2$$

the decomposition of $1 + n^2$ into prime factors. Todd's Theorem 9.3.2 assures us that $p_k \leq 2n$. We will prove that lifting this equality to $\mathbb{Z}[i]$,

$$1 + in = \Phi_1 \cdots \Phi_t,$$

gives an adapted factorization (and a Machin formula for the integer n).

Let Φ denote one of the factors $\Phi_1, \ldots, \Phi_t$.

• If $N(\Phi) = 2$, we know that $\Phi = \varepsilon(1 + i)$.

• If $N(\Phi) > 2$ is an odd integer, we use the Lemma 9.3.2: dividing n by $N(\Phi)$ gives rise to w_1 and $\Phi^{(1)}$ such that

$$1 + iw_1 = \Phi \cdot \Phi^{(1)}, \quad N(\Phi^{(1)}) \leq \tfrac{1}{2} N(\Phi).$$

If $N(\Phi^{(1)}) > 1$, we know that $1 + w_1^2 = N(\Phi) \cdot N(\Phi^{(1)}))$ is not a prime number. Use the Lemma again and divide w_1 by $N(\Phi^{(1)})$, which gives rise to w_2 and $\Phi^{(2)}$ such that

$$1 + iw_2 = \Phi^{(1)} . \Phi^{(2)}, \quad N(\Phi^{(2)}) \leq \tfrac{1}{2} N(\Phi)^{(1)}.$$

As $N(\Phi^{(1)}) = (1 + w_1^2)/N(\Phi)$, it is not necessary to find the explicit value of $\Phi^{(1)}$ to deduce w_2 from w_1. Starting anew with w_2 and $N(\Phi^{(1)})$ we obtain a *finite* sequence of integers

$$\begin{cases} 1 + iw_1 &= \Phi \cdot \Phi^{(1)}, \qquad N(\Phi^{(1)}) \leq \tfrac{1}{2} N(\Phi), \\ 1 + iw_2 &= \Phi^{(1)} . \Phi^{(2)}, \quad N(\Phi^{(2)}) \leq \tfrac{1}{2} N(\Phi)^{(1)}, \\ &\ldots \\ 1 + iw_r &= \Phi^{(r-1)} . \Phi^{(r)}, \; N(\Phi^{(r)}) = 1. \end{cases}$$

Solving these equations gives

$$\Phi = \frac{1+iw_1}{\Phi^{(1)}} = \frac{(1+iw_1)\Phi^{(2)}}{1+iw_2} = \cdots = \varepsilon \frac{(1+iw_1)(1+iw_3)\cdots}{(1+iw_2)(1+iw_4)\cdots}.$$

Therefore there exists an integer $M \geq 1$ such that

$$M\Phi = \varepsilon(1+iw_1)(1-iw_2)(1+iw_3)(1-iw_4)\cdots.$$

(The exact value $M = N(1+iw_2)\cdot N(1+iw_4)\cdots$ is irrelevant.)

Kern's algorithm is as follows.

- Start with $1+n^2 = p_1 \cdots p_t$.
- Initially, *list_factors* is empty. We collect factors $1+iw$ associated to each prime divisor d using

```
index := 1 ;
repeat
| w := centered_rem(n, d) ;     {n = dq + w,  |w| ≤ ½d}
| if index mod 2 = 1
| then list_factors := add(1 + wi, list_factors)
| else list_factors := add(1 − wi, list_factors) ;
| index := index + 1 ;
| d := (1 + w²)/d ;
| n := w
until w = 0
```

- The product of all factors $1+iw$ associated to $p_1, \ldots, p_t$ gives a Machin formula for the integer n.

Example

Choose $n = 1136$; since $1+n^2 = 1873 \cdot 53 \cdot 13$ and $1873 \leq 2n$, we know that n is decomposable and that there exists an adapted factorization $1+in = \Phi_{1873} \cdot \Phi_{53} \cdot \Phi_{13}$.

- Apply Kern's algorithm to the divisor 1873:

$1136 = 1 \times 1873 - 737$	$1+737^2 = 1873 \times 290$
$-737 = -3 \times 290 + 133$	$1+133^2 = 290 \times 61$
$133 = 2 \times 61 + 11$	$1+11^2 = 61 \times 2$
$11 = 5 \times 2 + 1$	$1+1^2 = 2 \times 1$

The centered remainders are $w_1 = -737$, $w_2 = 133$, $w_3 = 11$ and $w_4 = 1$. The norms necessary to compute the w_k are $N(\Phi^{(1)}) = (1+737^2)/1873 = 290$, $N(\Phi^{(2)}) = (1+133^2)/61 = 2$, $N(\Phi^{(3)}) = (1+11^2)/61 = 290$ and $N(\Phi^{(4)}) =$

$(1+1^2)/2 = 1$. Therefore, there exists an integer $M' \geq 1$ and a unit ε' such that

$$\begin{aligned} M'\Phi_{1873} &= \varepsilon'(1+w_1)(1-iw_2)(1+iw_3)(1-iw_4) \\ &= \varepsilon'(1-737i)(1-133i)(1+11i)(1-i) \end{aligned}$$

- Kern's algorithm applied to the two other factors gives

$1136 = 87 \times 13 + 5$	$1 + 5^2 = 13 \times 2$
$5 = 2 \times 2 + 1$	$1 + 1^2 = 2 \times 1$
$1136 = 21 \times 53 + 23$	$1 + 23^2 = 53 \times 10$
$23 = 2 \times 10 + 3$	$1 + 3^2 = 10 \times 1$

Thus we know that there are formulas

$$\begin{aligned} M''\Phi_{53} &= \varepsilon''(1+23i)(1-3i), \qquad M'' \geq 1, \\ M'''\Phi_{13} &= \varepsilon'''(1+5i)(1-i), \qquad M''' \geq 1. \end{aligned}$$

- From $M'M''M'''(1+in) = M'\Phi_{1873} \cdot M''\Phi_{53} \cdot M'''\Phi_{13}$, we deduce that there exists an integer $M \geq 1$ and a unit ε such that

$$M(1+in) = \varepsilon(1-i)^2(1-3i)(1+5i)(1+11i)(1+23i)(1-133i)(1-737i).$$

We use this to get the following Machin formula where the coefficent of Arctan(1) collects the unit ε and the factors $1 \pm i$:

$$\begin{aligned} \operatorname{Arctan}(1136) = h \operatorname{Arctan}(1) &- \operatorname{Arctan}(3) + \operatorname{Arctan}(5) + \operatorname{Arctan}(11) \\ &+ \operatorname{Arctan}(23) - \operatorname{Arctan}(133) - \operatorname{Arctan}(737). \end{aligned}$$

A calculator shows that $h = 2$.

Remarks

1) Kern's algorithm illuminates the condition $p \leq 2n$ of Todd's Theorem 9.3.2. As we have already remarked, the factors $1 + iw$ which appear satisfy $|w| \leq d$. When $d = p$ is odd, we have $w < \frac{1}{2}p \leq n$, which assures us that $1 + in$ is not among the collected factors.

2) We examine the behavior of Kern's algorithm when n is not decomposable: say, for example, when $n = 9$. The factorization $1 + n^2 = 2 \times 41$ gives $\Phi^{(1)} = 1+i$ and $2\Phi^{(2)} = (1+9i)(1-i)$. We get $2(1+9i) = (1+i)(1+9i)(1-i)$ and this equation is not a precursor of a Machin formula.

3) The first remark also shows that it is not necessary to completely factor $1 + n^2$ into primes. Any factorization which ensures that $d \leq 2n$ (where the inequalty is strict if d is even) will work. Take for example $n = 1136$; we can content ourselves with the factorization $1 + n^2 = 1873 \cdot 689$ which lifts

to $\mathbb{Z}[i]$ as $1+1136i = \Phi_{1873} \cdot \Phi_{689}$. Application of the algorithm to the factor $d = 689$ gives:

$1136 = 2 \cdot 689 - 242$	$1 + 242^2 = 689 \cdot 85$
$-242 = -3 \cdot 85 + 13$	$1 + 13^2 = 85 \cdot 2$
$13 = 6 \cdot 2 + 1$	$1 + 1^2 = 2 \cdot 1$

Thus there exists $M \geq 1$ and a unit ε such that

$$M\Phi_{689} = \varepsilon(1-242i)(1-13i)(1+i).$$

The Machin formula associated to the factorization $1+n^2 = 689 \cdot 1873$ is

$$\begin{aligned}\text{Arctan}(1136) = k\,\text{Arctan}(1) &+ \text{Arctan}(11) - \text{Arctan}(13)\\ &- \text{Arctan}(133) - \text{Arctan}(242) - \text{Arctan}(737).\end{aligned}$$

Numerical approximations shows that $k = 8$.

9.3.6. How to get rid of the Arctangent function

After a series of purely arithmetic calculations, it is frustrating to have to turn to numerical approximations to guess the precise value of the integer multiplying Arctan(1), thereby abandoning the absolute precision of arithmetic.

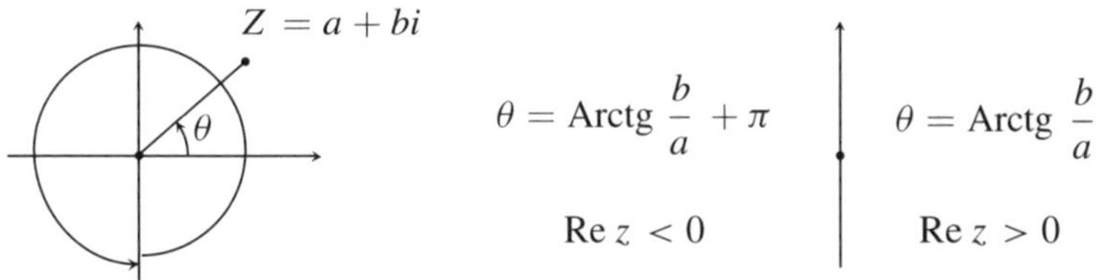

To avoid this false note, we choose a determination of the argument of a complex number and monitor the variation of the argument in the course of the various multiplications. Let

$$Z = a + ib, \quad ab \neq 0$$

be a complex number not on the axes. Choose the argument θ of Z to satisfy

$$-\frac{1}{2}\pi < \theta < \frac{3}{2}\pi.$$

(This unusual choice minimizes the number of cases we will have to handle.) If we put

$$a_+ = \begin{cases} 1 & \text{if } a > 0, \\ 0 & \text{if } a < 0, \end{cases}$$

we find that the argument of Z is

$$\theta(a+bi) = \operatorname{Arctg}\Big(\frac{b}{a}\Big) + (1-a_+)\pi. \tag{9.15}$$

Let $w \neq 0$ be a real number. Put $W = 1 + iw$ and

$$Z' = ZW = a' + ib' = (a - bw) + (b + aw)i.$$

Lemma 9.3.3. *If $ab \neq 0$ and $a'b' \neq 0$, then:*

$$\operatorname{Arctg}\Big(\frac{b'}{a'}\Big) = \operatorname{Arctg}\Big(\frac{b}{a}\Big) + \operatorname{Arctg}(w) + \operatorname{sgn}(b)(a'_+ - a_+)\pi. \tag{9.16}$$

Proof. The derivative of the function

$$w \in \mathbb{R} \longmapsto \operatorname{Arctg}\Big(\frac{b+aw}{a-bw}\Big) - \operatorname{Arctg}\Big(\frac{b}{a}\Big) - \operatorname{Arctg}(w)$$

with respect to w is zero. Consequently, this function is constant on every interval on which it is differentiable; that is, on every interval which does not contain a/b. Upon letting w tend to $\pm\infty$, we see that this constant equals

$$c = -\Big\{\operatorname{Arctg}\Big(\frac{a}{b}\Big) + \operatorname{Arctg}\Big(\frac{b}{a}\Big)\Big\} - \operatorname{sgn}(w)\frac{1}{2}\pi = -\frac{1}{2}\Big\{\operatorname{sgn}\Big(\frac{a}{b}\Big) + \operatorname{sgn}(w)\Big\}\pi$$

since

$$\operatorname{Arctan}(x) + \operatorname{Arctan}(x^{-1}) = \operatorname{sgn}(x)\tfrac{1}{2}\pi, \quad x \neq 0.$$

But $\operatorname{sgn}(a/b) = \operatorname{sgn}(a)\operatorname{sgn}(b)$ and

$$a' = a - bw \implies \operatorname{sgn}(a') = -\operatorname{sgn}(b)\operatorname{sgn}(w),$$

that is, $\operatorname{sgn}(w) = -\operatorname{sgn}(a')\operatorname{sgn}(b)$. Therefore

$$\begin{aligned} c &= -\tfrac{1}{2}\{\operatorname{sgn}(a)\operatorname{sgn}(b) - \operatorname{sgn}(a')\operatorname{sgn}(b)\}\pi \\ &= \operatorname{sgn}(b)\tfrac{1}{2}\{\operatorname{sgn}(a') - \operatorname{sgn}(a)\}\pi \\ &= \operatorname{sgn}(b)(a'_+ - a_+)\pi. \end{aligned}$$

□

Corollary 9.3.1. *If $ab \neq 0$ and $a'b' \neq 0$, then:*

$$\theta(Z') = \theta(Z) + \operatorname{Arctg}(w) + \big(1 - \operatorname{sgn}(b)\big)(a_+ - a'_+)\pi. \tag{9.17}$$

Proof. By definition

$$\theta(Z') = \operatorname{Arctg}\Big(\frac{b'}{a'}\Big) + (1 - a'_+)\pi.$$

Using (9.20) then (9.21), we get

$$\begin{aligned}\theta(Z') &= \operatorname{Arctan}\left(\frac{b}{a}\right) + \operatorname{Arctg}(w) + \operatorname{sgn}(b)(a'_+ - a_+)\pi + (1 - a'_+)\pi \\ &= \theta(Z) - (1 - a_+)\pi + \operatorname{Arctg}(w) + \operatorname{sgn}(b)(a_+ - a'_+)\pi + (1 - a'_+)\pi \\ &= \theta(Z) + \operatorname{Arctg}(w) + (a_+ - a'_+)\pi + \operatorname{sgn}(b)(a'_+ - a_+)\pi. \qquad \square\end{aligned}$$

9.3.7. Examples

1) Consider the adapted factorization where $M \geq 1$

$$\begin{aligned}M(1 + 1136i) = \varepsilon(1 - i)^2(1 - 3i)(1 + 5i)(1 + 11i) \\ (1 + 23i)(1 - 133i)(1 - 737i).\end{aligned}$$

As we have already remarked, the factor $\varepsilon(1 - i)^2$ does not interest us because it only modifies the coefficient of Arctan(1). We start then with

$$Z_1 = 1 - 3i$$

which has argument Arctan(−3), and we multiply it repeatedly by $1 + iw$ with $w = 5, 11, 23, -133, -737$, which gives the Gaussian numbers

$$Z_2, \ldots, Z_6 = 353800(1136 - i).$$

As the complex numbers $Z_2, \ldots, Z_5$ all have imaginary part $b > 0$, formula (9.17) tells us that we do not need any correction and, since Z_6 and $1136 - i$ have the same argument, we find that

$$\begin{aligned}\operatorname{Arg}(1136 - i) = &\operatorname{Arctan}(-3) + \operatorname{Arctan}(5) + \operatorname{Arctan}(11) \\ &+ \operatorname{Arctan}(23) + \operatorname{Arctan}(-133) + \operatorname{Arctan}(-737).\end{aligned}$$

To obtain $1 + 1136i$, we multiply $1136 - i$ by i, which increases the argument by $\frac{1}{2}\pi = 2\operatorname{Arctan}(1)$:

$$\begin{aligned}\operatorname{Arctan}(1136) &= \operatorname{Arg}(1 + 1136i) \\ &= 2\operatorname{Arctan}(1) + \operatorname{Arctan}(-3) + \operatorname{Arctan}(5) + \operatorname{Arctan}(11) \\ &\quad + \operatorname{Arctan}(23) + \operatorname{Arctan}(-133) + \operatorname{Arctan}(-737).\end{aligned}$$

2) Now consider the adapted factorization where $M \geq 1$:

$$M(1 + 1136i) = 2(1 + 11i)(1 - 13i)(1 - 133i)(1 - 242i)(1 - 737i).$$

We begin with

$$Z_1 = 1 + 11i$$

which has argument Arctan(11) and we multiply it successively by $1 + iw$ with $w = -13, -133, -242, -737$:

$$Z_2 = 2(72 - i), \qquad Z_3 = 122(-1 - 157i),$$
$$Z_4 = 10370(-447 + i), \qquad Z_5 = 3007300(1 + 1136i).$$

Thanks to formula (9.17), the corresponding arguments are:

$$\begin{aligned}
\theta_2 &= \text{Arctan}(11) + \text{Arctan}(-13),\\
\theta_3 &= \text{Arctan}(11) + \text{Arctan}(-13) + \text{Arctan}(-133) + 2\pi,\\
\theta_4 &= \text{Arctan}(11) + \text{Arctan}(-13) + \text{Arctan}(-133) + \text{Arctan}(-242) + 2\pi,\\
\theta_5 &= \text{Arctan}(11) + \text{Arctan}(-13) + \text{Arctan}(-133) + \text{Arctan}(-242)\\
&\qquad + \text{Arctan}(-737) + 2\pi.
\end{aligned}$$

Since $2\pi = 8\,\text{Arctan}(1)$, we have obtained the Machin formula:

$$\begin{aligned}
\text{Arctan}(1136) &= \text{Arg}(Z_5)\\
&= 8\,\text{Arctan}(1) + \text{Arctan}(11) + \text{Arctan}(-13)\\
&\qquad + \text{Arctan}(-133) + \text{Arctan}(-242) + \text{Arctan}(-737).
\end{aligned}$$

Exercise 3

Transform this theory into a program which calculates Machin formulas for decomposable integers $n \in [\![1, 100]\!]$.

Remark

The reader interested in an another approach to this subject might consult the book *Mathématiques et Informatique* by J. Berstel, J.-E. Pin and M. Pocchiola, Mc Graw-Hill (1991).

10. Polynomials

10.1. Definitions

For a mathematician, a polynomial A with coefficients in a ring k is an infinite sequence $(a_n)_{n\in\mathbb{N}}$ of elements which are all zero after some point (which depends on the sequence):

$$A = (a_0, a_1, a_2, \ldots, a_n, 0, 0, 0, \ldots).$$

Let $k[X]$ be the set of such sequences (the appearance of X will be justified a little later) One can give this set the structure of a ring by defining the operations of addition and multiplication as follows:

- for addition, let

$$A + B = (a_0 + b_0, \ldots, a_n + b_n, 0, \ldots)$$

- for multiplication $C = AB$, let the n-th element of C be:

$$c_n = \sum_{p+q=n} a_p b_q.$$

The ring k can be identified with the constant polynomials using the bijection:

$$a_0 \rightleftarrows (a_0, 0, \ldots).$$

With this identification, we can, in particular, multiply a polynomial by a *constant* so that $\lambda(a_0, \ldots, a_d, 0, \ldots) = (\lambda, 0, \ldots)(a_0, \ldots, a_d, 0, \ldots)$ is indeed the polynomial $(\lambda a_0, \ldots, \lambda a_d, 0, \ldots)$. If we now put

$$X = (0, 1, 0, \ldots),$$

a straightforward induction shows that for every integer $n \geq 1$,

$$X^n = (0, \ldots, 0, 1, 0, \ldots), \quad \text{the 1 being in the } n\text{-th place.}$$

Then, every polynomial can be written uniquely in the form

$$A = \sum_{i=0}^{n} a_i X^i$$

and one recovers the traditional presentation of a polynomial.

10.2. Degree of a Polynomial

If k is an *integral domain* (the product of two elements is zero if and only if one of the factors is zero) then it turns out that $k[X]$ is an integral domain.

Let A be a *nonzero* polynomial. Its *degree* is, by definition, the largest index i such that $a_i \neq 0$:

$$\deg(A) = \max\{i;\ a_i \neq 0\}.$$

For example the *nonzero* constant polynomials have degree 0.

This definition does not work if A is zero because the set of indexes i such that $a_i \neq 0$ is empty. What degree can we attribute to the zero polynomial? Several conventions are possible depending on what one wants to investigate. In general, one wants the formula

$$\deg(A \times B) = \deg A + \deg B$$

to continue to hold if A or B is zero (we suppose that k is an integral domain).

• *First convention*: one attributes degree $-\infty$ to the zero polynomial. This somewhat surprising convention is best from the point of view of the preceding criterion because:

$$\deg(A \times 0) = \deg A - \infty = -\infty = \deg(0).$$

• *Second convention*: the zero polynomial is considered as a constant polynomial of degree zero. With this choice,

$$\deg(A \times 0) = \deg(0) = 0.$$

So the formula $\deg(A \times B) = \deg A + \deg B$ does not continue to hold! It is necessary therefore to pay careful attention to the definition of degree that one uses.

10.3. How to Store a Polynomial

Recall that we use *basic* Pascal without pointers. Thus we can only use arrays whose size is fixed once and for all at the moment of compilation. Two natural solutions present themselves.

• We can consider a polynomial as a vector of *fixed dimension*

$$X^2 + 2X + 3 \rightleftarrows (3, 2, 1, 0, 0, 0, 0, 0, 0, 0) \tag{10.1}$$

using an array $A[0\,..\,deg\_max]$:

```
const deg_max = 10 ;
type poly = array[0 .. deg_max] of integer ;
var A : poly ;
```

We then store the polynomial $A = X^2 + 2X + 3$ using the code:

```
for i := 0 to deg_max do A[i] := 0 ;
A[0] := 3 ;  A[1] := 2 ;  A[2] := 1 ;
```

The first statement – the preliminary clearing of the coefficients in the array is vital.

- We can consider a polynomial as a pair (*degree*, *sequence of coefficients*):

$$X^2 + 2X + 3 \rightleftarrows \text{degree} = 2 \ \text{and} \ (3, 2, 1). \tag{10.2}$$

There is a subtlety here. We try to explain visually:

$$X^2 + 2X + 3 \rightleftarrows \begin{cases} (3, 2, 1, 0, 0, 0, 0, 0, 0, 0), \\ (3, 2, 1, ?, ?, ?, ?, ?, ?, ?) \ \text{and} \ \deg = 2. \end{cases}$$

(The question marks signal *undefined* values; that is, memory contents which have not been initialized and must be considered random. In other words, they are *litter.*)

Convention (10.2) requires that we store the degree at the same time as the coefficients. For polynomials with integer coefficients, one way to do this, inspired by the implementation of chains of characters, is for example:

```
const deg_max = 10 ;  deg = -1 ;
type poly = array[deg .. deg_max] of integer ;
var A : poly ;
```

The degree is thus $A[deg]$ and the coefficients are $A[0], \ldots, A[deg\_max]$. For example, we store the polynomial $A = X^2 + 2X + 3$ by typing:

$$A[0] := 3; \ \ A[1] := 2; \ \ A[2] := 1; \ \ A[deg] := 2.$$

There is no point specifying that $A[i] = 0$ for $i > \deg(A)$ since we have defined the degree of A.

Convention (10.2) functions much less well when the coefficients are real. In effect, although $A[deg]$ is an integer, it is treated as a real number. In this case, it is better to use a *record*:

```
const deg_max = 10 ;
type arr_coeff = array[0 .. deg_max] of real ;
poly =  record coeff : arr_coeff ;  deg : integer end ;
var A : poly ;
```

Now, the degree of A is $A.deg$ and the coefficient of X^k is $A.coef[k]$.

Convention (10.2) is seductive and furnishes *a priori* faster programs. But it is difficult to implement. In effect, when one calculates $C = A + B$, it is necessary to be very, very careful:

• If $\deg(A) = \deg(B)$, and if we work with real coefficients, one can use the code

```
for i := 0 to A.deg do C.coeff[i] := A.coeff[i] + B.coeff[i]
```

After that, we absolutely *must define* the degree of C while examining the $C.coeff[i]$ for $i \leq \deg(A)$.

• If $\deg(A) < \deg(B)$, it is necessary to use the code

```
C := B ;
for i := 0 to A.deg do C.coeff[i] := A.coeff[i] + C.coeff[i]
```

We do not need to find the degree of C because it is equal to that of B.

The product is still more difficult to write correctly.[1] Try it!

10.4. The Conventions we Adopt

• We store polynomials using convention (10.1).

• We assign the zero polynomial degree 0.

The degree function

If the coefficients are integers, we can use the test $A[i] \neq 0$ without any precaution:

```
function degree(A : poly) : integer ;
var i : integer ;
begin
  degree := 0 ;
  for i := 0 to deg_max do
      if A[i] ≠ 0 then degree := i
end ;
```

The statement *degree* := 0 in the preceding code is vital. If one omits it, the zero polynomial does not have a degree!

We must modify this code lightly in the case that the coefficients are not integers because the test $A[i] \neq 0$ is not satisfactory over the reals:

```
function degree(A : poly) : integer ;
const ε = 1E − 8 ;
var i : integer ;
begin
  degree := 0 ;
  for i := 0 to deg_max do
      if abs(A[i]) > ε then degree := i
end ;
```

[1] Here is a good illustration of the proverb *Above all, no tricks!!*

The procedure annul

This is an indispensable procedure. If we forget to initialize our polynomials, they will contain litter and their degree will not be correct.

```
procedure annul(var A : poly) ;
var i : integer ;
begin
| for i := 0 to deg_max do A[i] := 0
end ;
```

The add_poly and mult_poly procedures

To add and multiply two polynomials, it suffices to recopy the definitions of sum and product (we suppose that the coefficients are integers):

```
procedure add_poly(A, B : poly ;  var C : poly) ;  {returns C = A + B}
var i : integer ;
begin
| for i := 0 to deg_max do C[i] := A[i] + B[i]
end ;

procedure mult_poly(A, B : poly ;  var C : poly) ;  {returns C = A · B}
var i, k, temp : integer ;
begin
| if degree(A) + degree(B) > deg_max
| then writeln('error :  degree too large')
| else begin
| | annul(C) ;
| | for i := 0 to degree(A) + degree(B) do begin
| | | temp := 0 ;  for k := 0 to i do temp := temp + A[k] * B[i − k] ;
| | | C[i] := temp
| | end
| end
end ;
```

For beginners

Suppose that you know A is a polynomial and you want to translate $A = 6$ into code. Beginners often write $A := 6$, which elicits the error message *type mismatch* (*i.e.* the types are incompatible). The right code is:

$$annul(A)\,;\;\; A[0] := 6\,;$$

It is easy to understand why the compiler is perplexed: the objects "A" and "6" do not occupy the same place in memory.

We can profitably use this opportunity to reflect on the meaning of the inclusion $\mathbb{R} \subset \mathbb{R}[X]$. When a mathematician says: "*I identify* the real number a with the constant polynomial $(a, 0, 0, \ldots)$", what is being said is that henceforth he or she will *pretend* that these two objects are equal, thereby allowing him or her to write $A = 6$ without blushing. This *behavior* will not produce catastrophes because the canonical injection $\mathbb{R} \rightarrow \mathbb{R}[X]$ is a ring homomorphism. But *physically*, real numbers and constant polynomials are distinct objects, a distinction that is not lost on our computer because it is not able to "pretend".

Comfortable display of polynomials

Suppose that we want to display a polynomial in a visually comfortable manner respecting the following constraints:

- no more than five monomials are displayed on a line;
- zero monomials are not displayed;
- '$1X\hat{}n$' is displayed as '$X\hat{}n$' ;
- '$-1X\hat{}n$' is dispayed as '$-X\hat{}n$' ;

To display only five monomials on a line, we have to count the number of monomials, whence:

```
num_monomials := 0 ;
for i := deg_max downto 0 do begin
  if P[i] ≠ 0 then begin
    write(P[i], 'X^', i : 1) ;
    num_monomials := num_monomials + 1 ;
    if num_monomials mod 5 = 0 then writeln
  end
end ;
```

We now refine this code discussing whether $P[i]$ is positive, zero, or negative. Pascal does not display the '+' sign before a positive number. The constants make the code easy to read. Finally, several trials will show that one must not forget that a polynomial is sometimes zero.

```
procedure display_poly(P : poly) ;
const plus = ' + ' ; minus = ' - ' ; exponent = 'X' ;
var i, num_monomials : integer ;
begin
  num_monomials := 0 ;
  for i := deg_max downto 0 do begin
    if P[i] ≠ 0 then begin
      if P[i] > 1 then write(plus, P[i] : 1) else
      if P[i] = 1 then write(plus) else
      if P[i] = -1 then write(minus) else
      if P[i] < -1 then write(minus, -P[i] : 1) ;
```

```
      write(exponent, i : 1) ;
      num_monomials := num_monomials + 1 ;
      if num_monomials mod 5 = 0 then writeln
    end
  end ;
  if num_monomials = 0 then writeln('0') ; {case P = 0}
  if num_monomials mod 5 ≠ 0 then writeln
end ;
```

The last statement ensures the return to the correct line after displaying a polynomial. This precaution allows us to ask that several polynomials be displayed.

```
write('P = ') ; display_poly(P) ;
write('Q = ') ; display_poly(Q) ;
write('R = ') ; display_poly(R) ;
```

10.5. Euclidean Division

Let A and B be two polynomials with coefficents in a field. If B is not the zero polynomial, we know that there exists a unique pair (Q, R) of polynomials satisfying the conditions:

$$A = BQ + R, \quad \deg R < \deg B.$$

We refresh our memory by dividing $A = 2X^5 - X^4 + 3X^3 + 4X^2 - X + 1$ by $B = X^3 + 2X + 3$:

$2X^5 - X^4 + 3X^3 + 4X^2 - X + 1$	$X^3 + 2X + 3$
$-X^4 - X^3 - 2X^2 - X + 1$	$2X^2$
$-X^3 + 2X + 1$	$-X$
$4X + 4$	-1
stop	

We see that three sequences appear: the *remainders* R_i and partial *quotients* Q_i in which *monomials* M_i accumulate:

$R_0 = A$	B
$R_1 = R_0 - BM_1$	$Q_1 = Q_0 + M_1 \quad (Q_0 = 0)$
$R_2 = R_1 - BM_2$	$Q_2 = Q_1 + M_2$
$\vdots$	$\vdots$
$R_\ell = R_{\ell-1} - BM_\ell$	$Q_\ell = Q_{\ell-1} + BM_\ell$

The calculation is finished when the degree of the partial remainder is smaller than that of the divisor B. The three dots represent a loop which must be specified. Since we do not know the number of steps in advance, we cannot use a 'for' loop (if we divide $X^6+X^4+X^2$ by X^2, the division stops right away). Thus, we choose a 'while' loop, because we should definitely do nothing when $\deg A < \deg B$. A succinct mathematical description of the division algorithm is then:

```
R_0 := A ; Q_0 := 0 ; i := 0 ;
while deg(R_i) ≥ deg(B) do begin
| «calculate the monomial M_i» ;
| Q_{i+1} := Q_i + M_i ;
| R_{i+1} := R_i − B · M_i ;
| i := i + 1
end
```

Recall that the monomial M_i is the quotient of the highest degree monomials of the polynomials R_i and B. Suppressing the time index i and specifying the monomial M_i we have:

```
R := A ; Q := 0 ;
while deg(R) ≥ deg(B) do begin
| M := R[deg(R)]/B[deg(B)] X^(deg(R)−deg(B)) ;
| Q := Q + M ;
| R := R − B · M
end ;
```

Remark

If A and B are coefficients in an *integral domain* (for example, $\mathbb{Z}$), the quotient Q and remainder R have coefficients in the field of fractions of the ring. But if the coefficient of the highest degree monomial of B is *invertible* (as in the example on the preceding page), the algorithm shows that Q and R have coefficients in the ring because no fractions are introduced during the calculations. As a consequence, one remains in the integers (that is, one does not need recourse to fractions) when dividing by a monic polynomial.

```
procedure unitary_division(A, B : poly ; var Q, R : poly) ;
var i, coeff : integer ; monomial : poly ;
begin   {we suppose A, B, Q, R ∈ ℤ[X] and B monic}
| if B[degree(B)] ≠ 1
| then writeln('error : polyomial is not unitary')
| else begin
| | R := A ; annul(Q) ;
| | while degree(R) ≥ degree(B) do begin
| | | annul(monomial) ;
| | | monomial[degree(R) − degree(B)] := R[degree(R)] ;
| | | Q[degree(R) − degree(B)] := R[degree(R)] ; {Q := Q + monomial}
| | | mult_poly(monomial, B, monomial) ; {monomial := B · monomial}
```

```
      sub_poly(R, monomial, R)   {R := R − monomial}
    end
  end
end ;
```

We remark that there is no point using the general procedure *add_poly* to add a monomial to the quotient Q.

10.6. Evaluation of Polynomials: Horner's Method

The problem of economically calculating the value of a polynomial was resolved in the seventeenth century by Newton during a time when calculations were done by hand and techniques for economizing on additions and multiplications were much appreciated. The technique, however, is called Horner's method in honor of W.G. Horner who rediscovered and popularized it in 1819.

To calculate the value at x of the polynomial

$$A = 1 + 2X + 3X^2 - 4X^3 - 5X^4,$$

we could type in our program:

$$value := 1 + 2 * x + 3 * x * x - 4 * x * x * x - 5 * x * x * x * x \qquad (10.3)$$

But the situation gets more interesting if we want to calculate

$$A = a_0 + a_1 x + \cdots + a_n x^n.$$

We cannot type something like

$$value := a[0] + b[1] * x + \cdots + a[n] * x * \cdots * x$$

(which is, unfortunately, what some beginners do) because the compiler does not understand the three dots '...' which represent a repetition (*i.e.* a loop). We can fix this by defining a Pascal function $power(x, i)$ which returns the value of x^i and entering:

```
value := 0 ;
for i := 0 to n do value := value + A[i] * power(x, i)
```

This code is very clumsy and is a Penelope code: when we calculate x^i, we forget that we calculated x^{i-1} an instant earlier.

Let us return to the calculation of $v = 1 + 2x + 3x^2 - 4x^3 - 5x^4$. If we isolate the constant term we can treat x as a factor in what remains:

$$v = 1 + x(2 + 3x - 4x^2 - 5x^3).$$

Carrying out this transformation repeatedly on the polynomials in parentheses, we finally get

$$v = 1 + x(2 + x(3 + x(-4 + x(-5)))).$$

The parentheses that appear suggest a very natural *sequence* of calculations:

$$\begin{aligned} v_0 &= -5, & v_3 &= 2 + x v_2, \\ v_1 &= -4 + x v_0, & v_4 &= 1 + x v_3, \\ v_2 &= 3 + x v_1, & \mathit{value} &= v_4 \end{aligned} \tag{10.4}$$

This strategy presents considerable advantages:

- formula (10.3) requires 10 multiplications and 4 additions;
- in contrast, (10.4) uses only 4 multiplications and 4 additions !

More generally, the value of the polynomial $A = a_0 + a_1 X + \cdots + a_n X^n$ at x is the last term of either of the two sequences

$$\begin{array}{rcl|rcl} v_0 &=& a_n & v_n &=& a_n \\ v_1 &=& a_{n-1} + x\, v_0 & v_{n-1} &=& a_{n-1} + x\, v_n \\ v_2 &=& a_{n-2} + x\, v_1 & v_{n-2} &=& a_{n-2} + x\, v_{n-1} \\ &\vdots& & &\vdots& \\ v_n &=& a_0 + x\, v_{n-1} & v_0 &=& a_0 + x\, v_1 \end{array}$$

whose translations into code are

```
value := a_n                          |  value := a_n
for i := 1 to n do                    |  for i := n − 1 downto 0 do
    value := a_{n−i} + x · value      |      value := a_i + x · value
```

These two algorithms are called *Horner's method*. Experience shows that the one on the right (with the decreasing indices) is much more natural in practice.

```
function value(A : poly ; x : integer) : integer ;
var i, deg_A, temp : integer ;
begin
  deg_A := degree(A) ;
  temp := A[deg_A] ;
  for i := deg_A − 1 downto 0 do temp := A[i] + x ∗ temp ;
  value := temp
end ;
```

10.7. Translation and Composition

10.7.1. Change of origin

Let $A(X) = a_0 + a_1 X + \cdots + a_n X^n$ be a polynomial with coefficients in a ring and h an element of this ring. Let:

$$B(X) = A(X + h) = b_0 + b_1 X + \cdots + b_n X^n.$$

How can one calculate the b_i using the a_i and h?

Suppose for example that we have $A = 1 + 2X + 3X^2 - 4X^3 + 7X^4$ and $h = 1$. To calculate

$$B(X) = 1 + 2(X+1) + 3(X+1)^2 - 4(X+1)^3 + 7(X+1)^4,$$

it is necessary to resist the temptation to expand the terms $(X+1)^k$ because this strategy produces a very awkward code. Following the idea behind Horner's method:

$$\begin{aligned} B_4 &= 7 \\ B_3 &= -4 + (X+1)B_4 \\ B_2 &= 3 + (X+1)B_3 \end{aligned} \qquad \Bigg| \qquad \begin{aligned} B_1 &= 2 + (X+1)B_2 \\ B_0 &= 1 + (X+1)B_1 \\ B &= B_0 \end{aligned}$$

We have chosen decreasing indexes because they are more natural!

It is necessary to pay close attention to the change in context: we are not calculating here with *numbers* but with *polynomials*; we begin with the constant polynomial $B_4 = 7$, then we calculate the first degree polynomial B_3, etc.:

$$\begin{aligned} B_4 &= 7, \\ B_3 &= 3 + 7X, \\ B_2 &= 6 + 10X + 7X^2, \\ B_1 &= 8 + 16X + 17X^2 + 7X^3, \\ B_0 &= 9 + 24X + 33X^2 + 24X^3 + 7X^4. \end{aligned}$$

The adaptation of Horner's method to calculate $B(X) = A(X+h)$ is:

$$\begin{aligned} &B := a_n \; ; \\ &\textbf{for } i := n-1 \textbf{ downto } 0 \textbf{ do } B := a_i + (X+h) \cdot B \end{aligned} \tag{10.5}$$

We insist again: B is a *polynomial* that one modifies little by little, the operations addition and multiplication taking place in the ring of polynomials. To implement this algorithm, there are two possibilities.

- Translate algorithm (10.6) directly which leads to the code:

```
procedure translation(A : poly ; h : integer ; var B : poly) ;
var i, deg : integer ; temp : poly ; {returns B(X) = A(X + h)}
begin
  annul(B) ; deg := degree(A) ; B[0] := A[deg] ; {B = A[deg]}
  for i := deg - 1 downto 0 do begin
    annul(temp) ; temp[0] := h ; temp[1] := 1 ; {temp = X + h}
    mult_poly(B, temp, B) ; {B = temp · B}
    B[0] := B[0] + A[i]   {B = B + a_i}
  end
end ;
```

• We dwell a little further on algorithm (10.6) arguing that it is awkward to call a general procedure for multiplication of polynomials in order to multiply B by $X+h$. If we put

$$B^{(i)}(X) = b_0^{(i)} + b_1^{(i)} X + \cdots + b_n^{(i)} X^n,$$

the equality $B^{(i)} = a_i + (X+h)B^{(i+1)}$ gives

$$\begin{cases} b_n^{(i)} &= b_{n-1}^{(i+1)} + h\, b_n^{(i+1)}, \\ b_{n-1}^{(i)} &= b_{n-2}^{(i+1)} + h\, b_{n-1}^{(i+1)}, \\ \vdots & \\ b_1^{(i)} &= b_0^{(i+1)} + h\, b_1^{(i+1)}, \\ b_0^{(i)} &= a_i + h\, b_0^{(i+1)}. \end{cases} \tag{10.6}$$

If we agree to view the index i as representing time, we can consider $B^{(i)}$ as the *state* of the polynomial B at the instant i in a reverse count. If we use the formulas (10.6) in the order $b_n, \ldots, b_0$, we find that we can calculate the polynomial B on the spot which avoids using an array with two indexes:

```
procedure translation(A : poly ; h : integer ; var B : poly) ;
var i, k, deg : integer ;
begin
| annul(B) ; deg := degree(A) ;
| B[0] := A[deg] ;
| for i := deg - 1 downto 0 do begin
| | for k := deg downto 1 do B[k] := B[k - 1] + h * B[k] ;
| | B[0] := A[i] + h * B[0]
| end
end ;
```

This code is more compact and performs better than the preceding. However, it is totally incomprehensible unless accompanied by a description of how it was constructed. Moreover, it is very delicate to implement and very fragile (if one replaces the internal ‘downto’ loop with a ‘to’ loop, the calculations are completely false). This is the price one pays for resorting to a programming trick.

For beginners

A mathematician seldom resists this sort of of pleasure and indulges in all sorts of shortcuts at the outset. Nevertheless, experience shows that this attitude is a continual source of catastrophes and loss of time when one programs. One can never repeat enough: *to program is first of all to choose security*; tricks come later. First write a solid, “industrial” program which works; then one can fine tune it later by modifying certain procedures. Reserve the intellectual thrills for this time.

10.7.2. Composing polynomials

Let $A = a_0 + \cdots + a_n X^n$ and $B = b_0 + \cdots + b_m X^m$ be two polynomials. We want to calculate the polynomial $C = B \circ A$, that is:

$$C(X) = B\big(A(X)\big).$$

Suppose that we have $B = 1 + 2X + 3X^2 - 5x^3$ so that:

$$C(X) = 1 + 2A + 3A^2 - 5A^3.$$

This presentation suggests the following use of Horner's method:

$$\begin{aligned} C_3 &= -5, \\ C_2 &= 3 + A \cdot C_3, \\ C_1 &= 2 + A \cdot C_2, \\ C_0 &= 1 + A \cdot C_1, \end{aligned}$$

where, once again, the 'value' C_i is not a number but a polynomial, which means that the calculations take place in the ring of polynomials.

```
procedure composition_poly(A, B : poly ; var C : poly) ;
var i : integer ;                          {returns C = B ∘ A}
begin
 | if degree(A) * degree(B) > deg_max
 | then writeln('error : degree too high')
 | else begin
 | | annul(C) ; C[0] := B[degree(B)] ; {C = B[n]}
 | | for i := degree(B) − 1 downto 0 do begin
 | | | mult_poly(C, A, C) ; {C := C · B}
 | | | C[0] := C[0] + B[i]   {C := C + B[i]}
 | | end
 | end
end ;
```

10.8. Cyclotomic Polynomials

Let $n \geq 1$ be an integer and $\lambda = \mathrm{e}^{2i\pi/n}$. The *n-th cyclotomic polynomial* is the polynomial:

$$\Phi_n(X) \;=\; \prod_{\substack{1 \leq k \leq n \\ \mathrm{GCD}(k,n)=1}} (X - \lambda^k).$$

We then have

$$\Phi_n(X) = X^{\varphi(n)} + \cdots$$

where φ is the Euler phi-function. We shall show a little later that Φ_n is a polynomial with integer coefficients, which is not at all evident from the definition. While waiting, and to familiarize ourselves with the these objects, we calculate the first few cyclotomic polynomials from the definition.

- When $n = 1$, we have $\lambda = 1$ and $k = 1$, which gives:

$$\Phi_1(X) = X - 1.$$

- When $n = 2$, only $k = 1$ works, so that $\lambda = e^{i\pi} = -1$ and

$$\Phi_2(X) = X + 1.$$

- When $n = 3$, we have $\lambda = e^{2i\pi/3} = j$ and $k = 1, 2$ which gives:

$$\Phi_3(X) = (X - j)(X - j^2) = X^2 + X + 1.$$

With patience and a lot of care, one can calculate more cyclotomic polynomials. Happily, there are better ways.

10.8.1. First formula

Suppose, for simplicity, that $n = 12$. We have

$$X^{12} - 1 = \prod_{1 \leq k \leq 12} (X - e^{2i\pi k/12}).$$

The GCD of k and of 12 is one of the numbers $1, 2, 3, 4, 6, 12$. We partition $[\![1, 12]\!]$ as

$$[\![1, 12]\!] = I_1 \cup I_2 \cup I_3 \cup I_4 \cup I_6 \cup I_{12}$$

by placing in I_d the integers k which satisfy the condition $\mathrm{GCD}(k, 12) = d$ (in other words, the I_ℓ are the "level curves" of the function $k \mapsto \mathrm{GCD}(n, k)$). We can regroup the $(X - \lambda^k)$ as:

$$X^{12} - 1 = \prod_{d \mid 12} \prod_{k \in I_d} (X - e^{2i\pi k/12}).$$

If, for example, we examine the product associated to I_4, we recognize the cyclotomic polynomial

$$\Phi_3(X) = \prod_{k \in I_4} = (X - e^{2i\pi 4/12})(X - e^{2i\pi 8/12}) = (X - e^{2i\pi/3})(X - e^{4i\pi/3}).$$

The trick is simple: one musn't touch the $2i\pi$ when simplifying exponents. By proceeding the same way with the other subproducts, we encounter other cyclotomic polynomials and wind up with the formula:

$$X^{12} - 1 = \Phi_1(X)\Phi_2(X)\Phi_3(X)\Phi_4(X)\Phi_6(X)\Phi_{12}(X).$$

The generalization is immediate and gives the following result.

Theorem 10.8.1. *For all integers $n \geq 1$,*

$$X^n - 1 = \prod_{d \mid n} \Phi_d(X). \tag{10.7}$$

Corollary 10.8.1. *The cyclotomic polynomials have integer coefficients.*

Proof. We prove this by strong induction on n. First of all, the result is true if $n = 1$. Suppose that we have shown that the Φ_k have integral coefficients if $k < n$ and let $1 = d_1 < d_2 < \cdots d_k < n$ be the divisors of n which are strictly smaller than n. It follows from (10.7) that:

$$\Phi_n(X) = \frac{X^n - 1}{\Phi_{d_1}(X)\Phi_{d_2}(X)\cdots\Phi_{d_k}(X)}. \tag{10.8}$$

We conclude by remarking that Φ_n is the quotient of two polynomials with integral coefficients and the denominator is monic. □

Formula (10.8) is very valuable because it allows us to calculate Φ_n very rapidly a little at a time. We begin with $\Phi_1(X) = X - 1$.

- We now have without effort:

$$\begin{aligned}
\Phi_2(X) &= \frac{X^2-1}{\Phi_1(X)} = \frac{X^2-1}{X-1} = X + 1,\\
\Phi_3(X) &= \frac{X^3-1}{\Phi_1(X)} = \frac{X^3-1}{X-1} = X^2 + X + 1,\\
\Phi_4(X) &= \frac{X^4-1}{\Phi_1(X)\Phi_2(X)} = \frac{X^4-1}{X^2-1} = X^2 + 1.
\end{aligned}$$

- If p is a prime, the formula $\Phi_1\Phi_p = X^p - 1$ gives:

$$\Phi_p(X) = \frac{X^p - 1}{X - 1} = X^{p-1} + \cdots + X + 1.$$

Putting $Y = X^{p^{\ell-1}}$, one finds that $Y^p = X^{p^\ell}$ and induction on ℓ gives:

$$\Phi_{p^\ell}(X) = \frac{X^{p^\ell} - 1}{\Phi_1\Phi_p\Phi_{p^2}\cdots\Phi_{p^{\ell-1}}} = \frac{X^{p^\ell} - 1}{X^{p^{\ell-1}} - 1} = \frac{Y^p - 1}{Y - 1} = \Phi_p\big(X^{p^{\ell-1}}\big).$$

In a similar manner, if p does not divide m, one has:

$$\Phi_{pm}(X) = \frac{\Phi_m(X^p)}{\Phi_m(X)}.$$

If we want to calculate $\Phi_1, \Phi_2, \Phi_3, \ldots, \Phi_N$ in this order, then (10.8) suggests the algorithm:

```
Φ₁ := X − 1 ;
for ℓ := 2 to N do begin
| «calculate Prod := Φ_{d1} Φ_{d2}
| ... Φ_{dk}»
| Φ_ℓ := (X^ℓ − 1)/Prod
end
```

To find the divisors d_i, we sweep the interval $[\![1, \ell - 1]\!]$. We can limit the amplitude of the sweep by remarking that $\ell = dq$ and $d < \ell$ implies $d \leq \frac{1}{2}\ell$ since $q > 1$ means $q \geq 2$. The calculation of *Prod* can be effected as follows:

```
Prod := 1 ;
for d := 1 to ℓ div 2 do
    if ℓ mod d = 0 then Prod := Prod · Φ_d
```

Inserting this code into the preceding algorithm gives:

```
Φ₁ := X − 1 ;
for ℓ := 2 to N do begin
| Prod := 1 ;
| for d := 1 to ℓ div 2 do
|     if ℓ mod d = 0 then Prod := Prod · Φ_d ;
| Φ_ℓ := (X^ℓ − 1)/Prod
end
```

Exercise 1

Transform this algorithm into a Pascal program using the declaration:

```
type poly = array[0 .. deg_max] of integer ;
cyclotomic = array[0 .. deg_max] of poly ;
var Φ : cyclotomic ;
```

With this declaration, $\Phi[n]$ is the n-th cyclotomic polynomial.

10.8.2. Second formula

We know that the coefficients of Φ_p are equal to 1 when p is a prime number. If q is a prime distinct from p, one can prove that the coefficients of Φ_{pq} are equal to 0 or ± 1. This property also holds for the cyclotomic polynomials of index less than 105; in contrast, the coefficient of X^7 in Φ_{105} is equal to -2.

It is also possible to prove that there exist cyclotomic polynomials with arbitrarily large coefficients. The coefficients do not grow rapidly, however, since for $n < 385$ the coefficients of Φ_n are all less than or equal to 2 in absolute value.

If we want to inspect the results while calculating Φ_{105}, for example, the formula (10.8) is not so useful because it requires the calculation and storage of $\Phi_1, \ldots, \Phi_{104}$. Happily, the Mœbius inversion formula in multiplicative form tells us that

$$\Phi_n(X) = \prod_{d \mid n} (X^d - 1)^{\mu(n/d)}, \tag{10.9}$$

the product being taken over all divisors of n including the extremes 1 and n.

This formula immediately suggests an algorithm:

```
Num := 1 ; Den := 1 ;
for d := 1 to n do begin
  if n mod d = 0 then
  case μ(n div d) of
    +1 : Num := (X^d − 1) · Num ;
    −1 : Den := (X^d − 1) · Den ;
  end {case}
end ;
Φ_n := Num/Den
```

Exercise 2

Implement this algorithm.

• Beware, because the degree of the numerator or that of the denominator may exceed *deg_max*! Do not omit the error message 'degree too high' in the procedure for multiplication.

• Insert the calculation of the Euler function φ (see Chap. 8) into your program. This will at least allow you to inspect the degree of the displayed polynomial.

10.9. Lagrange Interpolation

Let $n \geq 1$ be an integer and consider $n+1$ points on the plane with distinct abcissas. Does there exsist a polynomial whose graph passes through these points?

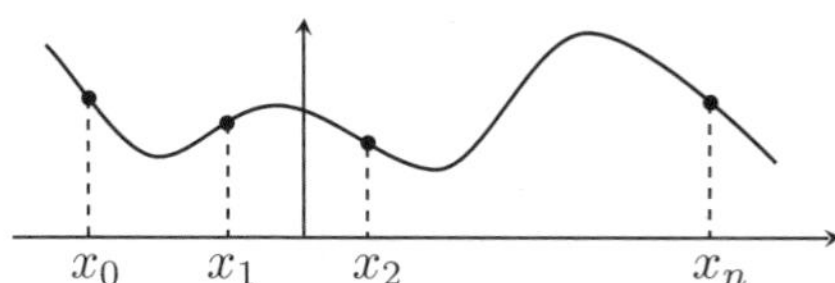

Theorem 10.9.1. *Let k be a commutative field, $x_0, x_1, \ldots, x_n$ distinct elements of k and $y_0, y_1, \ldots, y_n$ any elements of k. There exists a polynomial $A \in k[X]$, and only one, satisfying the conditions:*

$$\deg A \leq n \quad \textit{and} \quad A(x_i) = y_i \quad \textit{for} \quad i = 0, \ldots, n.$$

This unique polynomial is called the *Lagrange interpolating polynomial* associated to the data $x_0, \ldots, x_n$ and $y_0, \ldots, y_n$.

Proof. Consider the polynomial

$$\omega_i(X) = (X - x_0) \cdots (X - x_{i-1})(X - x_{i+1}) \cdots (X - x_n).$$

It is zero at $x_0, \ldots, x_{i-1}, x_{i+1}, \ldots, x_n$ and, in view of the absence of x_i, does not take the value 0 at x_i. Consequently, the polynomial

$$A(X) = \sum_{i=0}^{n} y_i \frac{\omega_i(X)}{\omega_i(x_i)} \tag{10.10}$$

is a solution. To prove uniqueness, suppose that A' and A'' are two solutions. Their difference $A'-A''$ vanishes at the $n+1$ points $x_0, \ldots, x_n$. Since the degree of the difference does not exceed n, it is necessarily the zero polynomial. □

Corollary 10.9.1. *Let $\omega(X) = (X - x_0)(X - x_1) \cdots (X - x_n)$. If A is the Lagrange interpolating polynomial, all the solutions of the system $P(x_i) = y_i$ for $i = 0, \ldots, n$ are given by the formula*

$$P = \omega \cdot Q + A, \quad Q \in k[X].$$

Proof. Divide P by ω to get

$$P = \omega \cdot Q + R, \quad \deg R < \deg \omega,$$

where the remainder satisfies the conditions $R(x_i) = P(x_i) = y_i$. Since its degree is less than or equal to n, we conclude that it is a Lagrange interpolating polynomial for A. □

Formula (10.10) is interesting. It tells us, for example, that the coefficients of the Lagrange interpolating polynomial are rational fractions, hence continuous functions in the x_i and y_i. It also leads to a (very clumsy) algorithm. Returning once again to the seventeenth century: Newton, who calculated without the knowledge or technique of Lagrange interpolating polynomials, used the following basis (now called the Newton basis) of the vector space of polynomials of degree $\leq n$:

$$\begin{aligned}
X^{(0)} &= 1, \\
X^{(1)} &= (X - x_0), \\
X^{(2)} &= (X - x_0)(X - x_1), \\
&\vdots \\
X^{(n)} &= (X - x_0)(X - x_1) \cdots (X - x_{n-1}).
\end{aligned}$$

Notice the absence of the monomial $X - x_n$ in this basis. If we put

$$A(X) = \alpha_0 + \alpha_1 X^{(1)} + \alpha_2 X^{(2)} + \cdots + \alpha_n X^{(n)},$$

the conditions $A(x_i) = y_i$ become:

$$\begin{aligned} y_0 &= \alpha_0, \\ y_1 &= \alpha_0 + \alpha_1(x_1 - x_0), \\ y_2 &= \alpha_0 + \alpha_1(x_2 - x_0) + \alpha_2(x_2 - x_0)(x_2 - x_1), \\ &\vdots \\ y_n &= \alpha_0 + \alpha_1(x_n - x_0) + \alpha_2(x_n - x_0)(x_n - x_1), \\ &\qquad + \cdots + \alpha_n(x_n - x_0)\cdots(x_n - x_{n-1}). \end{aligned}$$

The vector $(\alpha_0, \ldots, \alpha_n)$ is then the unique solution of the triangular Cramer system, which reproves the existence and uniqueness of the interpolating Lagrange polynomial. The solution of a triangular system is an exercise that we have already studied. Knowing that

$$\alpha_\ell = \frac{y_\ell - \alpha_0 - \alpha_1(x_\ell - x_0) - \cdots - \alpha_{\ell-1}(x_\ell - x_0)\cdots(x_\ell - x_{\ell-2})}{(x_\ell - x_0)\cdots(x_\ell - x_{\ell-1})}, \tag{10.11}$$

our first attempt at solving the system is:

$$\begin{aligned} &\alpha_0 := y_0 \; ; \\ &\textbf{for } \ell := 1 \textbf{ to } n \textbf{ do} \\ &\qquad \text{«}\textit{calculate } \alpha_\ell \textit{ using } (10.11)\text{»} \end{aligned} \tag{10.12}$$

When we try to program the numerator of formula (10.11) we encounter a difficulty

$$\begin{aligned} &S := y_\ell - \alpha_0 \; ; \\ &\textbf{for } k := 1 \textbf{ to } \ell - 1 \textbf{ do } S := S - \alpha_k * ??? \end{aligned}$$

because we must find *ourselves* the coefficients of the system using the x_i. If we delay addressing this problem by naming it

$$prod(\ell, k) = (x_\ell - x_0)\cdots(x_\ell - x_k), \tag{10.13}$$

the solution of the system becomes very simple:

$$\begin{aligned} &\alpha_0 := y_0 \; ; \\ &\textbf{for } \ell := 1 \textbf{ to } n \textbf{ do begin} \\ &\quad S := y_\ell - \alpha_0 \; ; \\ &\quad \textbf{for } k := 1 \textbf{ to } \ell - 1 \textbf{ do } S := S - \alpha_k * prod(\ell, k-1) \; ; \\ &\quad \alpha_\ell := S / prod(\ell, \ell - 1) \\ &\textbf{end} \end{aligned} \tag{10.14}$$

We only need to program the function *prod*, and this is mindless.

This code is certainly correct, but it is a Penelope code because the calculation of $prod(\ell, k)$ does not make use of that of $prod(\ell, k-1)$. However, we can "surf" on the wave of calculations using first order recurrences. Consider again the numerator of (10.11). To be certain to avoid stupidities, we first pass into trace mode using sequences:

$$\begin{aligned} S_0 &= y_\ell - \alpha_0, & P_1 &= x_\ell - x_0, \\ S_1 &= S_0 - \alpha_1 P_1, & P_2 &= P_1(x_\ell - x_1), \\ &\vdots & &\vdots \\ S_{\ell-1} &= S_0 - \alpha_{\ell-1} P_{\ell-1}, & P_\ell &= P_{\ell-1}(x_\ell - x_{\ell-1}). \end{aligned}$$

Next we get rid of the time index:

```
S := y_ℓ − α_0 ; P := x_ℓ − x_0 ;
for k := 1 to ℓ − 1 do begin
  | S := S − α_k ∗ P ;
  | P := (x_ℓ − x_k) ∗ P
end ;
α_ℓ := S/P
```

If we carry this code into (10.12) we obtain the following very nice algorithm:

```
α_0 := y_0 ;
for ℓ := 1 to n do begin
  | S := y_ℓ − α_0 ; P := x_ℓ − x_0 ;
  | for k := 1 to ℓ − 1 do begin
  |   | S := S − α_k ∗ P ; P := (x_ℓ − x_k) ∗ P                (10.15)
  | end ;
  | α_ℓ := S/P
end ;
```

Exercise 3

Transform algorithm (10.15) into a procedure (suppose that the data are real numbers). Then rewrite the procedure to work over $\mathbb{Q}$ supposing that the data are rational numbers.

Remark

The algorithm above is not the only one possible. If we put to simplfy

$$\pi_i = x_\ell - x_i,$$

and if we effect the division in (10.11) right away we obtain:

$$\alpha_\ell = \frac{y_\ell}{\pi_0\pi_1\cdots\pi_{\ell-1}} - \frac{\alpha_0}{\pi_0\pi_1\cdots\pi_{\ell-1}} - \frac{\alpha_1}{\pi_1\cdots\pi_{\ell-1}} - \frac{\alpha_2}{\pi_2\cdots\pi_{\ell-1}} - \cdots - \frac{\alpha_{\ell-1}}{\pi_{\ell-1}}.$$

This presentation suggests that we use Horner's method, which leads to the introduction of the sequence:

$$\begin{aligned}
S_0 &= y_\ell, \\
S_1 &= \frac{y_\ell - \alpha_0}{\pi_0} = \frac{S_0 - \alpha_0}{\pi_0}, \\
S_2 &= \frac{y_\ell - \alpha_0}{\pi_0 \pi_1} - \frac{\alpha_1}{\pi_1} = \frac{S_1 - \alpha_1}{\pi_1}, \\
S_3 &= \frac{y_\ell - \alpha_0}{\pi_0 \pi_1 \pi_2} - \frac{\alpha_1}{\pi_1 \pi_2} - \frac{\alpha_2}{\pi_2} = \frac{S_2 - \alpha_2}{\pi_2} \\
&\vdots \\
S_\ell &= \frac{S_{\ell-1} - \alpha_{\ell-1}}{\pi_\ell} = \alpha_\ell.
\end{aligned}$$

We obtain the celebrated method of *divided differences*:

```
α₀ := y₀ ;
for ℓ := 1 to n do begin
  | S := y_ℓ ;
  | for i := 0 to ℓ − 1 do S := (S − α_i)/(x_ℓ − x_i) ;
  | α_ℓ := S
end
```

The code is more compact. Nonetheless, it is slower and less precise because it contains many divisions which make it numerically unstable. Note finally that one can incorporate the statement $\alpha_0 := y_0$ into the loop by beginning the loop at $\ell = 0$.

10.10. Basis Change

When we calculate the Lagrange interpolating polynomial *à la Newton*, we obtain coordinates with respect to the Newton basis. However, we often need to know the coordinates in the canonical basis. Suppose, then, that we know the coefficients α_i of the polynomial

$$A(X) = \alpha_0 + \alpha_1 X^{(1)} + \alpha_2 X^{(2)} + \cdots + \alpha_n X^{(n)},$$

and that we want to calculate the coordinates a_i in the canonical basis

$$A(X) = a_0 + a_1 X + \cdots + a_n X^n.$$

Let us first consider the example:

$$A(X) = 2 + 3(X-2) - 4(X-2)(X-3) + 7(X-2)(X-3)(X-5).$$

Horner's method allows us to view $A(X)$ as the last term of a reverse sequence where the calculations take place in the polynomial ring:

$$\begin{aligned}
A_3 &= 7 &&= 7,\\
A_2 &= -4 + (X-5)A_3 &&= -39 + 7X,\\
A_1 &= 3 + (X-3)A_2 &&= 120 - 60X + 7X^2,\\
A_0 &= 2 + (X-2)A_1 &&= -238 + 240X + 74X^2 + 7X^3.
\end{aligned}$$

More generally, $A(X)$ is the last term of a reverse sequence

$$\begin{aligned}
A_n &= \alpha_n,\\
A_{n-1} &= \alpha_{n-1} + (X - x_{n-1})A_n,\\
A_{n-2} &= \alpha_{n-2} + (X - x_{n-2})A_{n-1},\\
&\vdots\\
A_0 &= \alpha_0 + (X - x_0)A_1.
\end{aligned}$$

If we view i as representing time, A_i becomes the *state* of the polynomial A at the instant i during a reverse count:

```
A := αn ;
for i := n − 1 downto 0 do A := αi + (X − xi)A
```

To translate this algorithm to Pascal, we can:

- use the general procedures for manipulation of polynomials:

```
procedure Newton_to_canonical_basis(α : poly ; x : data ; var A : poly) ;
var i, k : integer ; U : poly ;
begin
  annul(A) ; A[0] := α[n] ;                    {A := αn}
  for i := n − 1 downto 0 do begin
    annul(U) ; U[1] := 1 ; U[0] := −x[i] ;     {U := X − xi}
    mult_poly(A, U, A) ;                       {A := U · A}
    A[0] := A[0] + α[i]                        {A := A + αi}
  end
end ;
```

- Pull the algorithm apart a bit more and amuse ourselves once again with sequences. If we put

$$A_i(X) = a_0^{(i)} + a_1^{(i)}X + \cdots + a_n^{(i)}X^n,$$

the equation $A_i = \alpha_i + (X - x_i)A_{i+1}$ gives

$$\begin{cases} a_n^{(i)} &= a_{n-1}^{(i+1)} - x_i\, a_n^{(i+1)}, \\ a_{n-1}^{(i)} &= a_{n-2}^{(i+1)} - x_i\, a_{n-1}^{(i+1)}, \\ &\vdots \\ a_1^{(i)} &= a_0^{(i+1)} - x_i\, a_1^{(i+1)}, \\ a_0^{(i)} &= \alpha_i - x_i\, a_0^{(i+1)}. \end{cases} \tag{10.16}$$

If we use (10.16) to calculate $a_n, \ldots, a_0$ *in this order*, we can do the calculations on site *i.e.* 'staying inside' the vector $A = (a_0, \ldots, a_n)$:

```
procedure Newton_to_canonical_basis(α : poly ; x : data ; var A : poly) ;
var i, k : integer ;
begin
 | annul(A) ; A[0] := α[n] ;
 | for i := n − 1 downto 0 do begin
 | | for k := n downto 1 do A[k] := A[k − 1] − x[i] ∗ A[k] ;
 | | A[0] := α[i] − x[i] ∗ A[0]
 | end
end ;
```

The calculation of (10.16) from bottom to top provokes a catastrophe. Why?

Exercise 4

Find another algorithm for changing basis using the sequences

$$B_k = a_0 + a_1 P_1 + \cdots + a_k P_k, \qquad P_k = (X - x_0)\cdots(X - x_{k-1}).$$

10.11. Differentiation and Discrete Taylor Formulas

To simplify the exposition, we suppose henceforth that the interpolation points x_i are the integers $0, 1, \ldots, n$.

We know that the Lagrange interpolating polynomial defined by the conditions $P(x_i) = y_i$ has degree $\leq n$. The maximum degree is not necessarily attained: if, for example, $y_i = ax_i + b$ with $a \neq 0$, the degree is equal to 1. Is it possible to determine this degree *in advance*?

10.11.1. Discrete differentiation

Consider the linear map $\Delta : \mathbb{R}^{n+1} \to \mathbb{R}^n$ defined by

$$\Delta(y_0, y_1, \ldots, y_n) = (y_1 - y_0, y_2 - y_1, \ldots, y_n - y_{n-1}).$$

We call it *discrete differentiation.* Since the dimension of a vector decreases by one after each differentiation, we can differentiate a vector at most n times:

k	y_k	Δy_k	$\Delta^2 y_k$	$\Delta^3 y_k$	$\Delta^4 y_k$
0	2	3	4	−17	34
1	5	7	−13	17	
2	12	−6	4		
3	6	−2			
4	4				

If A is a polynomial and if we have $y_i = A(i)$ for $x = 0, 1, \ldots, n$, the derivatives of order greater than $\deg A$ are zero.

k	$y_k = A(k)$	Δy_k	$\Delta^2 y_k$	$\Delta^3 y_k$	$\Delta^4 y_k$
0	3	0	14	6	0
1	3	14	20	6	
2	17	34	26		
3	51	60			
4	111				

Discrete derivatives of the values of $A(X) = X^3 + 4X^2 - 5X + 3$

To explain this phenomenon, define a discrete derivative $\Delta : \mathbb{R}[X] \to \mathbb{R}[X]$ on the polynomial ring by putting

$$\Delta A(X) = A(X+1) - A(X).$$

If we evaluate the polynomial $\Delta A(X)$ at the point x, we get:

$$(\Delta A)(x) = A(x+1) - A(x).$$

Since $(\Delta A)(x)$ is the discrete derivative of the vector $(A(x), A(x+1)) \in \mathbb{R}^2$, we see that that the number $\Delta^k A(x)$ can be interpreted in two ways:

• We differentiate k times the polynomial $A(X)$, then take the value at x of the resulting polynomial.

• We differentiate k times the vector $(A(x), A(x+1), \ldots, A(x+k))$ of values of $A(X)$ at the points $x, x+1, \ldots, x+k$.

The first interpretation shows that $\Delta^k(A) = 0$ since $k > \deg A$.

Knowing that the Newton basis associated to the integers $0, \ldots, n$ is

$$X^{(0)} = 1,\ X^{(1)} = X,\ \ldots,\ X^{(n-1)} = X(X-1)\cdots X(n+1),$$

an immediate calculation shows that for $k > 0$,

$$\begin{aligned}\Delta X^{(k)} &= (X+1)^{(k)} - X^{(k)} \\ &= (X+1)\cdots(X+k-1)\big[X+k-X\big] \\ &= k\,X^{(k-1)}.\end{aligned}$$

Theorem 10.11.1. *The following discrete Taylor formula holds for each polynomial $A \in \mathbb{R}[X]$:*

$$A(X) = A(0) + \frac{\Delta A(0)}{1!}X^{(1)} + \frac{\Delta^2 A(0)}{2!}X^{(2)} + \cdots + \frac{\Delta^n A(0)}{n\,!}X^{(n)}.$$

Proof. Expand A in the Newton basis to get

$$A(X) = \alpha_0 + \alpha_1 X^{(1)} + \alpha_2 X^{(2)} + \cdots + \alpha_n X^{(n)}.$$

Then take the discrete derivative n times in succession:

$$\begin{aligned}A(X) &= \alpha_0 + \alpha_1 X^{(1)} + \alpha_2 X^{(2)} + \cdots + \alpha_n X^{(n)}, \\ \Delta A(X) &= \alpha_1 + 2\,\alpha_2 X^{(1)} + \cdots + n\,\alpha_n X^{(n-1)}, \\ \Delta^2 A(x) &= 2\,\alpha_2 + 3\cdot 2\,\alpha_3 X^{(1)} + \cdots + n(n-1)\alpha_n X^{(n-2)}, \\ &\vdots \\ \Delta^n A(X) &= n\,!\,\alpha_n.\end{aligned}$$

Upon putting $X^{(1)} = X = 0$ in these equations, we get $k\,!\,\alpha_k = \Delta^k A(0)$. □

Corollary 10.11.1. *Let $A = \alpha_0 + \alpha_1 X^{(1)} + \alpha_2 X^{(2)} + \cdots + \alpha_n X^{(n)}$ be a polynomial with real coefficients. Then:*

- $\alpha_k = \Delta^{(k)} A(0)/k\,!$;
- *the degree of $A \neq 0$ is the largest exponent k such that $\Delta^{(k)} A(0) \neq 0$;*
- *the coefficients of A are integers if and only if $\alpha_0, \ldots, \alpha_n$ are integers.*

Proof. The first two assertions follow directly from the discrete Taylor formula. To establish the third, note that if $\alpha_0, \ldots, \alpha_n$ are integers, then A clearly has integer coefficients. Conversely, suppose that A has integer coefficients and degree d:

$$A = a_0 + \cdots + a_d X^d, \quad a_d \neq 0.$$

We already know that $\alpha_i = a_i = 0$ for $i > d$. Comparing the monomials of d, we obtain $\alpha_d = a_d \in \mathbb{Z}$. We complete the proof by induction on degree. □

We can now determine the degree of the Lagrange interpolating polynomial. Suppose that we are given $x = (0, 1, 2, 3, 4, 5)$ and $y = (7, 4, 5, 10, 19, 32)$. The polynomial determined by the coefficients $A(x_i) = y_i$ for $i = 0, \ldots, 5$ has degree 2 because $\Delta^2 y \neq 0$ and $\Delta^3 y = 0$.

Remark

We have the following pretty corollary:

If $A \in \mathbb{R}[X]$, then $A(\mathbb{Z}) \subset \mathbb{Z}$ if and only if the $\Delta^{(k)}A(0)$ are all integers.

In effect, if A takes integer values on the integers, so does ΔA, and it follows by induction that the $\Delta^{(k)}A(0)$ are integers. Conversely, in the discrete Taylor formula, the $X^{(k)}/k!$ are polynomials with integral values on the integers (their values are 0 or a binomial coefficient).

10.12. Newton-Girard Formulas

Recall that $P(X_1, \ldots, X_n)$ is a *symmetric polynomial* if, for every permutation $s \in \mathfrak{S}_n$ of the indices,

$$P\big(X_{s(1)}, \ldots, X_{s(n)}\big) = P(X_1, \ldots, X_n).$$

The symmetric polynomials

$$\sigma_1 = \sum_{1 \le i \le n} X_i\,, \quad \sigma_2 = \sum_{1 \le i < j \le n} X_i X_j\,, \ldots, \quad \sigma_n = X_1 \cdots X_n$$

are called the *elementary symmetric functions* in the variables $X_1, \ldots, X_n$.

Theorem 10.12.1. *Let $P(X_1, \ldots, X_n)$ be a symmetric polynomial with coefficients in a ring k and $\sigma_1, \ldots, \sigma_n$ the elementary symmetric functions of the X_i. There exists a polynomial $Q(X_1, \ldots, X_n)$ with coefficients in the same ring k such that*

$$P(X_1, \ldots, X_n) = Q(\sigma_1, \ldots, \sigma_n).$$

We use this theorem, which is not difficult to prove. Note especially that P and Q have coefficients in the same ring, so, for example, $P \in \mathbb{Z}[X]$ implies $Q \in \mathbb{Z}[X]$.

Theorem 10.12.2 (Newton-Girard formulas). *Consider the Newton sums*

$$S_k(X_1, \ldots, X_n) = X_1^k + \cdots X_n^k, \quad k \ge 0,$$

which are manifestly symmetric in the X_i.

- *If $1 \le k \le n$,*

$$\begin{aligned}
&S_1 - \sigma_1 = 0,\\
&S_2 - S_1\sigma_1 + 2\sigma_2 = 0,\\
&S_3 - S_2\sigma_1 + S_1\sigma_2 - 3\sigma_3 = 0,\\
&\vdots\\
&S_n - S_{n-1}\sigma_1 + \cdots + (-1)^{n-1}S_1\sigma_{n-1} + (-1)^n n\sigma_n = 0;
\end{aligned}$$

- *If $k > n$, put $k = n + i$ with $i > 0$. Then*

$$S_{n+i} - S_{n+i-1}\sigma_1 + \cdots + (-1)^n S_i \sigma_n = 0.$$

Proof. We are going to use *formal series*, that is, series in which we are not concerned with convergence but only with algebraic operations on them.

For simplicity, suppose that $n = 3$. Consider the following polynomial in the indeterminates X_1, X_2, X_3 and T:

$$\varphi(T) = (1 - X_1 T)(1 - X_2 T)(1 - X_3 T). \tag{10.17}$$

We differentiate $\varphi(T)$ with respect to T in two ways.

- First expand (10.17), then differentiate to get:

$$\varphi'(T) = -\sigma_1 + 2\sigma_2 T - 3\sigma_3 T^2. \tag{10.18}$$

- Differentiate (10.17) directly to get:

$$\varphi'(T) = -\left\{\frac{X_1}{1 - X_1 T} + \frac{X_2}{1 - X_2 T} + \frac{X_3}{1 - X_3 T}\right\}\varphi(T). \tag{10.19}$$

Since we have the identity

$$\frac{1}{1 - XT} = 1 + XT + X^2T^2 + X^3T^3 + X^4T^4 + \cdots,$$

we can rewrite (10.19) as:

$$\begin{aligned}\varphi'(T) = \; & -\big(X_1 + X_1^2 T + X_1^3 T^2 + X_1^4 T^3 + \cdots\big)\varphi(T) \\ & -\big(X_2 + X_2^2 T + X_3^3 T^2 + X_2^4 T^3 + \cdots\big)\varphi(T) \\ & -\big(X_3 + X_3^2 T + X_3^3 T^2 + X_3^4 T^3 + \cdots\big)\varphi(T).\end{aligned}$$

Upon multiplying by T and regrouping vertically, we get:

$$T\varphi'(T) + \big(S_1 T + S_2 T^2 + S_3 T^3 + \cdots\big)\varphi(T) = 0.$$

Now replace φ' by (10.18) and φ by the expansion in (10.17) to get

$$\begin{aligned}& -\sigma_1 T + 2\sigma_2 T^2 - 3\sigma_3 T^3 \\ & \quad + (S_1 T + S_2 T^2 + S_3 T^3 + \cdots)(1 - \sigma_1 T + \sigma_2 T^2 - \sigma_3 T^3) = 0.\end{aligned}$$

It remains only to note that the coefficients of the T^k are zero to obtain the desired formulas. □

10.13. Stable Polynomials

Definition 10.13.1. *We say that a polynomial P with real coeficients is stable if all its zeroes belong to the half-plane* $\operatorname{Re} z < 0$.

This definition is essential for an engineer. When we design a wing of a plane, a shock absorber on a car, or an electric circuit, we want to know how to be certain that their oscillations decrease rapidly *no matter what* the initial conditions. In nice cases, one can show that the oscillations occur among the solutions of a differential equation with constant coefficients, say

$$y^{(n)} + a_1 y^{(n-1)} + \cdots + a_{n-1} y' + a_n y = 0. \tag{10.20}$$

Let $P(X) = X^n + a_1 X^{n-1} + \cdots + a_{n-1} X + a_n$ be the *characteristic equation* of the equation and $\alpha_k + i\beta_k$ its zeroes, so that the solutions of (10.20) are of the form

$$\sum p_k(t)\, \mathrm{e}^{(\alpha_k + i\beta_k)t} = \sum p_k(t)\big[\cos(\beta_k t) + i \sin(\beta_k t)\big]\mathrm{e}^{\alpha_k t},$$

where the $p_k(t)$ are polynomials. If the characteristic equation is stable, that is, if all α_k are less than 0, we can be certain that *all* the solutions are damped sinusoids which die as t tends to $+\infty$.

An immediate consequence of the definition of stability is the following.

Proposition 10.13.1. *If A and B are two polynomials with real coefficients, then*

$$A \text{ and } B \text{ are stable} \iff A \cdot B \text{ is stable.}$$

Recall that a polynomial with real coefficients is a product of irreducible polynomials of first and second degree:

- The polynomial $X + a$ is stable if and only if $a > 0$.
- The *irreducible* polynomial $X^2 + aX + b$ is stable if and only if both a and b are positive since the roots are $\frac{1}{2}(-a \pm i\sqrt{-\Delta})$ with $\Delta = a^2 - 4b < 0$.

If we combine these two remarks and the proposition, we see that

- the coefficients of a stable polynomial with real coefficients are either all positive, or all negative;
- the values of a stable polynomial at x are never zero for $x \geq 0$ and have the same sign as the coefficients of the polynomial.

Remark

The condition ‘all coefficients are positive’ *is not* sufficient for stability. For example, $x^3 + x^2 + x + 1$ is not stable because i is a root.

Is it possible to decide if a polynomial is stable without knowing the roots? The classical response to this celebrated problem uses Routh-Hurwitz determinants. In what follows, we present another method and program.

Theorem 10.13.1 (Sh. Strelitz [2]***).*** *Let $A = X^n + a_1X^{n-1} + \cdots + a_n$ be a polynomial with real coefficients and roots $\alpha_1, \ldots, \alpha_n$ and let*

$$B = X^m + b_1X^{m-1} + \cdots + b_m$$

be a monic polynomial with real coefficients, degree $m = \frac{1}{2}n(n-1)$, and roots $\alpha_i + \alpha_j$ for $1 \le i < j \le n$. Then:

$$A \text{ is stable} \iff \text{all coefficients of } A \text{ and } B \text{ are positive.}$$

Proof. Suppose first that A is stable. Since the zeros of A are in the half-plane $\operatorname{Re} z < 0$, we have $\operatorname{Re}(\alpha_i + \alpha_j) < 0$, which shows that B is stable with positive coefficients because it is monic.

Conversely, if A has positive coefficients, the real zeroes of A satisfy the condition $x < 0$. Let $\alpha = x + iy$ be a non-real zero of A, so that $y \neq 0$. Since $\overline{\alpha}$ is a zero of A different from α, we know that $\alpha + \overline{\alpha} = 2x$ is a real zero of B satisfying $x < 0$ since B has positive coefficients. □

Examples

- If $A = X^3 + aX^2 + bX + c$, the degree of B is $m = 3$ and

$$B = X^3 + 2aX^2 + (a^2 + b)X + (ab - c).$$

The polynomial A is therefore stable if and only if $a, b, c > 0$ and $ab > c$.

- If $A = X^4 + aX^3 + bX^2 + cX + d$, the degree of B is $m = 6$ and with much patience, one finds that:

$$B = X^6 + 3aX^5 + (3a^2 + 2b)X^4 + (a^3 + 4ab)X^3 + (2a^2b + b^2 + ac - 4d)X^2 \\ + (a^2c + ab^2 - 4ad)X + (abc - a^2d - c^2).$$

More generally, if we know how to express the coefficients of B using those of A without calculating the roots of A, we will obtain a test that tells us whether or not A is stable. Let S_k be the Newton sums of the roots of A:

$$S_k = \alpha_1^k + \cdots + \alpha_n^k.$$

[2] *On the Routh-Hurwitz Problem*, American Math. Monthly 84 (1977), pp. 542–544.

Knowing that we have $a_i = (-1)^i \sigma_i(\alpha_1, \dots, \alpha_n)$, the Newton-Girard formulas for the polynomial A are, for $k \le n$,

$$\begin{cases} S_1 + a_1 = 0, \\ S_2 + S_1 a_1 + 2a_2 = 0, \\ S_3 + S_2 a_1 + S_1 a_2 + 3a_3 = 0, \\ \vdots \\ S_n + S_{n-1} a_1 + \cdots + S_1 a_{n-1} + n a_n = 0, \end{cases} \tag{10.21}$$

and, when $k = n + i$ exceeds n:

$$S_{n+i} + S_{n+i-1} a_1 + \cdots + S_i a_n = 0, \quad i > 0. \tag{10.22}$$

Now consider the Newton sums T_ℓ of the roots of B:

$$T_\ell = \sum_{1 \le i < j \le n} (\alpha_i + \alpha_j)^\ell.$$

For $\ell \le m$, the Newton-Girard formulas of the polynomial B are:

$$\begin{cases} T_1 + b_1 = 0, \\ T_2 + T_1 b_1 + 2b_2 = 0, \\ T_3 + T_2 b_1 + T_1 b_2 + 3b_3 = 0, \\ \vdots \\ T_m + T_{m-1} b_1 + \cdots + T_1 b_{m-1} + m b_m = 0, \end{cases} \tag{10.23}$$

The formulas (10.21), (10.22) and (10.23) are linked by the following result.

Proposition 10.13.2. *For all $\ell \ge 1$:*

$$T_\ell = \tfrac{1}{2}\Big\{-2^\ell S_\ell + \sum_{i=0}^{\ell} C_\ell^i \, S_i \, S_{\ell-i}\Big\}. \tag{10.24}$$

(Recall that $S_0 = n$ by definition.)

Proof. If we expand the sum

$$(\mathrm{e}^{tz_1} + \cdots + \mathrm{e}^{tz_n})^2 = \mathrm{e}^{2tz_1} + \cdots + \mathrm{e}^{2tz_n} + 2 \sum_{1 \le i < j \le n} \mathrm{e}^{t(z_i + z_j)},$$

as a (formal) series, we obtain:

$$\Big\{\sum_{i=0}^{\infty} \frac{S_i}{i!} t^i\Big\}^2 = \sum_{i=0}^{\infty} \frac{2^i S_i}{i!} t^i + 2 \sum_{1 \le i < j \le n} \frac{T_i}{i!} t^i.$$

It suffices to equate the coefficients of t^ℓ. □

We can find the b_j in terms of the a_i by using the scheme:

$$(a_i) \xrightarrow{(10.21) \text{ and } (10.22)} (S_k) \xrightarrow{(10.24)} (T_\ell) \xrightarrow{(10.23)} (b_j). \tag{10.25}$$

If the a_i are integers, this scheme shows that the b_j are rational. Must we use the *rational* type to program the calculation of the b_j? Happily, we have the following result.

Lemma 10.13.1. *The coeffcients b_j are symmetric polynomials with integer coefficients in the variables a_i.*

Proof. We have $b_j = (-1)^j \overline{\sigma}_j$, where $\overline{\sigma}_j$ is the j-th elementary function in the $m = \frac{1}{2}n(n-1)$ variables $\alpha_i + \alpha_j$. Consequently, $\overline{\sigma}_j$ is a symmetric polynomial with integer coefficients in the n variables $\alpha_1, \ldots, \alpha_n$. Thus, b_j is also a polynomial with integer coefficients in the variables $a_i = (-1)^i \sigma_i(\alpha_1, \ldots, \alpha_n)$. □

The declarations of the program

Since we are not doing a polynomial operation (addition, multiplication, or division), we prefer to define a type *coefficient*. We will also need Newton sums.

```
program stable_polynomial ;
const deg_max = 10 ;
type coeff = array[1..deg_max] of integer ;
Newton_sum = array[0..deg_max] of integer ;
var a, b : coeff ; S, T : Newton_sum ; m, n : integer ;
```

Attention: with *degre_max* = 10, we can only test polynomials A of degree ≤ 5 since $C_5^2 = 10$. The maximum degree here is that of B, not of A!

The main body of the program

This follows word for word the scheme (10.26).

```
begin
| message ; choose(a, n) ;
| m := (n * (n - 1)) div 2 ;
| coeff_to_Newton_sum(a, S, n, m) ;
| change_Newton_sum(S, T, m) ;
| Newton_sum_to_coeff(T, b, m) ;
| display(b, m)
end .
```

The procedure coeff_to_Newton_sum

We want to implement (10.21) and (10.22); that is,

for $k := 1$ **to** n **do** «*calcultate* S_k *using*(10.21)» ;
for $k := n+1$ **to** m **do** «*calcultate* S_k *using*(10.22)» ;

the second loop picking up the results of the first. Being a little more explicit, we have:

$S_1 := -a_1$;
for $k := 2$ **to** n **do** «$S_k := -(S_{k-1}a_1 + \cdots + S_1 a_{k-1} + ka_k)$» ;
for $k := n+1$ **to** m **do** «$S_k := -(S_{k-1}a_1 + \cdots + S_1 a_{k-1})$»

Translation into Pascal is now a formality. It is worth taking the opportunity to initialize S_0, an indispensable precaution because (10.24), which expresses the T_j in terms of the S_i, explicitly involves S_0.

```
procedure coeff_to_Newton_sum(a : coeff ; var S : Newton_sum ;
                                                n, m : integer) ;
var i, k, temp : integer ;
begin
  S[0] := n ; {do not forget!}
  S[1] := -a[1] ;
  for k := 2 to n do begin
    temp := 0 ;
    for i := 1 to k - 1 do temp := temp + S[k - i] * a[i] ;
    S[k] := -(temp + k * a[k])
  end ;
  for k := n + 1 to m do begin
    temp := 0 ;
    for i := 1 to k - 1 do temp := temp + S[k - i] * a[i] ;
    S[k] := -temp
  end
end ;
```

The procedure change_Newton_sum

A first translation of the system (10.23) gives:

for $\ell := 1$ **to** m **do**
«$T_\ell := \frac{1}{2}(S_0 S_\ell + C_\ell^1 S_1 S_{\ell-1} + C_\ell^2 S_2 S_{\ell-2} + \cdots + C_\ell^\ell S_\ell S_0 - 2^\ell S_\ell)$»

We turn once more to the classical calculation of a sum which we turn into a loop:

temp $:= S_0 S_\ell$;
for $i := 1$ **to** ℓ **do** *temp* := *temp* $+ C_\ell^i S_i S_{\ell-i}$;
$T_\ell := \frac{1}{2}(temp - 2^\ell S_\ell)$

We could have written our procedure by programming the function $(i, \ell) \mapsto C_\ell^i$ which could calculate *any* binomial coefficient.[3] But this would be sloppy because we do not need the complete Pascal triangle: the line with the number ℓ suffices for our needs.

Once again, we try to surf on the wave of the calculations by seeking a recurrence relation relating C_ℓ^i and C_ℓ^{i-1}:

$$C_\ell^i = \frac{\ell(\ell-1)\cdots(\ell-i+1)}{1 \times 2 \times \cdots \times (i-1) \times i} = C_\ell^{i-1}\frac{(\ell-i+1)}{i}, \quad i \geq 1.$$

We benefit from the internal loop which increments i to insert this recurrence so as to determine the binomial coefficients as we go along:

```
C := 1 ; temp := S_0 S_ℓ ;
for i := 1 to ℓ do begin
|  C := C · (ℓ − i + 1)/i ; {now, C = C_ℓ^i}
|  temp := temp + C S_i S_{ℓ−i}
end ;
T_ℓ := ½(temp − 2^ℓ S_ℓ)
```

Similarly, to avoid programming the function $\ell \mapsto 2^\ell$, we introduce the recurrence $2^\ell = 2 \times 2^{\ell-1}$ in the external loop that increments ℓ:

```
P := 1 ;
for ℓ := 1 to m do begin
|  P := 2P ;                          {now, P = 2^ℓ}
|  C := 1 ; temp := S_0 S_ℓ ;
|  for i := 1 to ℓ do begin
|  |  C := C · (ℓ − i + 1)/i ; {now, C = C_ℓ^i}
|  |  temp := temp + C S_i S_{ℓ−i}
|  end ;
|  T_ℓ := ½(temp − P S_ℓ)
end
```

A little Pascal packaging where "/" becomes as usual "div" and our procedure is ready!

```
procedure change_Newton_sum(S : Newton_sum ;
                              var T : Newton_sum ; m : integer) ;
var i, ℓ, C, P, temp : integer ;
begin
|  P := 1 ;
|  for ℓ := 1 to m do begin
|  |  C := 1 ;
```

[3] For beginners: in view of the limits on integers in Pascal, the *worst* way to program this function would be to use the factorial function and the formula $C_\ell^i = \ell!/i!(\ell-i)!$. Formula $C_\ell^i = \ell(\ell-1)\cdots(\ell-i+1)/i!$ is better.

```
        P := 2 * P ;                          {now, P = 2^ℓ}
        temp := S[0] * S[ℓ] ;
        for i := 1 to ℓ do begin
          C := (C * (ℓ - i + 1)) div i ;  {now, C = C^i_ℓ}
          temp := temp + C * S[i] * S[ℓ - i]
        end ;
        T[ℓ] := (temp - P * S[ℓ]) div 2
      end
    end ;
```

The procedure Newton_sum_to_coeff

Writing this procedure is entirely similar to the procedure *coeff_to_Newton_sum*.

```
procedure Newton_sum_to_coeff(S : Newton_sum ;
                              var b : coeff ; m : integer) ;
var i, ℓ, temp : integer ;
begin
  b[1] := -T[1] ;
  for ℓ := 2 to m do begin
    temp := T[ℓ] ;
    for i := 1 to ℓ - 1 do temp := temp + T[ℓ - i] * b[i] ;
    b[ℓ] := -temp div i
  end
end ;
```

10.14. Factoring a Polynomial with Integral Coefficients

Let $P \in \mathbb{Z}[X]$ be a nonconstant polynomial whose coefficients are integers. We want to find all decompositions of P (if such exist) as a product of two nonconstant polynomials with integer coefficients. More precisely, we seek two polynomials $A, B \in \mathbb{Z}[X]$ satisfying the conditions:

$$P = AB, \quad \deg A > 0, \quad \deg B > 0.$$

10.14.1. Why integer (instead of rational) coefficients?

If P is irreducible in $\mathbb{Z}[X]$, there is no point seeking a factorization in $\mathbb{Q}[X]$ because the result is the same.

Theorem 10.14.1. *If U, V are two polynomials with rational coefficients such that UV has integral oefficents, then there exists a rational number $\kappa \neq 0$ such κU and $\kappa^{-1} V$ both have integral coefficients.*

To show this result, we introduce the following tool.

Definition 10.14.1. *The content of a polynomial $P \in \mathbb{Z}[X]$, denoted cont(P), is the largest integer $\kappa \geq 1$ such that $\kappa^{-1}P$ also has integral coefficients.*

It follows immediately from the definition that the content is the GCD of the coefficients of P and that that $\kappa^{-1}P$ has content equal to 1.

Lemma 10.14.1 (Gauss's lemma for contents). *If A and B are two polynomials with integer coefficients, then* $\operatorname{cont}(AB) = \operatorname{cont}(A) \cdot \operatorname{cont}(B)$.

Proof. Put $C = AB$. If α and β are the contents of A and B, respectively, the equation

$$C = \alpha\beta(\alpha^{-1}A)(\beta^{-1}B),$$

already shows that the content of C is a multiple of $\alpha\beta$. To finish, it suffices to show that the content of $(\alpha^{-1}A)(\beta^{-1}B)$ is equal to 1, which we reduce to proving the lemma in the case when the contents of A and B are equal to 1.

Put $A = \sum a_i X^i$, $B = \sum b_j X^j$ and let p be any prime number. Since the contents of A and B equal 1, we have the right to talk of the smallest indexes i_0 and j_0 such that a_i and b_j are not divisible by p. Knowing that the coefficient $c_{i_0+j_0}$ of $X^{i_0+j_0}$ in AB can be writen

$$c_{i_0+j_0} = a_{i_0}b_{j_0} + \sum_{\substack{i+j=i_0+j_0 \\ i<i_0 \text{ or } j<j_0}} a_i b_j \equiv a_{i_0}b_{j_0} \mod p,$$

we see that $c_{i_0+j_0}$ is not divisible by p. Since p was arbitrary, the GCD of the coefficients of AB is equal to 1. □

Proof of the theorem. Let $U, V \in \mathbb{Q}[X]$ be such that $UV \in \mathbb{Z}[X]$.

• If U has integral coefficients and if κ is its content, we can write $UV = (\kappa^{-1}U)(\kappa V)$ and $\operatorname{cont}(\kappa^{-1}U) = 1$. Since we do not know if κV has integer coefficients, introduce an integer $\ell \geq 1$ such that $\ell\kappa V$ has integral coefficients. From $\ell UV = (\kappa^{-1}U)(\ell\kappa V)$ and Gauss's lemma, we deduce:

$$\operatorname{cont}(\ell UV) = \operatorname{cont}(\kappa^{-1}U) \cdot \operatorname{cont}(\ell\kappa V) = \operatorname{cont}(\ell\kappa V).$$

On the other hand, $\operatorname{cont}(\ell UV) = \ell \operatorname{cont}(UV)$ since UV has integral coefficients. Thus ℓ divides the content of $\ell\kappa V$, *i.e.* κV has integral coefficients.

• If U does not have integral coefficients, let $\kappa > 1$ be the smallest integer such κU has integral coefficients. Then $UV = (\kappa U)(\kappa^{-1}V)$ reduces us to the preceding case. □

10.14.2. Kronecker's factorization algorithm

Consider the polynomial

$$P = 3X^4 + 15X^3 + 24X^2 + 21X + 9$$

and let *val_P* denote the following vector in $\mathbb{Z}^4$:

$$val\_P = \big((P(0), \ldots, P(4)\big) = (9, 72, 315, 936, 2205).$$

Letting $\mathrm{Div}(a)$ denote the set of positive divisors of the integer a, we have:

$$\begin{aligned}
\mathrm{Div}(9) &= \{1, 3, 9\};\\
\mathrm{Div}(72) &= \{1, 2, 3, 4, 6, 8, 9, 12, 18, 24, 36, 72\};\\
\mathrm{Div}(315) &= \{1, 3, 5, 7, 9, 15, 21, 35, 45, 63, 105, 315\};\\
\mathrm{Div}(936) &= \{1, 2, 3, 4, 6, 8, 9, 12, 13, 18, 24, 26, 36, 39, 52,\\
&\qquad 72, 78, 104, 117, 156, 234, 312, 468, 936\};\\
\mathrm{Div}(2205) &= \{1, 3, 5, 7, 9, 15, 21, 35, 45, 49, 63, 105, 147, 245,\\
&\qquad 315, 441, 735, 2205\}.
\end{aligned}$$

Consider the set:

$$\begin{aligned}
\mathcal{P} &= \mathrm{Div}\big(P(0)\big) \times \cdots \times \mathrm{Div}\big(P(4)\big)\\
&= \mathrm{Div}(9) \times \mathrm{Div}(72) \times \mathrm{Div}(315) \times \mathrm{Div}(936) \times \mathrm{Div}(2205),
\end{aligned}$$

This is a large set, since $\mathrm{Card}\,\mathcal{P} = 3 \times 12 \times 12 \times 24 \times 18 = 186\,624$.

Suppose that $P = AB$ is a factorization of P. Since

$$P(x) = A(x)B(x) \quad \text{for} \quad x = 0, \ldots, \deg(P),$$

the vectors

$$val\_A = \big(A(0), \ldots, A(\deg(P))\big) \text{ and } val\_B = \big(B(0), \ldots, B(\deg(P))\big)$$

are two elements of $\mathcal{P}$.

With the notation specified, we can find all factorizations of P using the following method due to B.A. Hausmann[4] (called *Kronecker's algorithm*): consider all the vectors $(a_0, \ldots, a_n) \in \mathcal{P}$, calculate the $b_i = P(i)/a_i$, construct

[4] B.A. Hausmann, *A new simplification of Kronecker's method of factorization of polynomials*, American Mathematical Monthly 47 (1937), pp. 574–576.

the Lagrange interpolating polynomials defined by the conditions $A(i) = a_i$ and $B(i) = b_i$ and verify that this is a factorization. For example:

$$\begin{cases} val\_A = (1, 2, 3, 4, 5), & A(X) = X + 1, \\ val\_B = (9, 36, 105, 234, 441), & B(X) = 3X^3 + 12X^2 + 12X + 9, \end{cases}$$

$$\begin{cases} val\_A = (1, 4, 7, 12, 21), & A(X) = \frac{1}{3}(X^3 - 3X^2 + 11X + 3), \\ val\_B = (9, 18, 45, 78, 105), & B(X) = -2X^3 + 15X^2 - 4X + 9. \end{cases}$$

The first pair is a factorization; the second isn't for two reasons: the coefficients of A are not integers and the degree of AB is too high. This last remark will put us on the way.

Theorem 10.14.2. *Let P, A, B be nonconstant polynomials with coefficients in $\mathbb{Z}$ satisfying the condition* $\deg(A) + \deg(B) \leq \deg(P)$. *Then, the following equivalence holds:*

$$P = AB \iff P(x) = A(x)B(x) \ \textit{for} \ x = 0, \dots, \deg(P).$$

Proof. The direction "$\Rightarrow$" is trivial. For the converse, remark that $\deg(P - AB) \leq \deg(P)$ and $(P - AB)(x) = 0$ for $x = 0, \dots, \deg(P)$ forces $P - AB = 0$. □

We can write the Kronecker algorithm very loosely as follows (put $p_k = P(k)$):

```
for (a_0, a_1, ..., a_n) := (1, 1, ..., 1) to (p_0, p_1, ..., p_n) do begin
  | Lagrange(A, a_0, a_1, ..., a_n) ;
  | (b_0, b_1, ..., b_n) := (p_0/a_0, p_1/a_1, ..., p_n/a_n) ;
  | Lagrange(B, b_0, b_1, ..., b_n) ;
  | if integer_coefficents(A) and integer_coefficents(B)
  |               and (degree(A) + degree(B) ≤ degree(P))
  | then begin display(A) ; display(B) end
end
```

A close analysis of this sketch raises the following questions:

• The use of 'for' loop tacitly assumes that we know how to run linearly over the set $\mathcal{P} = \text{Div } \mathcal{P}(0) \times \cdots \times \text{Div } \mathcal{P}(n)$.

• The definition of the b_i is not at all clear: it is incorrect when p_i is zero.

• We made a rather daunting implicit hypothesis: we only defined the set $\text{Div}(p_i)$ when $p_i > 0$ and we do not have the right to suppose that the values a_i and b_i of the factors A and B are positive.

10.14.3. Use of stable polynomials

It P is a stable polynomial, then A and B are also stable; moreover, since $P = AB = (-A)(-B)$, we know that we can restrict our attention to polynomials with positive coefficients.

Since the polynomial P that we wish to factor is not necessarily stable, we translate the coordinate axes horizontally and factor the polynomial $Q(X) = P(X + \mu)$ where μ is a strict upper bound on the moduli of the roots of P:

$$P(z) = 0 \implies |z| < \mu.$$

The polynomial Q is stable because $Q(z) = P(z+\mu) = 0$ implies $|z+\mu| < \mu$, whence:

$$\mathrm{Re}(z + \mu) \leq |z + \mu| < \mu.$$

If $P(X) = P_n X^n + \cdots + P_0$, we put:

$$\begin{aligned}\Phi(t) &= |P_n|\, t^n - \big(|P_{n-1}|\, t^{n-1} + \cdots + |P_0|\big)\\ &= |P_n|\, t^n \Big(1 - \frac{|P_{n-1}|}{t} - \cdots - \frac{|P_0|}{t^n}\Big).\end{aligned}$$

Since the function $t \mapsto 1/t^k$ is decreasing for $t > 0$, the rational function in parentheses varies from $-\infty$ to 1 as t grows from 0 to $+\infty$. Thus, the function Φ vanishes once and only once on $\mathbb{R}$.

Lemma 10.14.2. *If $\xi > 0$ is the unique real zero of the function $\Phi(t)$, then*

$$P(z) = 0 \implies |z| \leq \xi.$$

Proof. In effect, bounding $-P_n z^n = P_{n-1} z^{n-1} + \cdots + P_0$, gives

$$|P_n| \cdot |z|^n \leq |P_{n-1}| z|^{n-1} + \cdots + |P_0|,$$

so that $\Phi(|z|) \leq 0$, whence $|z| \leq \xi$. □

The case $P(X) = \Phi(X)$ shows that the result is best possible.

Translating these considerations into code is easy. Let $P \in \mathbb{Z}[X]$ be a polynomial of any degree $n > 0$. To determine the integral abscissa $x_0 > \xi$ that will become the new origin, we content ourselves with brute force.

```
procedure stable_abscissa(P : poly ; var x0 : integer) ;
var i : integer ;  Φ : poly ;
begin
  Φ[n] := abs(P[n]) ;
  for i := n − 1 downto 0 do Φ[i] := −abs(P[i]) ;
  x0 := 0 ;
  repeat x0 := x0 + 1 until value(Φ, x0) > 0
end ;
```

In view of the above, it is clear that $Q(X) = P(X + x_0)$ is stable.

Remark

The automatic search for $x_0 > \xi$ is not a universal panacea. When $P(X) = (X+1)^2$, this algorithm finds $x_0 = 3$ although $x_0 = 0$ would do because P is already stable. The result is annoying:

- If $x_0 = 3$, we have Card(Div(16) $\times$ Div(125) $\times$ Div(36)) $= 180$.
- If $x_0 = 0$, we have Card(Div(1) $\times$ Div(4) $\times$ Div(9) $= 9$.

Thus, the time for calculation grows when we replace P by Q ; moreover, since the coefficients of Q are bigger that those of P, we cannot factor polynomials whose degree is too big.

10.14.4. The program

We now know enough to begin to write our program. The declaration of types does not present any difficulty.

```
const deg_max = 5 ;
type poly = array[0 . . deg_max] of integer ;
values = array[0 . . deg_max] of integer ;
var P, Q : poly ;  val_P, val_Q : values ;  deg : integer ;
```

We could do with a single type *poly*. For clarity of exposition, it is preferable not to confound values and coefficients.

The main body of the program

We begin by entering the polynomial P. The procedure *choose* is also charged with eventually modifying the sign of P when the coefficient of the highest degree is negative. This done, we find the new origin $x_0 =$ *stable_abcissa* and translate axes, which amounts to replacing $P(X)$ by $P(X + x_0)$.

```
begin
  message ;
  choose(deg_P, P) ;
  find_stable_abcissa(P, stable_abcissa) ;
  translate(P, stable_abcissa, P, deg_P) ;
  initialize(val_P, val_A, deg_P) ;
  repeat
    decomposition_factors(val_P, val_A, deg_P) ;
    next(val_P, val_A, finish, deg_P) ;
  until finish
end .
```

Linearly traversing the set $\mathcal{P}$ (which is, recall, the product of the sets $\mathrm{Div}(P(x))$ for $x = 0, \dots, n$) is accomplished by repeated calls of the procedure *next* which returns, according to the case, the next vector *val_A* or the boolean *finish* which interrupts the “repeat” loop.

The procedure decomposition_factors

We now show that the stable polynomials have also some qualities. Since A and B have coefficients > 0, the functions $t \mapsto A(t)$ and $t \mapsto B(t)$ are strictly increasing on $\mathbb{R}_+$. Consequently, the vectors *val_A* and *val_B* must satisfy the draconian conditions

$$A(0) < A(1) < \cdots < A(n) \quad \text{and} \quad B(0) < B(1) < \cdots < B(n)$$

if they are to have a chance of being among the values of a divisor of P. Since Lagrange interpolations require rather lengthy calculations, we only launch them knowingly. For this, we ask the procedure *reconstruct*:

- to inform us if A has integral coefficients *via* the boolean *int_coeff_A*;
- to determine the degree *deg_A* of A, which can be done, as we shall see a little later, without entirely calculating A.

If the coordinates α_k of A *in the Newton basis* consisting of the $X^{(k)}$ are integers, the procedure *reconstruct* returns those in A; otherwise, A contains *indefinite* values which have no meaning.

```
procedure decomposition_factors(val_P, val_A : values ; deg_P : integer) ;
var A, B : poly ; val_B : values ; deg_A, deg_B : integer ;
begin
 | if increasing_seq(val_A, deg_P)
 | then begin
 |  | division_values(val_P, val_A, val_B, deg_P) ;
 |  | if increasing_seq(val_B, deg_P)
 |  | then begin
 |  |  | reconstruct(val_A, int_coeff_A, deg_A, A, deg_P) ;
 |  |  | reconstruct(val_B, int_coeff_B, deg_B, B, deg_P) ;
 |  |  | if int_coeff_A and int_coeff_B and (deg_A + deg_B ≤ deg_P)
 |  |  | then begin
 |  |  |  | return_canonical_basis(A, A, deg_A) ;
 |  |  |  | translate(A, −stable_abcissa, A, deg_A) ;
 |  |  |  | display(A) ;
 |  |  |  | return_canonical_basis(B, B, deg_B) ;
 |  |  |  | translate(B, −stable_abcissa, B, deg_B) ;
 |  |  |  | display(B) ;
 |  |  |  | verification(A, B)   {compares AB and P}
 |  |  |  end
 |  |  end
 |  end
end ;
```

The names of the other procedures or functions speak for themselves.

The procedure next

We use the algorithm detailed in Chapter 5 to linearly traverse the set $\mathcal{P} = \mathrm{Div}(\mathcal{P}(0)) \times \cdots \times \mathrm{Div}(\mathcal{P}(n))$ in the lexicographic order.

```
procedure next(val_P : values ; var val_A : values ;
                   var finish : boolean ; deg_P : integer) ;
var i, k : integer ;
begin
  finish := false ;
  k := -1 ;
  for i := 0 to deg_P do if val_A[i] < val_P[i] then k := i ;
  if k = -1 then finish := true
  else begin
    val_A[k] := next_divisor(val_P[k], val_A[k]) ;
    for i := k + 1 to deg_P do val_A[i] := 1
  end
end ;
```

We ask that the function *next_divisor* send us the smallest divisor of *val_P*$[k]$ which is strictly larger than *val_A*$[k]$. (When your program runs successfully, you can accelerate things spectacularly by pre-calculating once and for all all divisors of each $P(k)$.)

The procedure reconstruct

When the data are integers, we have seen that the Lagrange interpolating polynomial has rational coefficients. But since the interpolating points are consecutive integers $0, \ldots, n$ where $n = deg\_P$, we are going to use the discrete Taylor formula. Recall that if $\alpha_0, \ldots, \alpha_n$ are the coordinates of A in the Newton basis $\{1, X^{(1)}, \ldots, X^{(n)}\}$, we have:

$$\alpha_k = \frac{\Delta^k A(0)}{k\,!}. \tag{10.26}$$

We are going to use this formula to calculate the α_k because

- the largest integer k such that $\Delta^k A(0) \neq 0$ is the degree of A ;
- the α_k are obtained by dividing by $k\,!$ — if there were a nonzero remainder, then A would not be an element of $\mathbb{Z}[X]$.

Thus, we can avoid introducing rational numbers in our program, which simplifies writing a lot! The procedure *reconstruct*

- stores $\Delta^k A(0)$ in $A[k]$;
- deduces the degree of A from $A[k] = \Delta^k A(0)$;
- tries to divide $A[k]$ by $k\,!$ to make $\alpha_k = \Delta^k A(0)/k\,!$ If the remainder is zero, the variable $A[k]$ contains α_k; otherwise, the procedure leaves $A[k]$ alone and informs us via *int_coeff_A* that A does not have integral coefficients.

```
procedure reconstruct(val_A : values ; var int_coeff_A : boolean ;
                     var deg_A : integer ; var A : poly ; deg_P : integer) ;
var i, k, fact : integer ;
begin
 | annul(A) ; A[0] := val_A[0] ;
 | for k := 1 to deg_P do begin
 | | for i := 0 to deg_P - k do
 | | val_A[i] := val_A[i + 1] - val_A[i] ;
 | | A[k] := val_A[0]
 | end ;
 | for i := 0 to deg_max do if A[i] ≠ 0 then deg_A := i ;
 | int_coeff_A := true ; fact := 1 ; k := 1 ;
 | while (k ≤ deg_P) and int_coeff_A do begin
 | | fact := k * fact ;
 | | if A[k] mod fact = 0 then A[k] := A[k] div fact
 | | else int_coeff_A := false ;
 | | k := k + 1
 | end
end ;
```

Note the second internal loop: we translated a 'for' loop into a 'while' loop in order to insert the boolean *integer_coeff_A* which interrupts the loop as soon as we know that A does not have integer coefficients.

10.14.5. Last remarks

In our days, the algorithms employed by formal computational software are of a totally different nature (and are infinitely faster). Since these algorithms are very sophisticated, there is no question of programming them here.

We give, however, a glimpse of a more mature (and more difficult) algorithm. The idea is to reduce to factoring a polynomial with coefficients in a finite field.

Let p be a prime number and let $P \mapsto \overline{P}$ be the canonical map of $\mathbb{Z}[X]$ to $\mathbb{Z}_p[X]$ obtained by considering the coefficients of P as classes modulo p. Since the canonical map $\mathbb{Z}[X] \mapsto \mathbb{Z}_p[X]$ is a ring homomorphism, every factorization $P = AB$ in $\mathbb{Z}[X]$ gives rise to a factorization $\overline{P} = \overline{A}\,\overline{B}$ in $\mathbb{Z}_p[X]$.

• If $\overline{P}$ is irreducible in $\mathbb{Z}_p[X]$, there is no point of seeking a decomposition of P into factors in $\mathbb{Z}[X]$;

• By contrast, if $\overline{P}$ factors in $\mathbb{Z}_p[X]$, one can try to factor $\overline{P}$ in other prime fields. If one finds a prime number q for which $\overline{P}$ is irreducible, we are reduced to the preceding case. Otherwise, one collects the information and tries to lift the factorizations in different $\mathbb{Z}_p[X]$ to $\mathbb{Z}[X]$ using the Chinese Remainder Theorem. If this is impossible one knows that P is irreducible in $\mathbb{Z}[X]$.

Since the Chinese Remainder Theorem finds an infinity of solutions, one bounds the coefficients of A and B using the coefficients of P to choose a good couple (A, B).

This method also works with polynomials in several indeterminates.

11. Matrices

11.1. Z-Linear Algebra

In studying vector spaces, one shows that:

(i) every vector subspace V of $\mathbb{R}^n$ has a finite number of generators (one says that V is of *finite type*);
(ii) every vector subspace V of $\mathbb{R}^n$ has a basis and two bases always have the same cardinality (whence the notion of *dimension*) ;
(iii) if $e_1, \ldots, e_n$ generate a vector subspace V, one can always extract a *basis* of V from this set;
(iv) every linearly independent subset of $\mathbb{R}^n$ can be completed to a basis of $\mathbb{R}^n$.

We are going to examine what becomes of these results and the algorithms associated to them, when we replace R^n by $\mathbb{Z}^n$. For example, a subgroup of $\mathbb{Z}^2$ other than $\{0\}$ or $\mathbb{Z}^2$ resembles one of the following two subgroups:

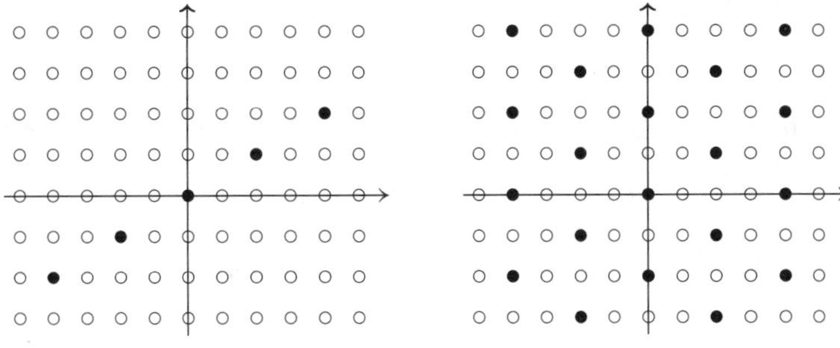

Fig. 11.1. *Subgroup* $x - 2y = 0$ *Subgroup* $x - 2y \equiv 0 \pmod 4$

Definition 11.1.1. *We say that* $x_1, \ldots, x_r$ *generate the subgroup* M *of* $\mathbb{Z}^n$ *if every* $x \in M$ *is of the form* $x = a_1x_1 + \cdots + a_rx_r$*, where all* a_i *belong to* $\mathbb{Z}$.

Definition 11.1.2. *We say that the vectors* $x_1, \ldots, x_r \in \mathbb{Z}^n$ *are* $\mathbb{Z}$*-linearly independent if* $a_1x_1 + \cdots + a_rx_r = 0$ *implies that* $a_1 = \cdots = a_r = 0$ *each time the* a_i *belong to* $\mathbb{Z}$.

Proposition 11.1.1. *There is an equivalence:*

$$\mathbb{Z}\text{-linearly independent} \iff \mathbb{Q}\text{-linearly independent.}$$

Proof. Let $x_1, \ldots, x_r$ be $\mathbb{Z}$-linearly independent vectors and suppose that $a_1x_1 + \cdots + a_rx_r = 0$ with $a_i \in \mathbb{Q}$. For appropriately chosen large $N > 1$, the linear combination $(Na_1)x_1 + \cdots + (Na_r)x_r = 0$ has integral coefficients. Thus $Na_i = 0$, which shows that the x_i are $\mathbb{Q}$-linearly independent. The converse is clear. □

This result is important because we can henceforth talk of *linear independence* without having to specify whether we are talking about it with respect to $\mathbb{Z}$ or to $\mathbb{Q}$.

Definition 11.1.3. *We say that $\varepsilon_1, \ldots, \varepsilon_r \in M$ form a basis of the additive subgroup M of $\mathbb{Z}^n$ if they both generate M and are linearly independent.*

Remarks

1) A basis $\varepsilon_1, \ldots, \varepsilon_r$ of a subgroup M of $\mathbb{Z}^n$ is also a vector space basis of the subspace $\text{Vect}_{\mathbb{Q}}(M) = \mathbb{Q}\varepsilon_1 + \cdots + \mathbb{Q}\varepsilon_r \subset \mathbb{Q}^n$ generated by M. Thus, two bases of M necessarily have the same cardinality. Hence, we have a notion of dimension (with respect to $\mathbb{Z}$ or $\mathbb{Q}$) once we succeed in establishing the existence of bases.

2) It is time for a counterexample: the results (iii) and (iv) *do not hold* for subgroups of $\mathbb{Z}^n$. To convince ourselves of this, consider the subgroup M of $\mathbb{Z}^2$ generated by the vectors

$$\varepsilon_1 = (2, 0), \quad \varepsilon_2 = (0, 3), \quad \varepsilon_3 = (5, 5).$$

It is easy to see that $M = \mathbb{Z}^2$ and $\text{Vect}_{\mathbb{Q}}(M) = \mathbb{Q}^2$. These three vectors are not linearly independent because $\dim_{\mathbb{Q}} \mathbb{Q}^2 = 2$; any two are a basis of $\text{Vect}_{\mathbb{Q}}(M)$, but do not generate M.

Definition 11.1.4. *We say that a matrix is unimodular if it has integral coefficients and if its inverse also has integer coefficients.*

Proposition 11.1.2. *Let A be a square matrix with integer coefficients. Then:*

$$A \text{ is unimodular} \iff \det A = \pm 1.$$

Proof. If A^{-1} has integer coefficients, then both $\det A$ and $\det(A^{-1})$ are integers, and $\det A \cdot \det(A^{-1}) = 1$ implies $\det A = \pm 1$. Conversely, if $\det A = \pm 1$, the classical formula $A^{-1} = (1/\det A)\,{}^t\text{Adj}\,(A)$ shows that the inverse of A has integer coefficients. □

Definition 11.1.5. *The unimodular matrices form a group, denoted* $\mathrm{Gl}(n, \mathbb{Z})$, *under matrix multiplication.*

Definition 11.1.6. *A matrix is said to be an elementary matrix if it is unimodular and one of the following three types:*

- $E_{i,j}(\lambda) = I + \lambda E_{i,j}$ *with* $i \neq j$ *; its inverse is* $E_{i,j}(-\lambda)$.
- $D_i = \mathrm{diag}(1, \ldots, -1, \ldots, 1)$ *which is its own inverse.*
- $T_{i,j} =$ *the matrix obtained by interchanging the* i*-th and* j*-th rows of the identity matrix; this matrix is also its own inverse.*

Manipulation of matrices

Let M be a matrix with integer coefficients and rows $L_1, \ldots, L_p$. A *row operation* on M consists of doing one of the following operations:

Matrix	*Operation*	*New matrix*
M	$L_i := L_i + \lambda L_j$	$M' = E_{i,j}(\lambda)M$
M	$L_i := -L_i$	$M' = D_i M$
M	$L_i \rightleftarrows L_j$	$M' = T_{i,j} M$

We will see a little later that one does not need to memorize these matrices; it is necessary only to note that performing row operations on a matrix M amounts to multiplying M on the left (that is, *premultiplying* M) by an elementary matrix.

Transposing the preceding equalities shows that *column operations* on M amount to multiplying on the right (or *postmultiplying*) by an elementary matrix:

Matrix	*Operation*	*New matrix*
M	$K_j := K_j + \lambda K_i$	$M' = M E_{i,j}(\lambda)$
M	$K_i := -K_i$	$M' = M D_i$
M	$K_i \rightleftarrows K_j$	$M' = M T_{i,j}$

For beginners

- It suffices to know that elementary matrices exist.
- A mnemonic device to remember the above is the *R-C rule*: replace the dash by the matrix M: to perform a row operation on M; one multiplies *before* by an elementary matrix; to perform a column operation on M, one multiplies *after*.

• Here are two classical errors to avoid. The first is to multiply a line (or column) by a rational number: the *only multiplication allowed* is changing the sign of a row (or column). (Since we are working exclusively with integers, we only have the right to multiply by the units of $\mathbb{Z}$, that is, by ± 1— otherwise, the inverse transformation would not be definited by a matrix with integer coefficients.) The second error consists of replacing the row L_i (resp. the column K_i) by the combination $aL_i + bL_j$ (resp. $aK_i + bK_j$), which is not allowed when $a \neq \pm 1$.

11.1.1. The bordered matrix trick

When one wants to know *explicitly* the row and column operations used, one uses a method that we already encountered in Chapter 8 when studying the Blankinship algorithm.

• Suppose that we want to perform row operations on M. Let

$$\widetilde{M} = (M, I) = \begin{pmatrix} m_{1,1} \dots & m_{1,s} & 1 & 0 & \dots & 0 \\ m_{2,1} \dots & m_{2,s} & 0 & 1 & \dots & 0 \\ \vdots & \vdots & \vdots & & \ddots & \vdots \\ m_{1,1} \dots & m_{r,s} & 0 & 0 & \dots & 1 \end{pmatrix}$$

be the matrix obtained from M by bordering it horizontally with the identity matrix with the same number of rows as M. To perform row operations on $\widetilde{M}$ amounts to multiplying it on the left by (unknown) elementary matrices $E_1, \dots, E_k$:

$$E_k \cdots E_1 \widetilde{M} = (E_k \cdots E_1 M, E_k \cdots E_1 I).$$

If we put $E = E_k \cdots E_1$, we obtain $E\widetilde{M} = (EM, E)$: the matrices E_i being multiplied together without knowing each separately!

• In a similar way, if we perform column operations on the bordered matrix

$$\widetilde{M} = \begin{pmatrix} M \\ I \end{pmatrix} = \begin{pmatrix} m_{1,1} & \dots & m_{1,s} \\ \vdots & & \vdots \\ m_{r,1} & \dots & m_{r,s} \\ 1 & \dots & 0 \\ \vdots & \ddots & \vdots \\ 0 & \dots & 1 \end{pmatrix},$$

the operations $E_1, \dots, E_k$ (in this order) automatically give:

$$\widetilde{M} E_1 \cdots E_k = \begin{pmatrix} M E_1 \cdots E_k \\ I E_1 \cdots E_k \end{pmatrix} = \begin{pmatrix} M E_1 \cdots E_k \\ E_1 \cdots E_k \end{pmatrix}.$$

11.1.2. Generators of a subgroup

Theorem 11.1.1. *Every subgroup M of $\mathbb{Z}^n$ is of finite type; that is, it is a set of linear combinations with integer coefficients of a finite number of vectors in M.*

Proof. When $n = 1$, the result is well-known: every subgroup of $\mathbb{Z}$ is of the form $d\mathbb{Z}$, hence generated by an element d of M. Now, suppose that the result is true for an integer $n \geq 1$ and let $M \subset \mathbb{Z}^{n+1}$ be a nonzero subgroup.

If M is contained in $\mathbb{Z}^n \times \{0\}$, the result holds (by the induction hypothesis). Otherwise, consider the projection $\varphi : M \to \mathbb{Z}$:

$$\varphi(x_1, \ldots, x_{n+1}) = x_{n+1}.$$

Since $\varphi(M)$ is a nonzero subgroup of $\mathbb{Z}$, we have $\varphi(M) = d\mathbb{Z}$ with $d > 0$ and there exists at least one vector $\varepsilon \in M$ such that $\varphi(\varepsilon) = d$. For $x \in M$, one can write $\varphi(x) = k(x)d$, so $x - k\varepsilon \in \ker\varphi$. But, by induction the additive subgroup $\ker\varphi = M \cap (\mathbb{Z}^n \times \{0\})$ is generated by $\varepsilon_1, \ldots, \varepsilon_r \in M$, which shows that x is a linear combination with integer coefficients of the vectors $\varepsilon, \varepsilon_1, \ldots, \varepsilon_r$. □

Theorem 11.1.2. *The vectors $\varepsilon_1, \ldots, \varepsilon_n$ are a basis of $\mathbb{Z}^n$ if and only if the matrix of these vectors in the canonical basis is a unimodular matrix.*

Proof. First suppose that $\varepsilon_1, \ldots, \varepsilon_n$ are a basis. Let $e_1, \ldots, e_n$ be the canonical basis of $\mathbb{Z}^n$ and put

$$x = x_1 e_1 + \cdots + x_n e_n = \xi_1 \varepsilon_1 + \cdots + \xi_n \varepsilon_n.$$

Consider the matrix $E = (\varepsilon_1, \ldots, \varepsilon_n)$ with columns $\varepsilon_1, \ldots, \varepsilon_n$ as well as the column matrices $\tilde{x} = {}^t(x_1, \ldots, x_n)$ and $\tilde{\xi} = {}^t(\xi_1, \ldots, \xi_n)$, so that we have the change of basis formula $\tilde{x} = E\,\tilde{\xi}$.

To say that $\varepsilon_1, \ldots, \varepsilon_n$ is a basis of $\mathbb{Z}^n$ means that $\tilde{\xi} = E^{-1}\tilde{x}$ has integral coordinates each time that $\tilde{x}$ does. Upon choosing $x = e_1$ (the first vector of the canonical basis), we obtain the vector with integer coordinates $\tilde{\xi} = E^{-1}e_1$, which shows that the first column of E^{-1} has integer coefficients, etc.

Conversely, the columns of a unimodular matrix are a basis of $\mathbb{Z}^n$ since the coordinate changes are made over the integers. □

11.1.3. The Blankinship algorithm

We consider how to adapt the method of Gaussian elimination and pivoting to the integers. We have already encountered this algorithm in the dimension two case in Chapter 8; its generalization is immediate. Let $a_1, \ldots, a_n$ be integers which are not all zero. Suppose that we want to calculate *simultaneously* the GCD of the a_i and a particular solution $u_1, \ldots, u_n$ of the Bézout equation:

$$u_1 a_1 + \cdots + u_n a_n = \mathrm{GCD}(a_1, \ldots, a_n).$$

We again content ourselves with presenting the generalization by example. Suppose that $n = 3$ and $a_1 = 9$, $a_2 = 5$ et $a_3 = 7$. Begin by bordering the vector ${}^t(9, 5, 7)$ by the identity matrix, obtaining a matrix M_0 on which we will perform row operations. The operations (see Table 11.1) are driven by the first column to which we apply the Euclidean algorithm.

The $(1, 1)$ entry of the matrix M_4 contains the desired GCD; across from it (on the first row) we find the solution of the Bézout equation. Thus, we obtain $\mathrm{GCD}(9, 5, 7) = 1$, which is not surprising since, for example, 9 and 5 are relatively prime. One checks directly that $(0, 3, -2)$ is a solution of the Bézout equation: $0 \cdot 9 + 3 \cdot 5 - 2 \cdot 7 = 1$.

Old matrix	*Operation*	*New matrix*
$M_0 = \begin{pmatrix} 9 & 1 & 0 & 0 \\ \boxed{5} & 0 & 1 & 0 \\ 7 & 0 & 0 & 1 \end{pmatrix}$	$L_1 := L_1 - L_2$ $L_3 := L_3 - L_2$	$M_1 = \begin{pmatrix} 4 & 1 & -1 & 0 \\ 5 & 0 & 1 & 0 \\ 2 & 0 & -1 & 1 \end{pmatrix}$
$M_1 = \begin{pmatrix} 4 & 1 & -1 & 0 \\ 5 & 0 & 1 & 0 \\ \boxed{2} & 0 & -1 & 1 \end{pmatrix}$	$L_1 := L_1 - 2L_3$ $L_2 := L_2 - 2L_3$	$M_2 = \begin{pmatrix} 0 & 1 & 1 & -2 \\ 1 & 0 & 3 & -2 \\ 2 & 0 & -1 & 1 \end{pmatrix}$
$M_2 = \begin{pmatrix} 0 & 1 & 1 & -2 \\ \boxed{1} & 0 & 3 & -2 \\ 2 & 0 & -1 & 1 \end{pmatrix}$	$L_3 := L_3 - 2L_2$	$M_3 = \begin{pmatrix} 0 & 1 & 1 & -2 \\ 1 & 0 & 3 & -2 \\ 0 & 0 & -7 & 5 \end{pmatrix}$
$M_3 = \begin{pmatrix} 0 & 1 & 1 & -2 \\ 1 & 0 & 3 & -2 \\ 0 & 0 & -7 & 5 \end{pmatrix}$	$L_1 \rightleftarrows L_2$	$M_4 = \begin{pmatrix} 1 & 0 & 3 & -2 \\ 0 & 1 & 1 & -2 \\ 0 & 0 & -7 & 5 \end{pmatrix}$

Table 11.1. *The Blankinship algorithm in practice*

Here is a very informal description of the algorithm:

$M := (a, I)$; $\{a \neq 0$ *is the column containing the* $a_i\}$
repeat
  «*seek* k *such that* $|a_k| = \min\{a_i\ ;\ a_i \neq 0\}$;
  if $k \neq 1$ **then** «*exchange lines* L_1 *and* L_k» ;
  if $a_1 < 0$ **then** $L_1 := -L_1$;
  for $k := 2$ **to** n **do**
    if $a_k \neq 0$ **then** $L_k := L_k - (a_k$ **div** $a_1)L_1$ $\{$*pivoting*$\}$
until $a_2 = \cdots = a_n = 0$

11.1.4. Hermite matrices

A matrix with integer coefficients is said to be in *row echelon form* if it resembles a staircase with steps of height 1:

$$A = \begin{pmatrix} \cdots & a_{1,j_1} & & & & \\ \cdots & \cdots & \cdots & a_{2,j_2} & & \\ \cdots & \cdots & \cdots & \cdots & \cdots & a_{3,j_3} \\ \cdots & \cdots & \cdots & \cdots & \cdots & \cdots \end{pmatrix}$$

The dots represent coefficients which are zero; in contrast, the "corners" of the steps are not zero:

$$a_{1,j_1} \neq 0, \quad a_{2,j_2} \neq 0, \quad a_{3,j_3} \neq 0, \ldots \qquad 1 \leq j_1 < j_2 < j_3 < \cdots.$$

Remarks

- The steps are of *height* equal to 1 and their *width* is ≥ 1.
- The nonzero rows are linearly independent.
- A row-echelon matrix is a triangular matrix in which the corners of the steps are not necessarily on the diagonal; they are above the diagonal as soon as a step of width ≥ 2 occurs.

$$\begin{matrix} 0 & 2 & 1 & 5 & 0 & 14 \\ 0 & 0 & 7 & 3 & 11 & 12 \\ 0 & 0 & 0 & 0 & 6 & 0 \\ 0 & 0 & 0 & 0 & 0 & 0 \end{matrix} \qquad \begin{matrix} 1 & 1 & 5 & 0 & 14 \\ 0 & 3 & 3 & 1 & 8 \\ 0 & 0 & 5 & 0 & 2 \\ 0 & 0 & 0 & 0 & 0 \end{matrix}$$

If the matrix ${}^tA$ is in row echelon form, we say that A is in *column echelon form*. Here are three examples:

$$\begin{matrix} 3 & 0 & 0 & 0 & 0 \\ 2 & 7 & 0 & 0 & 0 \\ 0 & 9 & 6 & 0 & 0 \\ 0 & 0 & 0 & 0 & 0 \end{matrix} \qquad \begin{matrix} 1 & 0 & 0 & 0 & 0 \\ 0 & 7 & 0 & 0 & 0 \\ 2 & 1 & 0 & 0 & 0 \\ 0 & 0 & 4 & 0 & 0 \end{matrix} \qquad \begin{matrix} 1 & 0 & 0 & 0 & 0 \\ 1 & 3 & 0 & 0 & 0 \\ 2 & 5 & 4 & 0 & 0 \\ 0 & 0 & 0 & 2 & 0 \end{matrix}$$

For such matrices, the steps are of width 1 and height ≥ 1. The columns which are not zero are linearly independent.

In what follows, we say that a matrix is a *Hermite matrix* if it is in row echelon or column echelon form.

Theorem 11.1.3 (Hermite). *Let A be any matrix with integer coefficients. There exist unimodular matrices E and F such that EA is in row echelon form and AF is in echelon form. (These are not, in general, unique.)*

Proof. We restrict ourselves to the case of rows. The key to the proof is the repeated application of Blankinship's algorithm to the columns of A. In applying this algorithm to the first column of A, we know how to explicitly construct a unimodular matrix E_1 such that:

$$E_1 A = \begin{pmatrix} d_1 & * \\ 0 & A' \end{pmatrix}, \quad d_1 = \mathrm{GCD}(a_{1,1}, \dots, a_{p,1}).$$

• If $d_1 \neq 0$, we apply Blankinship's algorithm to the first column of the matrix A', which shows the existence of a unimodular matrix E_2 such that:

$$E_2 E_1 A = \begin{pmatrix} d_1 & * & * \\ 0 & d_2 & * \\ 0 & 0 & A'' \end{pmatrix} \quad d_2 = \mathrm{GCD}(a'_{2,2}, \dots, a'_{p,2}).$$

• If $d_1 = 0$, we apply Blankinship's algorithm to the first column of the matrix $\begin{pmatrix} * \\ A' \end{pmatrix}$, which establishes the existence of a unimodular matrix E_2 such that:

$$E_2 E_1 A = \begin{pmatrix} 0 & d_2 & * \\ 0 & 0 & A'' \end{pmatrix}, \quad d_2 = \mathrm{GCD}(a'_{1,2}, \dots, a'_{p,2}).$$

It suffices to "clean out" the columns of the original matrix to obtain the desired result. □

Remark

When we want to obtain *explicitly and automatically* the unimodular matrix E such that EA is in row echelon form, it suffices to border A by the unit matrix and carry out the row operations on the bordered matrix $M = (A, I)$. The technique for columns is analogous.

Example

Put the following matrix in row echelon form.

$$A = \begin{pmatrix} 6 & 4 & -8 & -8 & 12 \\ 9 & 3 & -8 & -16 & 18 \\ 3 & 1 & -3 & -7 & 5 \\ -3 & -1 & 3 & 7 & -1 \end{pmatrix}.$$

Clean out the first column of $M_0 = (A, I)$:

$$M_0 = \left(\begin{array}{ccccc|cccc} 3 & 4 & -8 & -8 & 12 & 1 & 0 & 0 & 0 \\ 9 & 3 & -8 & -16 & 18 & 0 & 1 & 0 & 0 \\ \boxed{3} & 1 & -3 & -7 & 5 & 0 & 0 & 1 & 0 \\ -3 & -1 & 3 & 7 & -1 & 0 & 0 & 0 & 1 \end{array}\right) \quad \begin{array}{l} L_1 := L_1 - 2L_3 \\ L_2 := L_2 - 3L_3 \\ L_4 := L_4 + L_3 \\ L_1 \rightleftarrows L_3 \end{array}$$

We tidy up in the second column

$$M_1 = \left(\begin{array}{ccccc|cccc} 3 & 1 & -3 & -7 & 5 & 0 & 0 & 1 & 0 \\ 0 & 0 & 1 & 5 & 3 & 0 & 1 & -3 & 0 \\ 0 & \boxed{2} & -2 & 6 & 2 & 1 & 0 & -2 & 0 \\ 0 & 0 & 0 & 0 & 4 & 0 & 0 & 1 & 1 \end{array}\right) \qquad L_2 \rightleftarrows L_3$$

to get

$$M_2 = \left(\begin{array}{ccccc|cccc} 3 & 1 & -3 & -7 & 5 & 0 & 0 & 1 & 0 \\ 0 & 2 & -2 & 6 & 2 & 1 & 0 & -2 & 0 \\ 0 & 0 & 1 & 5 & 3 & 0 & 1 & -3 & 0 \\ 0 & 0 & 0 & 0 & 4 & 0 & 0 & 1 & 1 \end{array}\right).$$

If we put $M_2 = (H, E)$, we can check that $\det E = \pm 1$ and $H = EA$.

Exercise 1

Show that *any* unimodular matrix E is a product of elementary matrices. (Do this by performing row operations on E to obtain the identity.)

Exercise 2: Smith reduction *(Solution at end of chapter)*

Let A be a $p \times n$ matrix with integer coefficients. Show that there always exist two unimodular matrices P and Q, of dimensions $p \times p$ and $n \times n$, such that $S = PAQ$ is a diagonal matrix of the form

$$S = \begin{pmatrix} D & 0 \\ 0 & 0 \end{pmatrix} \quad \text{with} \quad D = \begin{pmatrix} d_1 & 0 & \cdots & 0 \\ 0 & d_2 & & \vdots \\ \vdots & & \ddots & 0 \\ 0 & \cdots & 0 & d_r \end{pmatrix} \quad \text{and } d_1, \ldots, d_r > 0.$$

To show this result, alternate row and column operations. To show that the algorithm terminates, consider the $(1, 1)$ entry after each operation.

Once the diagonal form is obtained, continue performing the operations so that d_i divides d_{i+1}. In this case, one says that S is the *Smith reduced* form of A. One can show this reduced form is unique; but P and Q are not unique (multiply S on the left and on the right by $\mathrm{diag}(1, \ldots 1, -1, 1 \ldots, 1)$).

11.1.5. The program Hermite

Let A be a $p \times n$ matrix with integer coefficients. We want to construct a unimodular matrix E such that $H = EA$ is a Hermite matrix in row echelon form.

The main part of the program

One encounters the three classical phases: entering the data, treatment of the data and output of results. The *verification* of the correctness of the program is a formality: it suffices to check that the matrix $EA - H$ is zero.

```
begin
  choose(A, row, col) ;
  H := A ;
  Hermite(H, E, row, col) ;
  display(E, row, col) ;  display(H, row, col) ;
  verification(A, E, H, row, col)
end .
```

The procedure Hermite

We require the function *zero_col*(A, ℓ, row, k) to tell us if a subcolumn of the $a_{i,k}$ for $\ell \le i \le row$ is zero.

• If the subcolumn is not zero, we "clean it out" using Blankinship's algorithm; this done, we go to the following column by incrementing ℓ (in other words, we go down a step);

• If the subcolumn is zero, we go to the next column without modifying ℓ because we are on a step of width > 1.

```
procedure Hermite(var A, E : matrix ;  row, col : integer) ;
var k, ℓ : integer ;
begin
  unit(E, row) ;  ℓ := 1 ;
  for k := 1 to col do
  if not zero_col(A, ℓ, row, k) then begin
    Blankinship(A, E, ℓ, row, k, col) ;
    ℓ := ℓ + 1
  end
end ;
```

The procedure Blankinship

This procedure "cleans" the k-th column out of the elements between rows ℓ and *row*; the subcolumn is cleaned out when $A[i, k] = 0$ for $i = \ell+1, \ldots, row$; that is, when *zero_col*$(A, \ell + 1, row, k)$ becomes true.

To clean out the subcolumn, we undertake the following actions:

• We begin by locating a pivot, that is, a nonzero element with smallest possible absolute value.

• We bring the pivot to the head of the column (that is, to the row ℓ) if it is on a lower row.

• We change the sign of the pivot if it is negative so that we do not have negative pivots.

• We "erode" the coefficients below the pivot by adding suitable multiples of the pivot to them.

Of course, each row operation on A is faithfully reproduced without delay on the rows of E.

```
procedure Blankinship(var A, E : matrix ;  ℓ, row, k, col : integer) ;
var p, j, coeff, pivot : integer ;
begin
| while not zere_col(A, ℓ + 1, row, k) do begin
| | p := row_pivot(A, ℓ, row, k) ;
| | if p > ℓ then begin
| | | swap_row(A, ℓ, p, col) ;
| | | swap_row(E, ℓ, p, row)
| | end
| end ;
| if A[ℓ, k] < 0 then begin
| | change_sign_row(A, ℓ, col) ;
| | change_sign_row(E, ℓ, col) ;
| end ;
| pivot := A[ℓ, k] ;
| for j := ℓ + 1 to row do begin
| | coeff := A[j, k] div pivot ;
| | if coeff ≠ 0 then begin
| | | add_row(A, j, ℓ, −coeff, col) ;
| | | add_row(E, j, ℓ, −coeff, row)
| | end
| end
end ;
```

Note that the extreme case where the subcolumn contains only a single pivot in a row of index $> \ell$ is treated correctly.

The function zero_col

This function is trivial modulo a subtle trap that some avoid without knowing it: the code of *Blankinship* functions correctly only if *zero_col*$(A, \ell + 1, row, k)$ answers that an empty subcolumn (*i.e.* when $\ell + 1 > row$) is declared *zero*. If we do not take this precaution, we can get get into an infinite loop because an empty column can only make an empty column.

```
function zero_col(A : matrix ; ℓ, row, k : integer) : boolean ;
var i : integer ;
begin
  zero_col := true   {to correctly treat the case ℓ + 1 > row}
  for i := ℓ + 1 to row do
      if A[i, k] ≠ 0 then zero_col := false
end ;
```

The function row_pivot

Here again, we must avoid the trap of believing that we can choose the head $a_{\ell,k}$ of the subcolumn as the first pivot-candidate (this is correct only if $a_{\ell,k} \neq 0$). This is why we run through the subcolumn twice.[1]

```
function row_pivot(A : matrix ; ℓ, row, k : integer) : integer ;
var i, place : integer ;
begin
  for i := ℓ to row do if A[i, k] ≠ 0 then place := i ;
  for i := ℓ to row do
      if (abs(A[i, k]) < abs(A[place, k])) and (A[i, k] ≠ 0)
      then place := i ;
  row_pivot := place
end ;
```

The other functions and procedures do not harbor any difficulties.

Exercise 3

Write a program that takes a matrix in echelon form and "scrambles" it up using elementary operations chosen at random. Then apply the procedure *Hermite* to the result. Is the form of the staircase the same? Are the corners of the steps the same?

Exercise 4: Inversion of a matrix with real coefficients

The pivoting strategy is far simpler when one works over a field since we can clean out a subcolumn in a single pass (speaking informally, we can immediately "kill" the coefficients instead of "eroding" them).

[1] Yes, this is heavy-handed; yes, we could do better and leave after a single run through. Never forget, however, that the absolute priority of a programmer is *security*. The code must be limpid. Heavy-handedness is not a fault; you can liven it up later, when everything works well. This industrial logic has nothing to do with the systematic search for tricks practiced in mathematics.

To invert a matrix with real coefficients, we can row reduce the matrix (A, I) by cleaning out columns to transform it to the matrix $E(A, I) = (I, A^{-1})$. Suppose that we are cleaning out the subcolumn k, and that up to this moment the matrix A has become:

$$EA = \begin{pmatrix} I_{k-1} & * \\ 0 & A' \end{pmatrix}.$$

We inspect the first column of A', that is $a_{k,k}, \ldots, a_{n,k}$, seeking a nonzero pivot of largest absolute value for reasons of numerical stability.

• If the pivot exists (which amounts to saying that the first column of A' is nonzero), we bring the pivot to (k, k) by exchanging rows. We then "kill" the $a_{i,k}$ for $i \neq k$ *in a single pass* using the linear combination:

$$L_i := L_i - \mu L_k, \quad \mu = a_{i,k}/a_{k,k}.$$

We end by dividing the row k by the pivot to make it equal to 1.

• If the pivot doesn't exist, we know that the first column of A' is zero, which means that A' is not invertible. But then A is not invertible because $\det(EA) = \det E \times \det A = \det I_{k-1} \times \det A' = \det A' = 0$.

```
procedure invert_matrix(var A, inv_A : matrix ; dim : integer ;
var invertible : boolean) ;
var i, k, place_pivot : integer ; value_pivot, coeff : real ;
begin
  k := 1 ; invertible := true ;
  while (k ≤ dim) and invertible do begin
    inspect_sub_column(A, k, dim, place_pivot, invertible) ;
    if invertible then begin
      exchange_rows(k, place_pivot, A) ;
      exchange_rows(k, place_pivot, invA) ;
      value_pivot := A[k, k] ;
      for i := 1 to dim do if i ≠ k then begin
        coeff := A[i, k]/value_pivot ;
        combination_rows(i, k, −coeff, A) ;
        combination_rows(i, k, −coeff, invA) ;
      end
      divide_row(k, value_pivot, A) ;
      divide_row(k, value_pivot, invA)
    end ;
    k := k + 1
  end
end ;
```

We remark that the principal loop is a "for $k := 1$ to *dim*" loop rewritten as a "while" loop so as to insert the interrupter *invertible* which interrupts the work of the procedure as soon as it is known that the matrix is not invertible.

Exercise 5

1) Suppose that A has coefficients in $\mathbb{Q}$ (resp. $\mathbb{C}$, resp. $\mathbb{Z}_p$ with p prime). Write a program which calculates its inverse.

2) Modify the procedure *invert_matrix* to calculate the determinant of the matrix A.

Exercise 6

Let k be a commutative field and $A \in \mathrm{Gl}(n, k)$ be an invertible matrix. There exists a lower triangular matrix L and an upper triangular matrix U such that $LAU = S$, where S is a permutation matrix; that is, a matrix obtained from the identiy matrix by permuting rows.

Proof. The idea is to operate on the rows of A *towards the bottom* and on the columns of A *towards the right*, which means that the only operations allowed are (we denote the rows of the matrix by the letter Λ):

- $\Lambda_j := \Lambda_j + \lambda\Lambda_i$ and $K_j := K_j + \lambda K_i$ with $i < j$ and $\lambda \in k$ whatever;
- $\Lambda_i := \lambda\Lambda_i$ and $K_i := \lambda K_i$ with $\lambda \in k^*$.

In particular, *exchanging* rows or columns is forbidden.

We start with the matrices $A_0 = A$, $L_0 = I_n$ and $U_0 = I_n$.

- We go down the first column of A_0 and stop at the first nonzero coefficient $a_{\ell,1}$, which we call the *pivot*. We *normalize* the pivot by dividing the ℓ-th rows of A_0 and L_0 by the pivot. We then clean out (towards the bottom) the first column of A using linear combinations of the form $\Lambda_j := \Lambda_j + \lambda\Lambda_\ell$ where $j > \ell$. As usual, each operation on the lines of A_0 is instantly reproduced on L_0.

$$A_0 \longmapsto \begin{pmatrix} 0 & * & \cdots & * \\ \vdots & * & \cdots & * \\ 0 & * & \cdots & * \\ 1 & a_{\ell,2} & \cdots & a_{\ell,n} \\ a_{\ell+1,1} & * & \cdots & * \\ \vdots & & & \\ a_{n,1} & * & \cdots & * \end{pmatrix} \longmapsto \begin{pmatrix} 0 & * & \cdots & * \\ \vdots & * & \cdots & * \\ 0 & * & \cdots & * \\ 1 & a_{\ell,2} & \cdots & a_{\ell,n} \\ 0 & * & \cdots & * \\ \vdots & * & \cdots & * \\ 0 & * & \cdots & * \end{pmatrix}.$$

- We then clean out the ℓ-th row of A_0 using linear combinations of the form $K_j := K_j + \lambda K_1$ with $j > 1$. Of course, these operations are reproduced right away on U_0.

$$\begin{pmatrix} 0 & * & \cdots & * \\ \vdots & * & \cdots & * \\ 0 & * & \cdots & * \\ 1 & a_{\ell,2} & \cdots & a_{\ell,n} \\ 0 & * & \cdots & * \\ \vdots & * & \cdots & * \\ 0 & * & \cdots & * \end{pmatrix} \longmapsto \begin{pmatrix} 0 & * & \cdots & * \\ \vdots & * & \cdots & * \\ 0 & * & \cdots & * \\ 1 & 0 & \cdots & 0 \\ 0 & * & \cdots & * \\ \vdots & * & \cdots & * \\ 0 & * & \cdots & * \end{pmatrix}.$$

After this first wave of operations, the matrices A_0, L_0 and U_0 become A_1, L_1 and U_1 and we have $A_1 = L_1 A_0 U_1$.

• We then handle the columns $k = 2, \dots, n$ and the associated rows in a similar manner. (We remark that the second pivot exists because A_1 is invertible and, by construction, it is on a row of index $\neq \ell$.)

• When the algorithm terminates, the matrix A_n is the desired matrix S and we have $S = L_n \cdots L_1 A U_1 \cdots U_n$. □

Transform this proof into a program which calculates L, U, S from A (assume, for simplicity, that the initial matrix is invertible). What modifications need to be made to the program to have it stop when A is not invertible?

Exercise 7: Reduction of quadratic Forms

Let S be a symmetric matrix with real coefficients. Then there exists an invertible matrix P such that ${}^tPAP$ is a diagonal matrix.

Proof. First suppose that the first column of A is not zero.

• If $a_{1,1} \neq 0$, we clean out the first column with operations towards the bottom, let $L_i := L_i + \lambda L_1$ with $i > 1$; if we take the precaution of following each row operation by the column operation $K_i := K_i + \lambda K_1$, the matrix A becomes ${}^tPAP$ with P being the product of the corresponding elementary matrices.

• If $a_{1,1} = 0$, we can be certain that there exists an $a_{\ell,1} \neq 0$ since the first column is not zero. We then perform the operations $L_1 := L_1 + L_\ell$ and $K_1 := K_1 + K_\ell$ on A which reduces us to the preceding case.

When we have finished cleaning out the first row and column, or if the first and row and column are zero, we pass to the submatrix consisting of the $a_{i,j}$ for which $2 \leq i, j \leq n$. □

Transform this proof into a program which takes the upper part of A and completes it to a symmetric matrix.[2]

[2] *Never* forget that the machine must serve humans; the user should only have to enter the minimum number of coefficients.

11.1.6. The incomplete basis theorem

Let $\varepsilon_1, \ldots, \varepsilon_r$ be vectors in $\mathbb{Z}^n$: do there exist vectors $\varepsilon_{r+1}, \ldots, \varepsilon_n$ such that $(\varepsilon_1, \ldots, \varepsilon_n)$ is a basis of $\mathbb{Z}^n$?

Definition 11.1.7. *A vector $\varepsilon \neq 0$ of $\mathbb{Z}^n$ is said to be visible from the origin if there does not exist a rational number $t \in]0, 1[$ such that $t\varepsilon$ has integer coordinates. (Think of trees in an orchard and look at Fig.* 11.2*)*

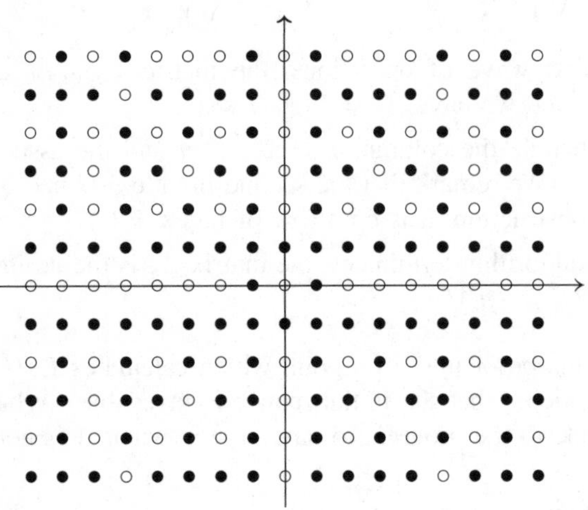

Fig. 11.2. *Visible points from the origin*

Lemma 11.1.1. *A vector with integer coordinates $\varepsilon \neq 0$ is visible from the origin if and only if its coordinates are relatively prime.*

Proof. Let $d > 0$ be the GCD of the coordinates of ε.

- If $d > 1$, the vector $d^{-1}\varepsilon$ has integer coordinates and hides ε.
- If $d = 1$ and if ε were hidden by the vector $\varepsilon' = t\varepsilon$ with $0 < t < 1$, then by considering a nonzero coordinate of ε, we obtain $t = p/q$ with $0 < p < q$ and $\mathrm{GCD}(p, q) = 1$. The equality $q\varepsilon' = p\varepsilon$ and Gauss's lemma shows that q divides all coordinates of ε, which is contrary to the hypothesis that $d = 1$. □

Proposition 11.1.3. *A vector $\varepsilon \neq 0$ in $\mathbb{Z}^n$ can be a member of a basis of $\mathbb{Z}^n$ if and only if it is visible from the origin.*

Proof. Let $(\varepsilon_1, \ldots, \varepsilon_n)$ be a basis of $\mathbb{Z}^n$ such that $\varepsilon_1 = \varepsilon$; we know that the matrix $E = (\varepsilon_1, \ldots, \varepsilon_n)$ is unimodular. If we expand $\det E$ along the first column, we obtain a linear combination with integer coefficients of the coordinates of ε which is equal to ± 1. By *Bézout's theorem*, the GCD of the coordinates of ε is equal to 1, and hence ε is visible.

Conversely, if ε is visible, Blankinship's algorithm *explicitly* gives us a unimodular matrix E such that $E\varepsilon = e_1$. The vector $\varepsilon = E^{-1}e_1$ then belongs to the basis of $\mathbb{Z}^n$ formed by the columns of E^{-1}. □

We can also state the proposition as follows: *a vector $\varepsilon \neq 0$ is the first column of a unimodular matrix if and only if it is visible from the origin.*

The case of r vectors is more technical and cannot be interpreted so intuitively. A *necessary* condition to belong to a basis of $\mathbb{Z}^n$ is — of course — to be visible from the origin. But this does not suffice.

Theorem 11.1.4. *Let $\varepsilon_1, \dots, \varepsilon_r$, with $1 \leq r \leq n$, be nonzero vectors in $\mathbb{Z}^n$ and put $M = (\varepsilon_1, \dots, \varepsilon_r)$. Let E be a unimodular matrix such that EM is in row echelon form:*

$$EM = \begin{pmatrix} T \\ 0 \end{pmatrix} \quad \textit{with} \quad T = \begin{pmatrix} d_1 & * & * \\ \vdots & \ddots & * \\ 0 & \dots & d_i \end{pmatrix}.$$

The vectors $\varepsilon_1, \dots, \varepsilon_r$ are part of a basis of $\mathbb{Z}^n$ if and only if the diagonal coefficients of T satisfy $d_i = \pm 1$ for $i = 1, \dots, r$.

Pay attention to the statement: the coefficients d_i in question are not *necessarily* the corners of the steps of the matrix EM in echelon form; they are the *diagonal coefficients* of T.

Proof. Suppose that one knows how to complete $\varepsilon_1, \dots, \varepsilon_r$ to a basis $(\varepsilon_1, \dots, \varepsilon_n)$ of $\mathbb{Z}^n$, so that the matrix $P = (\varepsilon_1, \dots, \varepsilon_n) = (M, \varepsilon_{r+1} \dots, \varepsilon_n)$ is unimodular. We can write:

$$EP = (EM, E\varepsilon_{r+1}, \dots, E\varepsilon_n) = \begin{pmatrix} T & * \\ 0 & Q \end{pmatrix}.$$

Taking the determinant, we obtain $\pm 1 = \det(EP) = \det T \times \det Q$ which first implies $\det T = d_1 \cdots d_i = \pm 1$ and then $d_i = \pm 1$ since the coefficients are integers.

Conversely, suppose that $d_i = \pm 1$ for $i = 1, \dots, r$. One round of pivoting towards the head of the rows of EM shows that there exists a unimodular matrix E' such that $E'EM = \begin{pmatrix} I_r \\ 0 \end{pmatrix}$, whence:

$$M = (E'E)^{-1}\begin{pmatrix} I_r \\ 0 \end{pmatrix}.$$

This formula says that the columns of M coincide with the r first columns of the unimodular matrix $(E'E)^{-1}$. □

The incomplete basis algorithm

Let $\varepsilon_1, \dots \varepsilon_r$ be vectors in $\mathbb{Z}^n$ and put $M = (\varepsilon_1, \dots \varepsilon_r)$. Blankinship's algorithm allows us to find a unimodular matrix E such that

$$EM = \binom{T}{0}.$$

The vectors $\varepsilon_1, \dots \varepsilon_r$ comprise the first r elements of a basis of $\mathbb{Z}^n$ if and only if the diagonal coefficients of T are ± 1. When this is the case, a second round of operations towards the top allows one to make a unimodular matrix E' such that

$$E'EM = \binom{I_r}{0}.$$

The desired basis is then the columns of A^{-1} if $A = E'E$.

To calculate the inverse of A, we can perform row operations on (A, I) to bring it to the form (I, A^{-1}). But this process seems a little strange: we start with I on which we perform row operations to obtain A, then we perform more operations on A to recover I.

Let us look a little more closely at what happens. Let $E_1, E_2, \dots, E_N$ be elementary matrices which transform I into A:

$$A = E_N \cdots E_1\, I.$$

Then,

$$A^{-1} = I\, E_1^{-1} \cdots E_N^{-1},$$

so we see that we can calculate A^{-1} *in parallel* by suitable column operations on the matrix I.

For this, we begin with the identity matrix I. Each time that we perform a row operation on M, we do a column operation on I using the following dictionary to translate between them:

	Row operation (matrix EM)	*Column operation (matrix ME^{-1})*
	$L_i := L_i + \lambda L_j$	$K_j := K_j - \lambda K_i$
$(\mathfrak{D})$	$L_i := -L_i$	$K_i := -K_i$
	$L_i \rightleftarrows L_j$	$K_j \rightleftarrows K_j$.

Notice the behavior of the indices and the change of sign in the first line.

With this specified, the incomplete basis algorithm goes as follows. Let

$$M = (\varepsilon_1, \dots \varepsilon_r) \quad \text{and} \quad N = I_n.$$

We do (row) operations on M to obtain a matrix in row echelon form; in parallel, we perform column opeartions on N using our dictionary $(\mathfrak{D})$. If, at some moment, we get a diagonal coefficient d_i of M which is not equal to ± 1, we stop because we know that the given vectors can not be extended to a basis of $\mathbb{Z}^n$. Otherwise, we continue our operations until M is transformed into the identity matrix. At this point, the first r columns of N are the vectors with which we began and the last $(n-r)$ columns, together with the first r, form a basis of $\mathbb{Z}^n$.

Example 1

Consider the vector $\varepsilon = {}^t(3,4,8)$. Since it is a vector visible from the origin, we know that can be extended to a basis. To complete the basis, we start with the matrices

$$M_0 = \begin{pmatrix} \boxed{3} \\ 4 \\ 8 \end{pmatrix}, \qquad N_0 = \begin{pmatrix} 1 & 0 & 0 \\ 0 & 1 & 0 \\ 0 & 0 & 1 \end{pmatrix}.$$

The choice of pivot leads to the operations

Row operation (on M)	*Column operation (on N)*
$L_2 := L_2 - L_1$	$K_1 := K_1 + K_2$
$L_3 := L_3 - 2L_1$	$K_1 := K_1 + 2K_3$

which gives the matrices:

$$M_1 = \begin{pmatrix} 3 \\ \boxed{1} \\ 2 \end{pmatrix}, \qquad N_1 = \begin{pmatrix} 1 & 0 & 0 \\ 1 & 1 & 0 \\ 2 & 0 & 1 \end{pmatrix}.$$

The new pivot having been chosen, we perform the operations

Row operation (on M)	*Column operation (on N)*
$L_1 := L_1 - 3L_2$	$K_2 := K_2 + 3K_1$
$L_3 := L_3 - 2L_2$	$K_2 := K_2 + 2K_3$

and we wind up with the matrices:

$$M_2 = \begin{pmatrix} 0 \\ \boxed{1} \\ 0 \end{pmatrix}, \qquad N_2 = \begin{pmatrix} 1 & 3 & 0 \\ 1 & 4 & 0 \\ 2 & 8 & 1 \end{pmatrix}.$$

It remains only to interchange rows $1, 2$ of M_2 and columns $1, 2$ of N_2 to finally obtain:

$$M_3 = \begin{pmatrix} 1 \\ 0 \\ 0 \end{pmatrix}, \qquad N_3 = \begin{pmatrix} 3 & 1 & 0 \\ 4 & 1 & 0 \\ 8 & 2 & 1 \end{pmatrix}.$$

The vector ε from the outset reappears in the first column of N_3 as the theeory predicts. The completed basis is formed by the three columns of N_3.

Example 2

Now consider the vectors:

$$\varepsilon_1 = {}^t(3, 2, 7, 9), \qquad \varepsilon_2 = {}^t(2, 5, 3, 6).$$

A first round of operations gives the matrices:

$$M = \begin{pmatrix} 1 & -3 \\ 0 & 1 \\ 0 & 0 \\ 0 & 0 \end{pmatrix}, \qquad N = \begin{pmatrix} 3 & 11 & 6 & 0 \\ 2 & 11 & 6 & 0 \\ 7 & 24 & 13 & 0 \\ 9 & 33 & 18 & 1 \end{pmatrix}.$$

Since $d_1 = d_2 = 1$, we know that the vectors ε_1 are ε_2 part of a basis of $\mathbb{Z}^4$. To extend them to a basis, we finish by doing row operations on M towards the top so as to "kill" the coefficient -3 (and we perform the corresponding column operations on N). We get:

$$M = \begin{pmatrix} 1 & 0 \\ 0 & 1 \\ 0 & 0 \\ 0 & 0 \end{pmatrix}, \qquad N = \begin{pmatrix} 3 & 2 & 6 & 0 \\ 2 & 5 & 6 & 0 \\ 7 & 3 & 13 & 0 \\ 9 & 6 & 18 & 1 \end{pmatrix}.$$

The desired basis is formed by the columns of N.

Exercise 8

Transform this algorithm into a program.

11.1.7. Finding a basis of a subgroup

Let $M = \langle g_1, \ldots, g_N \rangle$ be the subgroup of $\mathbb{Z}^n$ with generators $g_1, \ldots, g_N$ and consider the matrix whose columns are the generators:

$$G = (g_1, \ldots, g_N).$$

We perform *column* operations on G to obtain a matrix G' in column echelon form

$$G' = GF, \qquad F \in \mathrm{Gl}(n, \mathbb{Z}).$$

Theorem 11.1.5. *With the preceding notation, the nonzero columns of G' are a basis for the subgroup M.*

Proof. We already know that the nonzero columns $g'_1, \ldots, g'_r$ of G' are independent. Since F has integer coefficients, the g'_j are linear combinations with integer coefficients of the g_i. Conversely, since F^{-1} also has integer coefficients, the g_i are linear combinations of the g'_j with integer coefficients. □

Corollary 11.1.1. *Every subgroup of $\mathbb{Z}^n$ has a basis, since every subgroup is of finite type.*

Example 1

We return to the subgroup M of $\mathbb{Z}^2$ generated by the columns of the matrix

$$G = \begin{pmatrix} 2 & 0 & 5 \\ 0 & 3 & 5 \end{pmatrix}.$$

Performing column operations on the matrix $G = (g_1, g_2, g_3)$, we obtain:

$$G' = GF = \begin{pmatrix} 1 & 0 & 0 \\ 0 & 1 & 0 \end{pmatrix}$$

Thus a basis of M is formed from the canonical basis of $\mathbb{Z}^2$, which is not surprising since $M = \mathbb{Z}^2$.

Example 2

Consider the subgroup N of $\mathbb{Z}^3$ generated by the columns of the matrix

$$G = \begin{pmatrix} 3 & 5 & 2 & 0 \\ 1 & 2 & 4 & 7 \\ 4 & -6 & 8 & 2 \end{pmatrix}.$$

By performing column operations on G, we obtain the matrices:

$$G' = GF = \begin{pmatrix} 1 & 0 & 0 & 0 \\ -3 & 1 & 0 & 0 \\ 0 & -38 & 4 & 0 \end{pmatrix}, \quad F = \begin{pmatrix} 4 & -5 & 3 & -83 \\ -1 & 3 & -1 & 23 \\ -3 & 0 & -2 & 67 \\ 1 & 0 & 1 & -33 \end{pmatrix}.$$

Therefore, a basis of N consists of the first three columns of G':

Verifying these calculations is easy. We have

$$\begin{aligned} g_1' &= 4g_1 - g_2 - 3g_3 + g_4, \\ g_2' &= -5g_1 + 3g_2, \\ g_3' &= 3g_1 - g_2 - 2g_3 + g_4. \end{aligned}$$

Conversely, knowing that

$$F^{-1} = \begin{pmatrix} 3 & 5 & 2 & 0 \\ 10 & 17 & 10 & 7 \\ 96 & 160 & 97 & 67 \\ 3 & 5 & 3 & 2 \end{pmatrix},$$

we have

$$\begin{aligned} g_1 &= 3g_1' + 10g_2' + 96g_3', & g_3 &= 2g_1' + 10g_2' + 97g_3', \\ g_2 &= 5g_1' + 17g_2' + 160g_3', & g_4 &= 0g_1' + 7g_2' + 67g_3'. \end{aligned}$$

Exercise 9

Convert this algorithm to a program. But be careful because very large integers can easily appear in the intermediate calculations.

11.2. Linear Systems with Integral Coefficients

Consider the general linear system with p equations in n unknowns

$$Ax = b, \quad i.e. \quad \begin{cases} a_{1,1}x_1 + \cdots + a_{1,n}x_n = b_1, \\ \ldots \\ a_{p,1}x_1 + \cdots + a_{p,n}x_n = b_p. \end{cases} \tag{11.1}$$

Suppose the $a_{i,j}$ and b_i are *integers* and that we want to find integer solutions of this system; that is, *all* vectors $x \in \mathbb{Z}^n$ satisfying the system.

11.2.1. Theoretical results

We begin by recalling a result which is as indispensable as it is trivial to prove.

Lemma 11.2.1. *If $\tilde{x} \in \mathbb{Z}^n$ is a particular solution of* (11.1), *then all other solutions are of the form $x = \tilde{x} + \xi$, where $\xi \in \mathbb{Z}^n$ is the general solution of the associated homogeneous system $A\xi = 0$.*

Students traditionally learn to solve linear systems by performing *row* operations because this does not change the value of the x_i. But when one programs, it turns out to be preferable to perform *column operations* which *do change the unknowns*. The following result justifies this practice.

Proposition 11.2.1. *Let F be any unimodular matrix and put $B = AF$. The map*

$$y \longmapsto x = Fy$$

is a bijection between solutions $y \in \mathbb{Z}^n$ of the system $By = b$ and solutions $x \in \mathbb{Z}^n$ of the system $Ax = b$.

Proof. It follows from $Ax = b$ that $AFy = b$ and conversely. Moreover, $x = Fy$ and $y = F^{-1}x$ show that the coefficients of x are integers if and only if the coefficients of y are integers. □

11.2.2. The case of a matrix in column echelon form

Solving the system

$$\begin{cases} 2x & = b_1, \\ 7x & = b_2, \\ 3x + 2y & = b_3, \\ x - 2y + 5z & = b_4. \end{cases} \tag{11.2}$$

in integers is especially easy because we are dealing with a matrix in column echelon form. If we solve the equations successively, working first over the field of rational numbers, we obtain:

$$\begin{cases} x = \frac{1}{2}b_1 = \frac{1}{7}b_2\,, \\ y = \frac{1}{2}\left(b_3 - \frac{3}{2}b_1\right), \\ z = \frac{1}{5}\left(b_4 - \frac{1}{2}b_1 + (b_3 - \frac{3}{2}b_1)\right). \end{cases} \tag{11.3}$$

These calculations show that the system has a solution $(x, y, z) \in \mathbb{Z}^3$ if and only if the following conditions hold:

- $\frac{1}{2}b_1$ and $\frac{1}{7}b_2$ are equal integers;
- $\frac{1}{2}\left(b_3 - \frac{3}{2}b_1\right)$ is an integer;
- $\frac{1}{5}\left(b_4 - \frac{1}{2}b_1 + (b_3 - \frac{3}{2}b_1)\right)$ is an integer.

Since our goal is to program the solution of the system, we present the calculations somewhat differently. To simplify, suppose that the second member is the vector $b = {}^t(4, 14, 8, 15)$ so the solution is the $x = {}^t(2, 1, 3)$.

We write K_1, K_2, K_3 the columns of the matrix of the system; below we write the identity matrix. This done, we write the vector $\beta_0 = -b$ bounded below by the zero column (see Table 11.2). We then try to make the vector $\beta_0 = -b$ zero by adding suitable integer multiples of the columns of the system:

$$\beta_1 = \beta_0 + 2K_1, \quad \beta_2 = \beta_1 + K_2, \quad \beta_3 = \beta_2 + 3K_3 = 0.$$

Note that these linear combinations are unique because the K_i are independent: the first linear combination is the only one capable of killing $-b_1$, etc.

system			$\beta_0=-b$	$\beta_1=\beta_0+2K_1$	$\beta_2=\beta_1+K_2$	$\beta_3=\beta_2+3K_3$
2	0	0	-4	0	0	0
7	0	0	-14	0	0	0
3	2	0	-8	-2	0	0
1	-2	5	-15	-13	-15	0
1	0	0	0	2	2	2
0	1	0	0	0	1	1
0	0	1	0	0	0	2
identity matrix						solution

Table 11.2. *Practical resolution of the system* (11.2)

• If we succeed in making $\beta_0 = -b$ vanish, it is clear that the corresponding linear combinations of the columns of the identity matrix (below the horizontal line) are a particular solution of the system.

• If we fail, the system does not possess an integer solution because b does not belong to the subgroup generated by the columns of the sytem.

11.2.3. General case

Blankinship's algorithm allows us to explicitly compute a unimodular matrix F such that $B = AF$ is in column echelon form. We then solve the system

$$By = b. \tag{11.4}$$

Finally, we return to the solutions of the original equation by the transformation $x = Fy$. But, as we will see, this last step can be done *automatically* if we organize our calculations well.

Consider for example the system defined by

$$A = \begin{pmatrix} 1 & 2 & 0 & 1 & 3 \\ 0 & 1 & -1 & 1 & 2 \\ 1 & -1 & 2 & 0 & -1 \end{pmatrix}, \quad b = \begin{pmatrix} 5 \\ 4 \\ -5 \end{pmatrix}.$$

• To find B and F, we border A below by the identity matrix. We then perform column operations on this large matrix until A is in column echelon form:

$$\begin{pmatrix} A \\ I \end{pmatrix} \xmapsto{\text{column operations}} \begin{pmatrix} B \\ F \end{pmatrix}, \quad B = AF, \quad F \in \mathrm{Gl}(n, \mathbb{Z}).$$

In our example, we find the following matrices F and $B = AF$.

						β_0	β_1	β_2	β_3
B	1	0	0	0	0	−5	0	0	0
	0	1	0	0	0	−4	−4	0	0
	1	−3	1	0	0	5	10	−2	0
F	1	−2	−1	3	3	0	5	−3	−5
	0	1	0	−1	−2	0	0	4	4
	0	0	1	−2	−2	0	0	0	2
	0	0	1	−1	−2	0	0	0	2
	0	0	0	0	1	0	0	0	0

• We write the vector $\beta_0 = -b$ to the right of B and the zero vector to the right of F. To this big vector, we add appropriate multiples of the columns of B and of F, which gives the vectors $\beta_1, \beta_2, \beta_3 = 0$. We find therefore:

▷ a particular solution $\tilde{x}$ of the initial system $Ax = b$ which is the vector $\tilde{x} = {}^t(-5, 4, 2, 2, 0)$ situated below β_3;

▷ a basis of the subgroup of solutions of the homogeneous system which is formed by the vectors $\xi_1 = {}^t(4, -1, -2, -1, 0)$ and $\xi_2 = {}^t(0, 1, -2, 1, 1)$ situated below the zero columns of B.

The solutions in integers of the system $Ax = b$ are thus the vectors

$$x = \tilde{x} + \lambda\xi_1 + \mu\xi_2, \quad \lambda, \mu \in \mathbb{Z}.$$

To explain this minor miracle, we let r denote the rank of A and write out the columns of our matrices:

$$\begin{pmatrix} B \\ F \end{pmatrix} = \begin{pmatrix} B_1 & B_2 & \dots & B_r & 0 & \dots & 0 \\ F_1 & F_2 & \dots & F_r & F_{r+1} & \dots & F_n \end{pmatrix}.$$

If the system $By = b$ has a solution in integers, there exist integers y_i satisfying

$$y_1 B_1 + \cdots + y_r B_r - b = 0; \tag{11.5}$$

in other words, $By = b$ has integer solutions if and only if it is possible to zero out the vector $\beta_0 = -b$ by adding to it integer multiples of the columns $B_1, \dots, B_r$.

We now turn to the columns of F. The vector which appears under the last vector β_r is $y_1 F_1 + \cdots + y_r F_r$, where the y_i satisfy (8.5). This allows us to write

$$A(y_1 F_1 + \cdots + y_r F_r) = y_1 A F_1 + \cdots + y_r A F_r = y_1 B_1 + \cdots + y_r B_r = b,$$

which shows that the vector $\tilde{x} = y_1 F_1 + \cdots + y_r F_r$ is indeed a particular solution of the complete system $Ax = b$.

Finally, it is clear that $By = 0$ is equivalent to $y = {}^t(0, \dots 0, \lambda_{r+1}, \dots, \lambda_n)$. Returning to $x = Fy$, we find that the solutions of $Ax = 0$ are the vectors

$$\xi = \lambda_{r+1} F_{r+1} + \cdots + \lambda_n F_n. \qquad \square$$

11.2.4. Case of a single equation

Suppose that we want to solve the equation:

$$12\,x_1 - 6\,x_2 + 9\,x_3 - 21\,x_4 = b. \tag{11.6}$$

Since the matrix of the system has a single row, pivoting is very rapid:

$$\begin{matrix} 12 & -6 & 9 & -21 \\ 1 & 0 & 0 & 0 \\ 0 & 1 & 0 & 0 \\ 0 & 0 & 1 & 0 \\ 0 & 0 & 0 & 1 \end{matrix} \quad \longmapsto \quad \begin{matrix} 0 & -6 & 3 & 3 \\ 1 & 0 & 0 & 0 \\ 2 & 1 & 1 & -4 \\ 0 & 0 & 1 & 0 \\ 0 & 0 & 0 & 1 \end{matrix}$$

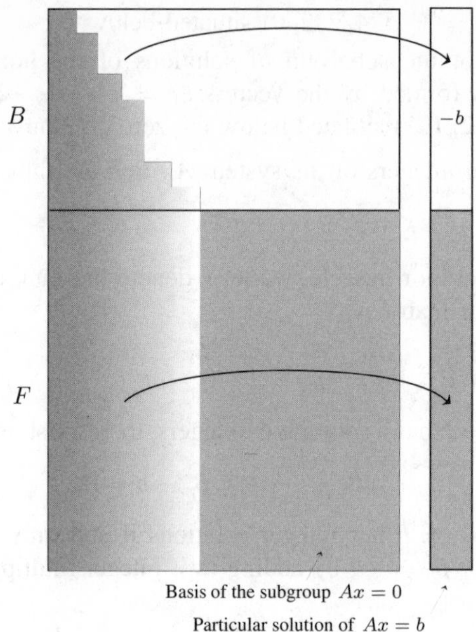

Fig. 11.3. Algorithm for solving the system $Ax = b$ in integers

$$\begin{matrix} & 0 & 0 & 3 & 0 & & 3 & 0 & 0 & 0 \\ & 1 & 0 & 0 & 0 & & 0 & 0 & 1 & 0 \\ \longmapsto & 2 & 3 & 1 & -5 & \longmapsto & 1 & 3 & 2 & -5 \\ & 0 & 2 & 1 & -1 & & 1 & 2 & 0 & -1 \\ & 0 & 0 & 0 & 1 & & 0 & 0 & 0 & 1 \end{matrix}$$

Thus, we have GCD$(12, -6, 9, -21) = 3$. In view of the simplicity of the situation, it is pointless to border the matrix with the vector $\beta_0 = -b$: the given equation has solutions in integers if and only if the condition $b \equiv 0 \pmod 3$ is satisfied. When this is the case, all the solutions are:

$$x = \frac{b}{3}\begin{pmatrix}0\\1\\1\\0\end{pmatrix} + \lambda\begin{pmatrix}0\\3\\2\\0\end{pmatrix} + \mu\begin{pmatrix}1\\2\\0\\0\end{pmatrix} + \nu\begin{pmatrix}0\\-5\\-1\\1\end{pmatrix}, \qquad \lambda, \mu, \nu \in \mathbb{Z} \text{ arbitrary.}$$

11.3. Exponential of a Matrix: Putzer's Algorithm

Let A be an $n \times n$ matrix with complex coefficients. One defines the exponential e^{At} in the classical way by the series

$$e^{At} = \sum_{k=0}^{\infty} \frac{A^k t^k}{k!}. \tag{11.7}$$

This pretty formula wreaks much havoc because it is often presented to students who have not yet mastered ordinary series.

Because of the Cayley-Hamilton theorem, we know that – in a certain sense – series of matrices do not *exist*! For there is an equation of the form $A^n = p(A)$, where p is a polynomial of degree less than n, so that A^n, A^{n+1}, etc., are polynomials in A of degree not exceeding n. If we replace these matrices in (11.7) by the corresponding polynomials we find that the exponential of A is of the form

$$e^{At} = \alpha_0(t)I + \alpha_1(t)A + \cdots + \alpha_{n-1}(t)A^{n-1},$$

where the α_i are functions defined by *series* (these are the remnants of the initial series). The explicit determination of the α_i is due Putzer.[3] We fix the following notation:

- Let $\lambda_1, \ldots, \lambda_n$ denote the eigenvalues of A. We do not make any restrictive hypotheses: the eigenvalues may be multiple, and we do not suppose that A is diagonalizable.
- Put (notice the shift in the indices):

$$\begin{aligned} B_1 &= I, \\ B_2 &= (A - \lambda_1 I)B_1, \\ B_3 &= (A - \lambda_2 I)B_2, \\ &\vdots \\ B_n &= (A - \lambda_{n-1} I)B_{n-1}. \end{aligned} \tag{11.8}$$

With this notation, we can write the Cayley-Hamilton theorem as:

$$B_{n+1} = (A - \lambda_n I)B_n = 0. \tag{11.9}$$

[3] E.J. Putzer, *Avoiding the Jordan canonical form in the discussion of linear systems with constant coefficients*, American Mathematical Monthly 73 (1966), pp. 2–7. This appeared over thirty years ago... The diffusion of this algorithm into the teaching world has been very, very slow.

- Let $y_1, \ldots, y_n$ be the solutions of the following differential equations:

$$\begin{aligned} y_1' &= \lambda_1 y_1, & y_1(0) &= 1; \\ y_2' &= \lambda_2 y_2 + y_1, & y_0(0) &= 0; \\ &\vdots & &\vdots \\ y_n' &= \lambda_n y_n + y_{n-1}, & y_n(0) &= 0. \end{aligned} \tag{11.10}$$

Here, there is no shift in indices; on the contrary, the first equation keeps to itself (it does not have a second term and the initial condition is different from the others).

Theorem 11.3.1 (Putzer, 1966). *With the notation above,*

$$e^{At} = y_1(t)B_1 + y_2(t)B_2 + \cdots + y_n(t)B_n. \tag{11.11}$$

Proof. Let $E(t)$ be the term on the right of (11.11). It suffices to show that $E(0) = I$ and $E'(t) = AE(t)$ since these two conditions characterise the matrix (11.7). It is clear that $E(0) = I$, Diferentiating E gives:

$$E' = y_1' B_1 + \cdots + y_n' B_n.$$

Use (11.10) and then (11.7) to get:

$$\begin{aligned} E' &= \lambda_1 y_1 B_1 \\ &\quad + \lambda_2 y_2 B_2 + y_1 B_2 \\ &\qquad + \lambda_3 y_3 B_3 + y_2 B_3 \\ &\qquad\quad \ddots \\ &\qquad\qquad + \lambda_{n-1} y_{n-1} B_{n-1} + y_{n-2} B_{n-1} \\ &\qquad\qquad\quad + \lambda_n y_n B_n + y_{n-1} B_n \\ &= \lambda_1 y_1 B_1 \\ &\quad + \lambda_2 y_2 B_2 + y_1 (A - \lambda_1 I) B_1 \\ &\qquad + \lambda_3 y_3 B_3 + y_2 (A - \lambda_2 I) B_2 \\ &\qquad\quad \ddots \\ &\qquad\qquad + \lambda_{n-1} y_{n-1} B_{n-1} + y_{n-2} (A - \lambda_{n-2} I) B_{n-2} \\ &\qquad\qquad\quad + \lambda_n y_n B_n + y_{n-1} (A - \lambda_{n-1} I) B_{n-1} \end{aligned}$$

After simplifying, we are left with $E' = y_1 AB_1 + \cdots + y_{n-1} AB_{n-1} + \lambda_n y_n B_n$; that is,

$$E' = y_1 AB_1 + \cdots + y_{n-1} AB_{n-1} + y_n AB_n$$

since (11.9) is written $\lambda_n B_n = AB_n$.

□

Example

If A is a 3×3 matrix with a triple eigenvalue λ, its exponential is:

$$e^{At} = e^{\lambda t} I + t\, e^{\lambda t}(A - \lambda I) + \tfrac{1}{2} t^2 e^{\lambda t}(A - \lambda I)^2.$$

(Whether or not A is diagonalizable doesn't matter.)

Remark

Formula (11.11) is an algorithm in a *theoretical and academic* sense only; it is poorly behaved numerically[4] because it requires knowledge of the eigenvalues of the matrix. (The main difficulty is the *precise* calculation of these numbers.)

Exercise 10

Write a Pascal program which calculates the matrices B_i knowing the matrix A and its eigenvalues. The calculation can be done over $\mathbb{Z}$, $\mathbb{R}$ or $\mathbb{C}$.

Remarks

1) You can try to calculate the eigenvalues of the matrix from its characteristic polynomial (Chapter 5), but you should not expect miracles because getting good precision in all circumstances is very delicate.

2) To make a matrix with integer coefficients and integer eigenvalues, we begin with a triangular matrix with integer coefficients

$$A = \begin{pmatrix} \lambda_1 & * & * \\ 0 & \ddots & * \\ 0 & 0 & \lambda_n \end{pmatrix}.$$

and complicate it using row and column operations of the form $A \mapsto EAE^{-1}$. For translation, you can use the algorithm of the incomplete basis (§1.6) as a dictionary.

3) If you have become enchanted by integers, you might wonder if the preceeding method can manufacture all matrices $A \in M(n, \mathbb{Z})$ whose integers are eigenvalues.

Theorem 11.3.2 (Leavitt and Whaples, 1948). *Let A be an $n \times n$ matrix with integer coefficients. The eigenvalues of A are integers if and only if there exists a unimodular matrix E such that $U = E^{-1}AE$ is an upper triangular matrix.*

[4] C. Moler and C. Van Loan, *Nineteen dubious ways to compute the exponential of a matrix*, SIAM Review, 20 (1978), pp. 801–836.

Proof. Let A be a matrix with integer coefficients and integer eigenvalues. Since the eigenvalues are rational, we already know how to find a matrix $P \in \mathrm{Gl}(n, \mathbb{Q})$ such that $U = P^{-1}AP$ is upper triangular. We write $AP = PU$ and multiply P by a suitable integer to ensure that this matrix has integer coefficients.

By performing row operations on P in accord with Blankinship's algorithm, we know that we will eventually find a unimodular matrix E such that $U' = EP$ is upper triangular. We get

$$(EAE^{-1})EP = EPU.$$

The identity $EAE^{-1} = U'UU'^{-1}$ then shows that EAE^{-1} is upper triangular. On the other hand, EAE^{-1} has integer coefficients since E is unimodular. The converse is clear. □

11.4. Jordan Reduction

Jordan's theory always intimidates apprentice mathematicians and many others as well. The origin of this malaise is easy to diagnose: needlessly abstract explanations culminating in deceptive exercises in which the transition matrix *is not* in general made explicit. We are going to detail here an elementary algorithm[5] which:

- *proves* the existence of the Jordan canonical form;
- *explicitly furnishes* the Jordan canonical form *as well as* the transition matrix.

Like the Putzer algorithm, this algorithm is of *academic* interest only; it is not adapted to numerical calculation because it requires that one know in advance the eigenvalues of a matrix (which is, we recall again, the numerically difficult part).

11.4.1. Review

Let $f : E \to E$ be an endomorphism of an n-dimensional vector space E over a field k which has all of its eigenvalues in k. Let $\lambda_1, \dots, \lambda_p$ be the distinct eigenvalues of f, with multiplicities $m_1, \dots, m_p$, so that $m_1 + \cdots + m_p = n$. The *characteristic spaces* of f are the vector subspaces of E:

$$E_i = \ker(f - \lambda_i \,\mathrm{id})^{m_i}.$$

We accept without proof the following elementary results (which are not at all difficult to prove):

[5] U. Pittelkow and H.-J. Runckel, *A short and constructive approach to the Jordan canonical form of a matrix*, Serdica 7 (1981), pp. 348–359. Added in proof: for an even simpler algorithm, see also A. Bujosa, R. Criado, C. Vega, *Jordan normal form via elementary transformations*, SIAM Review, 40 (1998), pp. 947–956.

(i) $E = E_1 \oplus \cdots \oplus E_p$;
(ii) $\dim E_i = m_i$;
(iii) E_i is mapped to itself by f;
(iv) the map $u_i = f - \lambda_i \,\mathrm{id} : E_i \to E_i$ is *nilpotent.*

Thus, the characteristic subspace E_i contains the eigenspace asociated with the eigenvalue λ_i. It results from (i) that A is not diagonalizable when some characteristic subspace strictly contains an eigenspace.

The Pittelkow-Runckel algorithm finds a basis $e_1^{(i)}, \dots, e_{m_i}^{(i)}$ of each characteristic subspace E_i. In this basis the matrix of the nilpotent endomorphism u_i is:

$$\begin{pmatrix} 0 & \varepsilon_1 & & & \\ & 0 & \varepsilon_2 & & \\ & & \ddots & \ddots & \\ & & & 0 & \varepsilon_{m_i} \end{pmatrix}, \qquad \varepsilon_i = 0, 1. \tag{11.12}$$

Consequently, in the basis $e_1^{(1)}, \dots, e_{m_1}^{(1)}, \dots, e_1^{(p)}, \dots, e_{m_p}^{(p)}$ of E (see (i)), the matrix of f is:

$$\begin{pmatrix} J_1 & & & \\ & J_2 & & \\ & & \ddots & \\ & & & J_p \end{pmatrix} \quad \text{with} \quad J_i = \begin{pmatrix} \lambda_i & \varepsilon_1 & & & \\ & \lambda_i & \varepsilon_2 & & \\ & & \ddots & \ddots & \\ & & & \lambda_i & \varepsilon_{m_i} \end{pmatrix}. \tag{11.13}$$

11.4.2. Reduction of a nilpotent endomorphism

Let $u : E \to E$ be a nilpotent endomorphism of a finite dimensional vector space. When $x \in E$ is not zero, we consider the iterates $x^{(k)} = u^k(x)$ of x under u:

$$x^{(0)} \neq 0, \quad x^{(1)} \neq 0, \quad \dots, \quad x^{(\ell)} \neq 0, \quad x^{(\ell+1)} = 0.$$

Definition 11.4.1. *The last exponent of a vector $x \neq 0$ is the greatest integer $\ell \geq 0$ such that $x^{(\ell)} \neq 0$.*

For example, $\ell = 0$ if and only if $x \in \ker u$.

These iterates are a natural tool in Jordan theory: when $x^{(\ell)}, \dots, x^{(0)}$ are a basis (note the decreasing indices), the matrix of the endomorphism u in this basis is a Jordan matrix (11.12) with $\varepsilon_i = 1$.

Definition 11.4.2. *Let $x_1, \dots, x_r$ be nonzero vectors and $\ell_1, \dots, \ell_r$ their last exponents. Put:*

$$\Sigma = \left\{x_1^{(\ell_1)}, \dots, x_r^{(\ell_r)}\right\},$$
$$\widetilde{\Sigma} = \left\{x_1^{(\ell_1)}, \dots, x_1^{(0)}, \dots, x_r^{(\ell_r)}, \dots, x_r^{(0)}\right\}.$$

We say the the system $\widetilde{\Sigma}$ of vectors is deployed over the system Σ.

Proposition 11.4.1. *The system* Σ *is linearly independent if and only if the deployed system* $\widetilde{\Sigma}$ *is linearly independent.*

Proof. To avoid a deluge of indices which will teach us nothing, we content ourselves with a particular case:

$$\Sigma = \{x_1^{(1)}, x_2^{(3)}, x_3^{(2)}, x_4^{(2)}, x_5^{(0)}\}.$$

Consider the following linear combination of the vectors of $\widetilde{\Sigma}$

$$\begin{aligned} 0 = {} & a_0 x_1^{(0)} + a_1 x_1^{(1)} \\ & + b_0 x_2^{(0)} + b_1 x_2^{(1)} + b_2 x_2^{(2)} + b_3 x_2^{(3)} \\ & + c_0 x_3^{(0)} + c_1 x_3^{(1)} + c_2 x_3^{(2)} \\ & + d_0 x_4^{(0)} + d_1 x_4^{(1)} + d_2 x_4^{(2)} \\ & + e_0 x_4^{(0)} \end{aligned} \tag{11.14}$$

and associate to it the array on the left below:

(3)		b_3			
(2)		b_2	c_2	d_2	
(1)	a_1	b_1	c_1	d_1	
(0)	a_0	b_0	c_0	d_0	e_0
	x_1	x_2	x_3	x_4	x_5

(3)		β			
(2)			γ	δ	
(1)	α				
(0)					ε
	x_1	x_2	x_3	x_4	x_5

The condition that Σ be independent tells us that each time that we have a linear combination symbolized by the tableau on the right, the coefficients $\alpha, \beta, \gamma, \delta, \varepsilon$ are necessarily zero.

However, if we apply u^3 to the linear combination (11.14), we move the coefficients towards the top in the left tableau: those that leave disappear and the "holes" which appear are filled with zeroes. Thus, we obtain the right hand tableau with $\alpha = 0$, $\beta = b_0$, $\gamma = \delta = \varepsilon = 0$, which rquires that $b_0 = 0$. Applying u^2 to (11.14), we get $\alpha = 0$, $\beta = b_1$, $\gamma = c_0$, $\delta = d_0$, $\varepsilon = 0$, from whence $b_1 = c_0 = d_0 = 0$. If we apply u, we likewise obtain $a_0 = b_2 = c_1 = d_1 = 0$. We are left with $\alpha = a_1$, $\beta = b_3$, $\gamma = c_2$, $\delta = d_2$ and $\varepsilon = e_0$ which are necessarily zero. □

Proposition 11.4.2. *Let* $x_1, \ldots, x_r$ *and* y *be nonzero vectors whose last exponents* $\ell_1, \ldots, \ell_r$ *and* ℓ *satisfy* $\ell_i \geq \ell$ *for* $i = 1, \ldots, r$ *and*

$$y^{(\ell)} = a_1 x_1^{(\ell_1)} + \cdots + a_{r-1} x_r^{(\ell_{r-1})} + a_r x_r^{(\ell_r)}.$$

If the vector

$$y' = y^{(0)} - \left(a_1 x_1^{(\ell_1 - \ell)} + \cdots + a_r x_r^{(\ell_r - \ell)}\right),$$

is not zero, its last exponent ℓ' satisfies $\ell' < \ell$.

Proof. In effect, $y'^{(\ell)}$ is zero by construction. □

11.4.3. The Pittelkow-Runckel algorithm

The Pitttelkow-Runckel algorithm finds a Jordan basis of a nilpotent endomorphism. To do this, it manipulates the system Σ in a loop (steps 2 to 4) until the system becomes linearly independent.

1) *Initialization:* Choose nonzero vectors $x_1, \dots, x_r$ such that the deployed system associated to

$$\Sigma = \left\{x_1^{(\ell_1)}, \dots, x_r^{(\ell_r)}\right\}$$

generates E (for want of better, one can choose the x_i to be generators of E since the deployed system $\widetilde{\Sigma}$ contains the x_i).

2) *Exit test of the loop:* If Σ is linearly independent, the algorithm teminates and the deployed system $\widetilde{\Sigma}$ is the desired Jordan basis.

3) *Body of the loop:* If Σ is not independent, we can suppose, upon renumbering the vectors, tha $\ell_1 \geq \ell_2 \geq \cdots \geq \ell_r$. Thus, there exists an index $k \in [\![1, r]\!]$ such that:

$$x_k^{(\ell_k)} + a_1 x_1^{(\ell_1)} + \cdots + a_{k-1} x_{k-1}^{(\ell_{k-1})} = 0.$$

Set:

$$y = x_k^{(0)} + a_1 x_1^{(\ell_1 - \ell_k)} + \cdots + a_{k-1} x_{k-1}^{(\ell_{k-1} - \ell_k)}.$$

If $\ell_k = 0$ or if y is zero, remove the vector $x_k^{(\ell_k)}$ from Σ; otherwise, replace $x_k^{(\ell_k)}$ by $y^{(\ell)}$, where ℓ is the last exponent of y. (In practice, remember that it is the vector with the smallest exponent that disappears.)

4) *End of the loop:* return to 2).

Example

Consider the 5×5 nilpotent matrix:

$$A = \begin{pmatrix} 2 & 6 & -3 & 5 & -2 \\ 1 & 3 & 16 & 9 & -1 \\ 1 & 3 & 5 & 5 & -1 \\ -2 & -6 & -13 & -11 & 2 \\ -1 & -3 & 3 & -2 & 1 \end{pmatrix}, \quad A^2 = \begin{pmatrix} -1 & -3 & 4 & -2 & 1 \\ 4 & 12 & 5 & 15 & -4 \\ 1 & 3 & 2 & 4 & -1 \\ -3 & -9 & -6 & -12 & 3 \\ 1 & 3 & -1 & 3 & -1 \end{pmatrix}$$

$$A^3 = \begin{pmatrix} 2 & 6 & 4 & 8 & -2 \\ -1 & -3 & -2 & -4 & 1 \\ 0 & 0 & 0 & 0 & 0 \\ 0 & 0 & 0 & 0 & 0 \\ -1 & -3 & -2 & -4 & 1 \end{pmatrix}, \quad A^4 = \begin{pmatrix} 0 & 0 & 0 & 0 & 0 \\ 0 & 0 & 0 & 0 & 0 \\ 0 & 0 & 0 & 0 & 0 \\ 0 & 0 & 0 & 0 & 0 \\ 0 & 0 & 0 & 0 & 0 \end{pmatrix}.$$

- The algorithm begins[6] with the system:

$$\Sigma_0 = \{e_1^{(3)}, e_2^{(3)}, e_3^{(3)}, e_4^{(3)}, e_5^{(3)}\} = \begin{pmatrix} 2 & 6 & 4 & 8 & -2 \\ -1 & -3 & -2 & -4 & 1 \\ 0 & 0 & 0 & 0 & 0 \\ 0 & 0 & 0 & 0 & 0 \\ -1 & -3 & -2 & -4 & 1 \end{pmatrix}.$$

- Since $e_5^{(3)} + e_1^{(3)} = 0$, we put $x = e_5^{(0)} + e_1^{(0)}$ and replace e_5 by x:

$$\Sigma = \{e_1^{(3)}, e_2^{(3)}, e_3^{(3)}, e_4^{(3)}, x^{(0)}\} = \begin{pmatrix} 2 & 6 & 4 & 8 & 1 \\ -1 & -3 & -2 & -4 & 0 \\ 0 & 0 & 0 & 0 & 0 \\ 0 & 0 & 0 & 0 & 0 \\ -1 & -3 & -2 & -4 & 1 \end{pmatrix}.$$

- Since $e_4^{(3)} = 4e_1^{(3)}$, put $y = e_4^{(0)} - 4e_1^{(0)}$ and replace e_4 by y:

$$\Sigma_2 = \{e_1^{(3)}, e_2^{(3)}, e_3^{(3)}, y^{(2)}, x^{(0)}\} = \begin{pmatrix} 2 & 6 & 4 & 2 & 1 \\ -1 & -3 & -2 & -1 & 0 \\ 0 & 0 & 0 & 0 & 0 \\ 0 & 0 & 0 & 0 & 0 \\ -1 & -3 & -2 & -1 & 1 \end{pmatrix}.$$

- Since $e_3^{(3)} = 2e_1^{(3)}$, put $z = e_3^{(0)} - 2e_1^{(0)}$ and replace e_3 by z:

$$\Sigma_3 = \{e_1^{(3)}, e_2^{(3)}, z^{(2)}, y^{(2)}, x^{(0)}\} = \begin{pmatrix} 2 & 6 & 6 & 2 & 1 \\ -1 & -3 & -3 & -1 & 0 \\ 0 & 0 & 0 & 0 & 0 \\ 0 & 0 & 0 & 0 & 0 \\ -1 & -3 & -3 & -1 & 1 \end{pmatrix}.$$

- Since $z^{(2)} = e_2^{(3)}$, put $t = z^{(0)} - e_2^{(1)}$ and replace z by t:

$$\Sigma_4 = \{e_1^{(3)}, e_2^{(3)}, t^{(1)}, y^{(2)}, x^{(0)}\} = \begin{pmatrix} 2 & 6 & -4 & 2 & 1 \\ -1 & -3 & 2 & -1 & 0 \\ 0 & 0 & 0 & 0 & 0 \\ 0 & 0 & 0 & 0 & 0 \\ -1 & -3 & 2 & -1 & 1 \end{pmatrix}.$$

- Since $y^{(2)} = e_1^{(3)}$, put $u = y^{(0)} - e_1^{(1)}$ and replace y by u:

$$\Sigma_5 = \{e_1^{(3)}, e_2^{(3)}, t^{(1)}, u^{(1)}, x^{(0)}\} = \begin{pmatrix} 2 & 6 & -4 & -2 & 1 \\ -1 & -3 & 2 & 1 & 0 \\ 0 & 0 & 0 & 0 & 0 \\ 0 & 0 & 0 & 0 & 0 \\ -1 & -3 & 2 & 1 & 1 \end{pmatrix}.$$

[6] We deliberately chose a redundant system to lengthen the algorithm. But it is possible to do better by choosing the vectors so that $\widetilde{\Sigma}$ generates $\mathbb{R}^5$.

- Since $u^{(1)} = -e_1^{(3)}$, put $v = u^{(0)} + e_1^{(2)}$ and replace u by v:

$$\Sigma_6 = \{e_1^{(3)}, e_2^{(3)}, t^{(1)}, v^{(0)}, x^{(0)}\} = \begin{pmatrix} 2 & 6 & -4 & -7 & 1 \\ -1 & -3 & 2 & 3 & 0 \\ 0 & 0 & 0 & 0 & 0 \\ 0 & 0 & 0 & 0 & 0 \\ -1 & -3 & 2 & 2 & 1 \end{pmatrix}.$$

- Since $-e_1^{(3)} + e_2^{(3)} + t^{(1)} = 0$, put $w = t^{(0)} - e_1^{(2)} + e_2^{(2)}$ and replace t by w

$$\Sigma_7 = \{e_1^{(3)}, e_2^{(3)}, w^{(0)}, v^{(0)}, x^{(0)}\} = \begin{pmatrix} 2 & 6 & -10 & -7 & 1 \\ -1 & -3 & 5 & 3 & 0 \\ 0 & 0 & 0 & 0 & 0 \\ 0 & 0 & 0 & 0 & 0 \\ -1 & -3 & 5 & 2 & 1 \end{pmatrix}.$$

- Since $e_2^{(3)} + v^{(0)} + x^{(0)} = 0$, eliminate v:

$$\Sigma_8 = \{e_1^{(3)}, e_2^{(3)}, w^{(0)}, x^{(0)}\} = \begin{pmatrix} 2 & 6 & -10 & 1 \\ -1 & -3 & 5 & 0 \\ 0 & 0 & 0 & 0 \\ 0 & 0 & 0 & 0 \\ -1 & -3 & 5 & 1 \end{pmatrix}.$$

- Since $2e_1^{(3)} + e_2^{(3)} + w^{(0)} = 0$, eliminate w:

$$\Sigma_9 = \{e_1^{(3)}, e_2^{(3)}, x^{(0)}\} = \begin{pmatrix} 2 & 6 & 1 \\ -1 & -3 & 0 \\ 0 & 0 & 0 \\ 0 & 0 & 0 \\ -1 & -3 & 1 \end{pmatrix}.$$

- Since $e_2^{(3)} = 3e_1^{(3)}$, put $p = e_2^{(0)} - 3e_1^{(0)}$ and replace e_2 by p:

$$\Sigma_{10} = \{e_1^{(3)}, p^{(0)}, x^{(0)}\} = \begin{pmatrix} 2 & -3 & 1 \\ -1 & 1 & 0 \\ 0 & 0 & 0 \\ 0 & 0 & 0 \\ -1 & 0 & 1 \end{pmatrix}.$$

- Since $e_1^{(3)} + p^{(0)} + x^{(0)} = 0$, we eliminate p:

$$\Sigma_{11} = \{e_1^{(3)}, x^{(0)}\} = \begin{pmatrix} 2 & 1 \\ -1 & 0 \\ 0 & 0 \\ 0 & 0 \\ -1 & 1 \end{pmatrix}.$$

• The last system is manifestly linearly independent, and the desired Jordan base is the deployed system

$$P = \widetilde{\Sigma}_{11} = \left\{e_1^{(3)}, e_1^{(2)}, e_1^{(1)}, e_1^{(0)}, x^{(0)}\right\} = \begin{pmatrix} 2 & -1 & 2 & 1 & 1 \\ -1 & 4 & 1 & 0 & 0 \\ 0 & 1 & 1 & 0 & 0 \\ 0 & -3 & -2 & 0 & 0 \\ -1 & 1 & -1 & 0 & 1 \end{pmatrix}$$

and the Jordan reduction is:

$$J = \begin{pmatrix} 0 & 1 & 0 & 0 & 0 \\ 0 & 0 & 1 & 0 & 0 \\ 0 & 0 & 0 & 1 & 0 \\ 0 & 0 & 0 & 0 & 0 \\ 0 & 0 & 0 & 0 & 0 \end{pmatrix}$$

One can check that $P^{-1}AP = J$ as well (to avoid a painful calculation of the inverse of P, it is preferable to check that $AP = PJ$).

11.4.4. Justification of the Pittelkow-Runckel algorithm

The algorithm does not loop indefinitely because the cardinality of $\widetilde{\Sigma}$ decreases by at least once each time through the loop.

Note that if the vector x belongs to the subspace $\mathrm{Vect}(\widetilde{\Sigma})$, then *all* its iterates belong to this subspace by definition of the deployed system.

With the notation above:

• if the vector y is zero or if $\ell_k = 0$ (which means $u(x_k) = 0$), let Σ' be the new system obtained by removing $x_k^{(\ell_k)}$ from Σ;

• if y is not zero, let Σ' be the system obtained by replacing the vector $x_k^{(\ell_k)}$ in Σ by $y^{(\ell)}$.

In each of the two cases, since $x_k^{(0)}$ belongs to the subspace $\mathrm{Vect}\{\widetilde{\Sigma}'\}$, one deduces that $\mathrm{Vect}\{\widetilde{\Sigma}'\} = \mathrm{Vect}(\widetilde{\Sigma})$, which shows that the assertion "the deployed system $\widetilde{\Sigma}$ generates E" is an invariant of the loop.

The algorithm stops when Σ becomes independent. It follows from Proposition 11.4.1 that $\widetilde{\Sigma}$ is also independent. But since $\widetilde{\Sigma}$ never stops spanning E, we consclude that it is a basis. By virtue of the preceding remarks, $\widetilde{\Sigma}$ is a Jordan basis.

11.4.5. A complete example

Consider the 7×7 matrix

$$A = \begin{pmatrix} 4 & 14 & 2 & 4 & 7 & 3 & 0 \\ 1 & 3 & 1 & 2 & 1 & 0 & 0 \\ 0 & 4 & 1 & 2 & 2 & 0 & -1 \\ -1 & -6 & -1 & -1 & -3 & -1 & 0 \\ -1 & 2 & -1 & -2 & 2 & 1 & 0 \\ -2 & -14 & -1 & -4 & -7 & -1 & 1 \\ 1 & 2 & 1 & 0 & 1 & 1 & 2 \end{pmatrix}$$

whose eigenvalues are $\lambda = 2$ (with multiplicity 3) and $\mu = 1$ (with multiplicity 4).

• Put $B = A - 2I$, so that the characteristic subspace E_2 associated to the eigenvalue $\lambda = 2$ is the kernel of B^3:

$$B = \begin{pmatrix} 2 & 14 & 2 & 4 & 7 & 3 & 0 \\ 1 & 1 & 1 & 2 & 1 & 0 & 0 \\ 0 & 4 & -1 & 2 & 2 & 0 & -1 \\ -1 & -6 & -1 & -3 & -3 & -1 & 0 \\ -1 & 2 & -1 & -2 & 0 & 1 & 0 \\ -2 & -14 & -1 & -4 & -7 & -3 & 1 \\ 1 & 2 & 1 & 0 & 1 & 1 & 0 \end{pmatrix},$$

$$B^2 = \begin{pmatrix} 1 & -2 & 2 & 2 & -1 & 0 & 1 \\ 0 & 9 & -1 & 0 & 4 & 2 & -1 \\ -1 & -10 & 0 & -4 & -5 & -1 & 1 \\ 0 & 2 & 0 & 1 & 1 & 0 & 0 \\ 0 & -18 & 2 & 0 & -8 & -4 & 2 \\ 0 & 8 & -2 & 0 & 4 & 1 & -2 \\ 1 & 8 & 1 & 4 & 4 & 1 & 0 \end{pmatrix},$$

$$B^3 = \begin{pmatrix} 0 & 8 & -2 & 0 & 4 & 1 & -2 \\ 0 & -17 & 3 & 0 & -8 & -3 & 3 \\ 0 & 6 & -1 & 2 & 3 & 0 & -1 \\ 0 & -2 & 0 & -1 & -1 & 0 & 0 \\ 0 & 34 & -6 & 0 & 16 & 6 & -6 \\ 0 & -10 & 3 & 0 & -5 & -1 & 3 \\ 0 & -4 & 0 & -2 & -2 & 0 & 0 \end{pmatrix}.$$

We recall[7] that $B : E_2 \to E_2$ is nilpotent and we seek a Jordan basis of this endomorphism.

---

[7] The same letter is used to denote a matrix and the associated endomorphism in the canonical basis.

Examining the powers of B shows that the vector e_1 belongs to E_2 and that its last exponent is 2. Consequently the vectors $e_2^{(2)}, e_2^{(1)}, e_2$ are independent because they form a deployment of the system $\Sigma' = \{e_2^{(2)}\}$.

This remark allows us to take a shortcut. Since E_2 is of dimension 3, the deployed system

$$\widetilde{\Sigma}' = \{e_2^{(2)}, e_2^{(1)}, e_2\}$$

is a basis of E_2. Now, we are perfectly within our rights to begin the algorithm with the system Σ'. But since this system is linearly independent, the algorithm terminates immediately and tells us that $\widetilde{\Sigma}'$ is desired Jordan basis.

• Put $C = A - I$, so that the characteristic subspace E_1 associated to the eigenvalue $\mu = 1$ is the kernel of C^4:

$$C = \begin{pmatrix} 3 & 14 & 2 & 4 & 7 & 3 & 0 \\ 1 & 2 & 1 & 2 & 1 & 0 & 0 \\ 0 & 4 & 0 & 2 & 2 & 0 & -1 \\ -1 & -6 & -1 & -2 & -3 & -1 & 0 \\ -1 & 2 & -1 & -2 & 1 & 1 & 0 \\ -2 & -14 & -1 & -4 & -7 & -2 & 1 \\ 1 & 2 & 1 & 0 & 1 & 1 & 1 \end{pmatrix},$$

$$C^2 = \begin{pmatrix} 6 & 26 & 6 & 10 & 13 & 6 & 1 \\ 2 & 12 & 1 & 4 & 6 & 2 & -1 \\ -1 & -2 & -1 & 0 & -1 & -1 & -1 \\ -2 & -10 & -2 & -4 & -5 & -2 & 0 \\ -2 & -14 & 0 & -4 & -7 & -2 & 2 \\ -4 & -20 & -4 & -8 & -10 & -4 & 0 \\ 3 & 12 & 3 & 4 & 6 & 3 & 1 \end{pmatrix},$$

$$C^3 = \begin{pmatrix} 10 & 44 & 10 & 18 & 22 & 10 & 1 \\ 3 & 14 & 3 & 6 & 7 & 3 & 0 \\ -3 & -12 & -3 & -4 & -6 & -3 & -1 \\ -3 & -14 & -3 & -6 & -7 & -3 & 0 \\ -3 & -14 & -3 & -6 & -7 & -3 & 0 \\ -6 & -28 & -6 & -12 & -14 & -6 & 0 \\ 6 & 26 & 6 & 10 & 13 & 6 & 1 \end{pmatrix},$$

$$C^4 = \begin{pmatrix} 15 & 66 & 15 & 28 & 33 & 15 & 1 \\ 4 & 18 & 4 & 8 & 9 & 4 & 0 \\ -6 & -26 & -6 & -10 & -13 & -6 & -1 \\ -4 & -18 & -4 & -8 & -9 & -4 & 0 \\ -4 & -18 & -4 & -8 & -9 & -4 & 0 \\ -8 & -36 & -8 & -16 & -18 & -8 & 0 \\ 10 & 44 & 10 & 18 & 22 & 10 & 1 \end{pmatrix}.$$

A basis of this subspace is formed by the vectors

$$\begin{aligned} f_1 &= e_1 - e_3, && \text{last exponent} = 2, \\ f_2 &= 2e_3 - e_4 - 2e_7, && \text{last exponent} = 0, \\ f_3 &= e_3 - e_6, && \text{last exponent} = 2, \\ f_4 &= e_2 - 2e_5, && \text{last exponent} = 0. \end{aligned}$$

The first system examined by the algorithm is

$$\Sigma_0'' = \left\{f_1^{(2)}, f_2^{(0)}, f_3^{(2)}, f_4^{(0)}\right\} = \begin{pmatrix} 0 & 0 & 0 & 0 \\ 1 & 0 & -1 & 1 \\ 0 & 2 & 0 & 0 \\ 0 & -1 & 0 & 0 \\ -2 & 0 & 2 & -2 \\ 0 & 0 & 0 & 0 \\ 0 & -2 & 0 & 0 \end{pmatrix}.$$

Since $f_4^{(0)} = f_1^{(2)}$, we can already eliminate the vector $f_4^{(0)}$:

$$\Sigma_1'' = \left\{f_1^{(2)}, f_2^{(0)}, f_3^{(2)}\right\}.$$

Since $f_1^{(2)} + f_3^{(2)} = 0$, we put $f_5 = f_1^{(1)} + f_3^{(1)}$ and we replace f_3 by f_5 in Σ_1'', which gives the system

$$\Sigma_2'' = \left\{f_1^{(2)}, f_2^{(0)}, f_5^{(0)}\right\} = \begin{pmatrix} 0 & 0 & 0 \\ 1 & 0 & 1 \\ 0 & 2 & 0 \\ 0 & -1 & 0 \\ -2 & 0 & -2 \\ 0 & 0 & 0 \\ 0 & -2 & 0 \end{pmatrix}.$$

Since $f_1^{(2)} = f_5^{(0)}$, we can eliminate f_5:

$$\Sigma_3'' = \left\{f_1^{(2)}, f_2^{(0)}\right\}$$

Since Σ_3'' is independent, the desired Jordan base is $\widetilde{\Sigma}_3''$.

By taking the union of the bases $\widetilde{\Sigma}'$ of E_2 and $\widetilde{\Sigma}_3''$ of E_1, we define the transition matrix

$$P = \begin{pmatrix} 1 & 2 & 1 & 0 & 1 & 1 & 0 \\ 0 & 1 & 0 & 1 & 0 & 0 & 0 \\ -1 & 0 & 0 & 0 & 0 & -1 & 2 \\ 0 & -1 & 0 & 0 & 0 & 0 & -1 \\ 0 & -1 & 0 & -2 & 0 & 0 & 0 \\ 0 & -2 & 0 & 0 & -1 & 0 & 0 \\ 1 & 1 & 0 & 0 & 0 & 0 & -2 \end{pmatrix}.$$

and the desired Jordan form is then:

$$J = P^{-1}AP = \begin{pmatrix} 2 & 1 & 0 & 0 & 0 & 0 & 0 \\ 0 & 2 & 1 & 0 & 0 & 0 & 0 \\ 0 & 0 & 2 & 0 & 0 & 0 & 0 \\ 0 & 0 & 0 & 1 & 1 & 0 & 0 \\ 0 & 0 & 0 & 0 & 1 & 1 & 0 \\ 0 & 0 & 0 & 0 & 0 & 1 & 0 \\ 0 & 0 & 0 & 0 & 0 & 0 & 1 \end{pmatrix}.$$

We check the validity of these calculations by verifying that $PJ = AP$.

11.4.6. Programming

Everything depends on your level.

• If you are a beginner, you can write a program to carry out the matrix operations.

• If you are seasoned, you can introduce a procedure to solve the linear system in the preceding program so as to make carrying out the algorithm more automatic.

We recall once again that the Pittelkow-Ruckel algorithm is not an algorithm for numerical calculation; it is only a demystification of Jordan's theory. Do not try to automate the entire program. Program instead some *instructional software* which reserves the thoughtful part (that is, the oversight of the calculations and the decisions to take) for the user of the program; if the user remains passive, he or she will learn nothing and carry away the impression that the theory is difficult.

• In order to obtain matrices with a given Jordan form prescribed in advance, start with the prescribed form J and complicate it with succesive elementary operations $A \mapsto E^{-1}AE$.

12. Recursion

12.1. Presentation

Certain mathematical objects are inherently fascinating. Consider, for example, the integers or differential equations. What can be more banal than the integers? Yet, what riches they hide! Differential equations are genuine "black holes": with the seven characters $x'' + x = 0$, we define the number π and all of trigonometry; with one character more, the equation $x'' + x^3 = 0$ defines periodic functions on $\mathbb{R}$ with distinct periods!

Recursion is another "black hole": several lines of code can lead to a procedure seemingly impossible to describe iteratively. It is also a very fruitul programming discipline: a recursive procedure contains its own proof. When necessary, standard techniques of derecursifying[1] allow one to *automatically* transform a recursive code into an otherwise inaccessible iterative code.

Despite these qualities, recursion frequently inspires fear in beginners.

- It seems mysterious: what does the machine do? How can one understand and execute a code that refers only to itself, so that the least imprecision leads to a crash?

- Beginners forget that recursion is simply reasoning by induction adapted to a computer, and requires only that one know some very simple techniques.

12.1.1. Two simple examples

On can translate the definition of the function $n!$ that mathematicians use *directly* into Pascal:

```
function fact(n : integer) : integer ;
begin
| if n ≤ 1 then fact := 1 else fact := n * fact(n − 1)
end ;
```

[1] See D. Krob, *Algorithmes et structures de données*, Ellipses (1989). Some individuals seek to introduce recursion into every new problem, even if it seems artificial at first; one can also explain formidable algorithmes in this manner. If this manner of thought also fascinates you, I recommend the very original J. Arsac, *Foundations of programming*, Academic Press (1985)

The first time that one sees code like this, one is is incredulous.[2]

How can a machine – a set of condensers, interrupters and a clock – *understand* this summit of the human spirit that is induction, this stupefying means that we conceived to master the infinite? We shall explain this later at length. In the interim, we remark on some features of the syntax.

- Consider the statement

$$fact := n * fact(n - 1)$$

and note the following nuances:

- The left of the assignment symbol is concerned with the definition of the value of the function; thus one finds only the *name* of the function;
- To the right of the assignment symbol, one finds an arithmetic expression. An arithmetic expression can contain one or more function calls, *including calls to a function which is in the process of being defined*: the compiler sees nothing inconvenient here. Note that here the name of the function is necessarily followed by its argument in parentheses.
- The following test is essential

$$\textbf{if } n \le 1 \textbf{ then} \ldots$$

If you forget it, to find the value of $fact(3)$, your program will first try to calculate $fact(2), fact(1), fact(0), fact(-1), fact(-2)$, etc., and will crash when its memory is entirely filled by the incessant recursive calls.

- The call parameter n is passed by value, that is, "without var". In fact, if we put n in "var", the compiler will not accept $fact(n-1)$ since $(n-1)$ is not the address of a variable in memory.[3]

A procedure can be recursive; that is, it can call itself. If we like, although it is of no practical interest, we could calculate $n!$ using a procedure:

```
procedure factorial(var y : integer ; n : integer) ;
begin   {returns n! in y}
  if n ≤ 1 then y := 1
  else begin
    factorial(y, n − 1) ;     {y = (n − 1)!}
    y := n ∗ y                {y = n!}
  end
end ;
```

[2] Examine your memories: didn't you feel the same uneasiness when you first encountered the definition of the factorial function in mathematics?

[3] But you could put everything back in order by declaring a local variable *temp* and replacing the faulty code $fact := n * fact(n-1)$ by $temp := n - 1$; $fact := n * fact(temp)$.

The body of the procedure contains the statements; now a call to a procedure *which itself consists of the procedure which is in the process of being defined* is a statement like any other! This is the reason that this code is accepted.

We can define the Fibonacci series as a function

```
function Fib(n : integer) : integer ;
begin
| if n ≤ 1 then Fib := n   {not Fib := 1, because Fib(0) = 0}
| else Fib := Fib(n − 1) + Fib(n − 2)
end ;
```

or as a procedure:

```
procedure Fibonacci(n : integer ; var u : integer) ;
var y, z : integer ; {returns F_n in u}
begin
| if n ≤ 1 then u := n
| else begin
| | Fibonacci(n − 1, y) ; {y = F_{n−1}}
| | Fibonacci(n − 2, z) ; {z = F_{n−2}}
| | u := y + z
| end
end ;
```

12.1.2. Mutual recursion

Suppose that we want to write two procedures A and B each of which calls the other, meaning that the code for A contains a call to B and that of B contains a call to A. How do we type this? Knowing that A contains one or more calls to B, we ought to type the code for B before that of A. But the same rule requires that the code for A appear before that of B. We find ourselves in the computer science version of the classical chicken and egg paradox.[4]

To resolve this dilemma, we use the statement "forward" which allows us to *detach* the declarative part of a procedure from the body of the procedure. So, we can write

```
procedure A(var x, y : integer) ; forward ;
procedure B(u : real ; var v : integer) ; forward ;
```

Having made the declaration, you can type the following in the reserved parts of the procedures.

```
procedure A{(var x, y : integer)} ;
{constants, types, local variables, etc.}
begin
```

[4] Which appeared first?

```
  | code which may eventually contain
  | calls to B or to procedures calling B
end ;
procedure B{(u : real ; var v : integer)} ;
{constants, types, local variables, etc.}
begin
  | code which may eventually contain
  | calls to A or to procedures calling A
end ;
```

• The code for *A* need not precede that of *B*. You could, if you wish, write the body of *B* before that of *A*.

• Note the comments which allow one to keep in sight the parameters of both procedures. This is a very useful technique when the heading of the procedure is several screens distant from the body! (Some compilers are tolerant and allow repetition of arguments.)

• You can use the "forward" procedure with any procedure or function, even if there are no mutual calls: some programmers, in fact, systematically use forwards so as to never have to move code.

We shall see soon a magnificent example of mutual calls (the *time-waster*). In Chapter 13, we will encounter much more elaborate examples.

12.1.3. Arborescence of recursive calls

To understand what a program does during a recursive call, we construct a tree which represents successive procedure calls. Its root is the calling procedure and its branches are the procedures called. For example Figure 12.1 displays the tree diagramming the recursive calls associated to the statement Fib(4): to execute Fib(4), the program first calculates Fib(3) and Fib(2) before adding them. But calling Fib(3) and Fib(2) starts the calculation of Fib(2), Fib(1) and Fib(0).

Precisely what these calculations are does not matter: the tree of recusive calls and the route (computer scientists speak of a *visit*) allow us to understand the *history* of the calulations.

12.1.4. Induction and recursion

What follows is the code for the aptly named procedure, *mystery*. What does it do? Do not cheat and look up the answer that follows! Try to discover yourself (without turning to a computer or looking at Fig. 12.3) what *mystery*(4) does.

Also observe your own behavior: one of the goals of the exercise is to observe your own reactions.

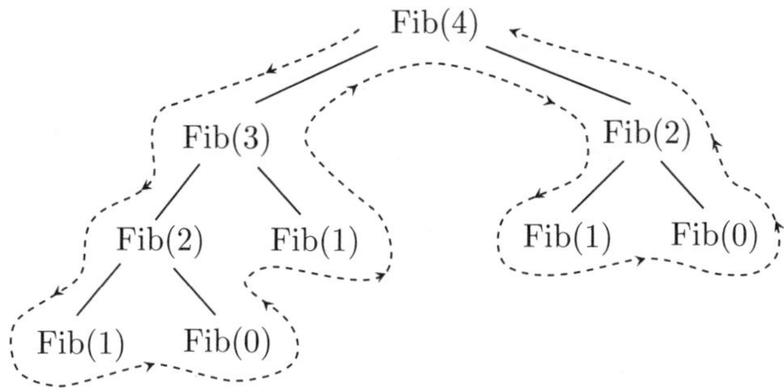

Fig. 12.1. *Recursive calls associated to the statement* Fib(4)

```
procedure mystery(n : integer) ;
begin
| if n = 0
| then writeln(n : 3)
| else begin write(n : 3) ; mystery(n − 1) ; writeln(n : 3) end
end ;
```

When one poses this question to a beginner, one discovers that he or she will reason as follows: "Let's see, *mystery*(4) writes 4 then calls *mystery*(3), which writes 3, then calls *mystery*(2), etc." Our beginner then guesses that the procedure begins by writing 4, 3, 2, 1, 0 on the screen. A careful beginner will even specify that these numbers are written on a line because the statement *write* comes into play and not *writeln*. But then, the intellectual mechanics screech to a halt.

To understand the problem, we sketch the tree of recursive calls (using the notation "*m*" for *mystery*, "*w*" for *write* and "*wln*" for *writeln*).

We find that the naïve method of the beginner consists of plunging into the left half of the tree. Unhappily, this error is easy to make: the human mind, unlike a program, does not easily remember statements which remain in wait: the climb back to the root of the tree is difficult, even impossible! Experience shows that simply reading the code will not suffice: most of the time, the tree is too complicated to be sketched …

What moral should we draw from this experience? *One should refuse to plunge into a recursive call tree and replace this suicidal plunge with an induction hypothesis* so as to *never* leave the code of the procedure.

To imagine the induction hypothesis, it suffices to simulate (by hand!) the calls *mystery*(0), *mystery*(1), *mystery*(2) and *mystery*(3) in this order.

• The first call writes 0 on the screen and leaves the cursor on the following line.

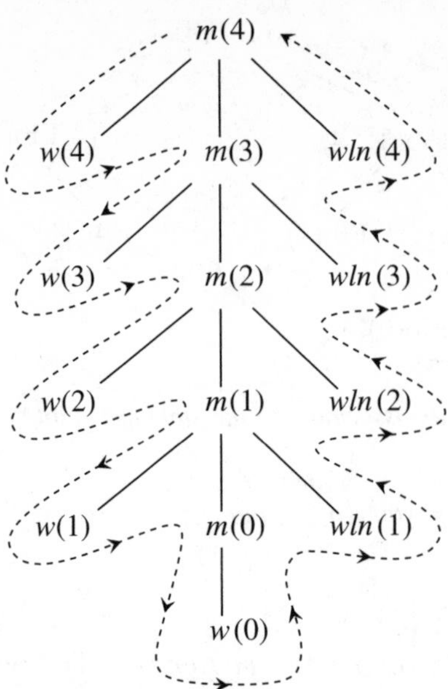

Fig. 12.2. Traversing the tree of recursive calls in the case mystery(4)

• Knowing this, it is not at all difficult to convince oneself that the second call writes 1, 0 on the line, then a 1 on the following line, and leaves the cursor on the second line.

• When we pass to *mystery*(2), we must execute the statements *write*(2), *mystery*(1) and *writeln*(2), which writes 2, 1, 0 on the first line, 1 on the second line, 2 on the third line and leaves the cursor on the third line.

We are now in known territory: induction.

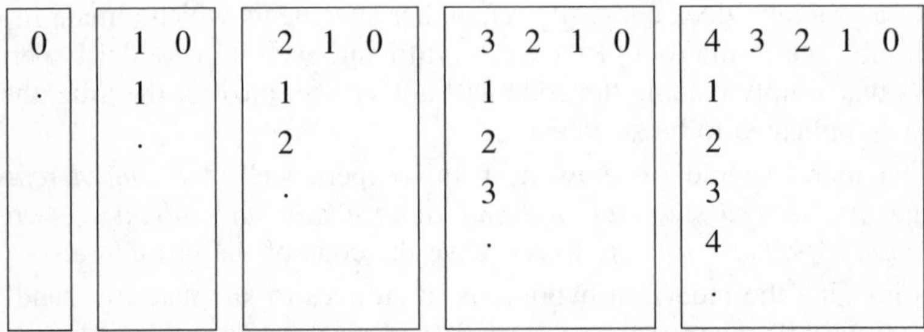

Fig. 12.3. *Solution: What the calls mystery*(0)*, mystery*(1)*, mystery*(2)*, mystery*(3) *and mystery*(4) *do. The dot indicates the position of the cursor.*

Exercise 1

Try to discover without cheating — that is without using your computer — what the following procedures do.

```
procedure mystery1(n : integer) ;
begin
  if n ≤ 0 then writeln
  else begin
    write(n : 3) ;
    mystery2(n − 1) ;
    writeln(n : 3)
  end
end ;
```

```
procedure mystery2(n : integer) ;
begin
  if n ≤ 0 then writeln
  else begin
    write(n : 3) ;
    mystery1(n) ;
  end
end ;
```

Exercise 2

Same question with the procedures:

```
procedure mystery3(n : integer) ;
begin
  if n ≤ 0
  then writeln
  else begin
    write(n : 3) ;
    mystery3(n − 1) ;
    mystery4(n − 1)
    writeln(n : 3)
  end
end ;
```

```
procedure mystery4(n : integer) ;
begin
  if n ≤ 0
  then writeln(n : 3)
  else begin
    write(n : 3) ;
    mystery4(n − 1) ;
  end
end ;
```

12.2. The Ackermann function

Another celebrated classic of recursion theory is the Ackermann function. This is a function or two variables $x, y \in \mathbb{N}$ and a parameter $n \in \mathbb{N}$ which controls its complexity:

- $A(0, x, y) = x + 1$;
- $A(n, x, 0) = \begin{cases} x & \text{if } n = 1, \\ 0 & \text{if } n = 2, \\ 1 & \text{if } n = 3, \\ 2 & \text{if } n \geq 4 ; \end{cases}$
- $A(n, x, y) = A\big(n - 1, A(n, x, y - 1), x\big) \quad \text{if } n > 0 \text{ and } y > 0.$

The translation into Pascal code follows the definition step by step and presents no difficulty. The interest of this function lies elsewhere: it is barely calculable in a sense that we will not try to make precise.

```
function Ackermann(n, x, y : integer) : integer ;
begin
  if n = 0 then Ackermann := x + 1 else
  if y = 0 then
  case n of
    1 : Ackermann := x ;
    2 : Ackermann := 0 ;
    3 : Ackermann := 1 ;
    else Ackermann := 2
  end   {case}
  else Ackermann := Ackermann(n − 1, Ackermann(n, x, y − 1), x)
end ;
```

If you program this function, expect surprises as your computer goes nuts very rapidly.

Theorem 12.2.1. *The Ackermann function is defined on all of* $\mathbb{N}^3$ *and:*

$$A(1,x,y) = x + y, \quad A(2,x,y) = x \cdot y,$$
$$A(3,x,y) = x^y, \quad A(4,x,y) = 2^{(x^y)}.$$

Proof. Before showing that this function is defined on $\mathbb{N}^3$ (which is not at all evident), let us explicitly work out the cases $n = 1, 2, 3, 4$.

- $A(1,x,0) = x$ by definition. Thus,

$$\begin{aligned} A(1,x,1) &= A(0, A(1,x,0), x) = (x+1) + 0 = x + 1, \\ A(1,x,2) &= A(0, A(1,x,1), x) = (x+1) + 1 = x + 2, \\ A(1,x,3) &= A(0, A(1,x,2), x) = (x+2) + 1 = x + 3. \end{aligned}$$

Induction on y then shows that $A(1,x,y) = x + y$.

- $A(2,x,0) = 0$ by definition. Thus,

$$\begin{aligned} A(2,x,1) &= A(1, A(2,x,0), x) = A(1,0,x) = x, \\ A(2,x,2) &= A(1, A(2,x,1), x) = A(1,x,x) = 2x, \\ A(2,x,3) &= A(1, A(2,x,2), x) = A(1,2x,x) = 3x. \end{aligned}$$

Induction on y then shows that $A(2,x,y) = xy$.

- $A(3,x,0) = 1$ by definition. Thus:

$$\begin{aligned} A(3,x,1) &= A(2, A(3,x,0), x) = A(2,1,x) = x, \\ A(3,x,2) &= A(2, A(3,x,1), x) = A(2,x,x) = x \cdot x, \\ A(3,x,3) &= A(2, A(3,x,2), x) = A(2,x^2,x) = x \cdot x \cdot x. \end{aligned}$$

Induction on y then shows that $A(3,x,y) = x^y$.

- $A(4, x, 0) = 1$ by definition. Thus:

$$
\begin{aligned}
A(4, x, 1) &= A(3, A(4, x, 0), x) = A(3, 2, x) = 2^x, \\
A(4, x, 2) &= A(3, A(4, x, 1), x) = A(3, 2^x, x) = (2^x)^x = 2^{x^2}, \\
A(4, x, 3) &= A(3, A(4, x, 2), x) = A(3, x^2, x) = (2^{x^2})^x = 2^{x^3}.
\end{aligned}
$$

We end with an induction on y.

Now let us determine the domain of definition: do the recursive calls stop at each triple? We will prove this using transfinite induction (Chap. 2) on $\mathbb{N}^3$ endowed with the lexicographic order.

Let $\mathbb{D}$ be the domain of definition of the Ackermann function, that is the set of triples at which the recursive calls stop. Choose a triple (N, X, Y) and suppose that $A(n, x, y)$ is defined for all triples $(n, x, y) < (N, X, Y)$. We want to prove that $A(N, X, Y)$ exists.

- If $N = 0$ or $Y = 0$, we know that $A(N, X, Y)$ exists because there is no recursive call. In other words $\mathbb{D}$ already contains the triples $(0, X, Y)$ and $(N, X, 0)$.

- If $N > 0$ and $Y > 0$, the induction hypothesis assures us that the number $\alpha = A(N, X, y - 1)$ exists because $(N, X, y - 1) < (N, X, Y)$. Since we also have $(N - 1, \alpha, X) < (N, X, Y)$, the induction hypothesis now implies that $A(N, X, Y) = A(N - 1, \alpha, X)$ is defined.

12.3. The Towers of Hanoi

Consider a board on which three equidistant pegs, called A, B, C have been stood vertically. At the outset, n disks with decreasing radii are positioned on rod A so as to form a pyramid. We want to move disks from rod A towards C respecting the following rule: one can take a disk from the top of a rod in order to put it on another rod subject to the condition that it does not cover a disk of smaller radius (in other words, the disks must always form pyramids.

This problem appears difficult, but is very simply solved when one reasons inductively. Let us call *Hanoi*(A, B, C, n) the operation which consists of moving the n upper disks of A to C using B as an intermediate rod.

- If $n = 1$, it suffices to move disk A to C.

- If $n > 1$, we can begin by moving the $n - 1$ upper disks of A to B (see Fig. 12.4) using C as an intermediate peg. We then move the largest disk of A to C, Next, we begin again and move the $n - 1$ top disks of peg B to C this time using A as an intermediate peg.

The translation into a program is now child's play.

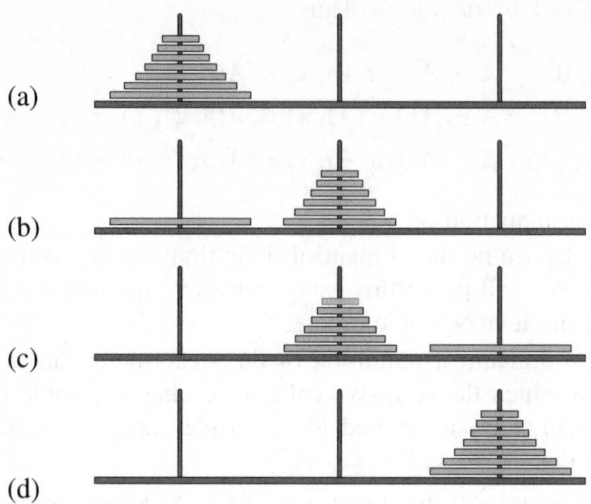

Fig. 12.4. Tower of Hanoi: (a) *initial situation,* (b) *after Hanoi*$(A, C, B, n-1)$, (c) *after move*(A, C) (d) *after Hanoi*$(B, A, C, n-1)$

```
program towers_of_Hanoi ;
var A, B, C : char ;  n : integer ;
procedure move(X, Y : char) ;
begin
| writeln('move disk from peg ', X, ' to peg ', Y)
end ;
procedure Hanoi(A, B, C : char ;  n : integer) ;
begin
| if n = 1 then move(A, C)
| else begin
| | Hanoi(A, C, B, n − 1) ;
| | move(A, C) ;
| | Hanoi(B, A, C, n − 1)
| end
end ;
begin
| write('number of disks = ') ;  readln(n) ;
| A := 'A' ;  B := 'B' ;  C := 'C' ;  {this is not a joke!}
| Hanoi(A, B, C, n)
end .
```

If we sketch (Fig. 12.4) the tree of recursive calls for *Hanoi*$(A, B, C, 3)$, we can *foresee* what the computer will do. But, as we have already mentioned, the limits are quickly attained: try, for instance, to completely sketch the tree of calls for *Hanoi*$(A, B, C, 5)$!

Never forget: it is an *induction hypothesis* which allows us to write the procedure. The tree of recursive calls simply allows us to understand the Pandora box that we have opened ...

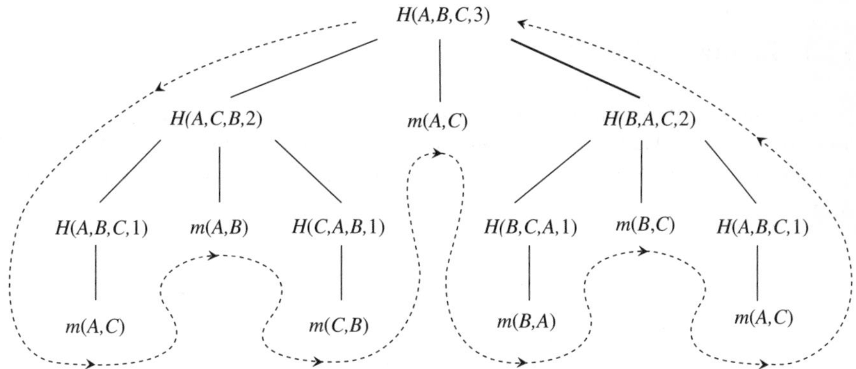

Fig. 12.5. *Tree of recursive calls of Hanoi*$(A, B, C, 3)$

Exercise 3

Sketch the tree of recursive calls for *Hanoi*$(A, B, C, 4)$.

Exercise 4

Our program is nevertheless frustrating because it indicates only how one needs to move the disks. Why not show the movements of the disks on the screen? To do this, we must know the *state* at each time of the system: that is the composition of each of the pyramids on each of the pegs A, B, C. Hence, the declaration:

```
type peg = record
  | name : char ;
  | ht : integer ;
  | ray : array[1..10] of integer
end ;
var A, B, C : peg ;
```

You initalize the towers with the code:

```
with A do begin
  | name := 'A' ;  ht := n ;  for i := 1 to ht do ray[i] := i
end ;
with B do begin name := 'B' ;  ht := 0 end ;
with C do begin name := 'C' ;  ht := 0 end ;
```

It is not necessary to modify the procedure *Hanoi*, other than the types. In contrast, it is necessary to entirely rewrite the procedure *move* which *modifies* the composition of the pegs (do not forget the "var"!) and *animates* the screen.

12.4. Baguenaudier

Bauguenaudier is a centuries old puzzle consisting of interlaced rings and a looped double rod which one wants to remove. By pulling the rod to the left and passing the rightmost ring into onto one side of the rod, one can free the ring on the right.

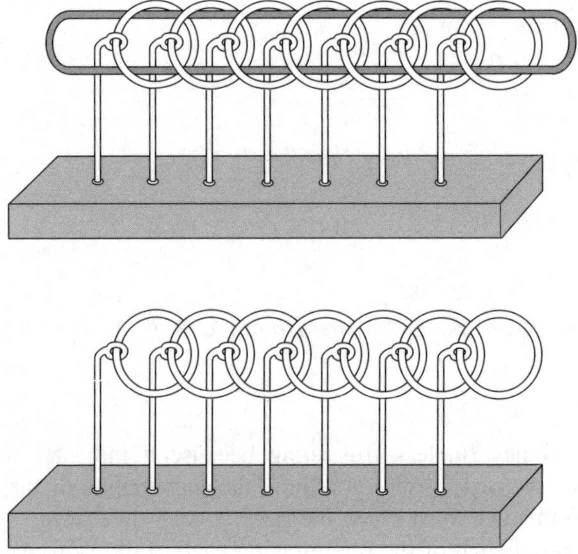

Fig. 12.6. *A baguenaudier*

More generally, on numbering the rings from right to left, one finds that to free (or interlace) the ring k, it suffices that the $k-2$ first rings are free and that the $(k-1)$-th is in place. This remark allows us to formalize this puzzle simply using a ruler with holes and n balls. The holes are numbered from 1 to n. One has to fill the n holes respecting the following rules (where *to play* means placing or removing a ball):

- the ruler is empty at the outset;
- a hole can contain only a single ball;
- one can always play hole number 1;
- one can always play the hole that follows the first filled hole.

To better grasp the nuances of the game, we detail the passage from the last row of the array on the left to the first row of the array on the right where each row has five holes (see below). At the bottom of the array on the left, the first occupied hole is hole 2. Thus, we either can play hole 1 (adding a ball) or hole 3 (removing a ball), Since playing hole 1 leads us backwards we remove a ball from hole 3.

To summarize, we alternately use the two rules since using the same ones twice in a row does nothing.

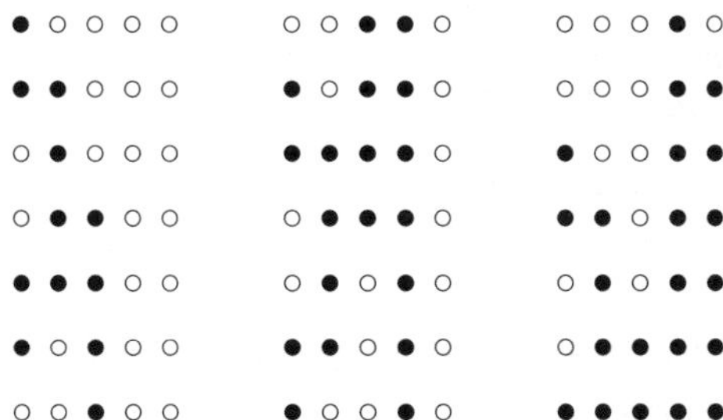

Fig. 12.7. *How to play baguenaudier with five holes*

When one plays baguenaudier with three, then four, then five holes, etc., one discovers (involuntary) strategies for filling or emptying the segment formed by holes 1 to p. We formulate an induction hypothesis by supposing that we know:

- how to *fill* the segment consisting of the first p holes (supposed empty);
- how to *empty* the segment consisting of the first p holes (supposed full).

If $p \geq 3$ and if the segment of holes 1 to p is empty (we make no assumptions about the other holes), we can fill our segment by:

(i) filling the segment of holes 1 to $p - 1$;
(ii) emptying the segment of holes 1 to $p - 2$;
(iii) playing hole p;
(iv) filling the segment of holes 1 to $p - 1$.

To empty the same segment — supposed full this time — we can:

(i) empty the segment of holes 1 to $p - 2$;
(ii) play hole p;
(iii) fill the segment of holes 1 to $p - 2$;
(iv) empty the segment of holes 1 to $p - 1$.

The programming is immediate. This is a splendid example of mutual recursion(where several procedures call one another).

This example also shows that it is practically impossible to sketch the tree of recursive calls once the situation is complicated. We are forced to rely on induction.

Declarations

We represent the baguenaudier by an array of booleans. The constants *empty* and *full* make the program more readable and make it unnecessary to memorize conventions.

```
const empty = true ;  full := false ;  n = 5 ;
type table = array[1..n] of boolean ;
var baguenaudier : table ;
```

The main body of the pogram is the simplest part.

```
begin
  message ;
  for i := 1 to n do baguenaudier[i] := empty ;
  fill_segment(n)
end .
```

The procedures fill_segment and empty_segment

These procedures faithfully translate the strategy we have written. Since the procedures mutually call one another, we use the “forward” statement.

```
procedure fill_segment(p : integer ;  var baguenaudier : table) ;
forward ;
procedure empty_segment(p : integer ;  var baguenaudier : table) ;
forward ;
```

We now write the bodies of the procedures.

```
procedure fill_segment ;
begin {the holes 1 to p are empty ;  afterwards, one doesn't know}
  case p of
    1 : play_hole(1, baguenaudier) ;
    2 :  begin
           play_hole(1, baguenaudier) ;
           play_hole(2, baguenaudier) ;
         end
    else {now,  p ≥ 3}
    fill_segment(p − 1, baguenaudier) ;
    empty_segment(p − 2, baguenaudier) ;
    play_hole(p, baguenaudier) ;
    fill_segment(p − 2, baguenaudier) ;
  end {case}
end ;
```

The separation of cases $n = 1, 2$ from the general case $n \geq 3$ is neither capricious nor happenstance; rather it is a consequence of attentively examining the strategy employed. Since it needs a segment that contains at least three holes, we are obliged to treat the cases with one or two holes separately.

```
procedure empty_segment ;
begin {the holes 1 to p are full ; afterwards, one doesn't know}
  case p of
    1 : play_hole(1, baguenaudier) ;
    2 : begin
          play_hole(2, baguenaudier) ;
          play_hole(1, baguenaudier) ;
        end
    else {now, p ≥ 3}
    empty_segment(p − 2, baguenaudier) ;
    play_hole(p, baguenaudier) ;
    fill_segment(p − 2, baguenaudier) ;
    empty_segment(p − 1, baguenaudier) ;
  end {case}
end ;
```

Programming is straightforward: it remains to write the procedure *play_hole* (several lines of code to modify and display the new baguenaudier).

12.5. The Hofstadter Function

The Hofstadter function[5] is defined as follows:

$$G(n) = \begin{cases} 0 & \text{if } n = 0, \\ n - G\big(G(n-1)\big) & \text{if } n \geq 1. \end{cases}$$

Here are its first values; it is not at all clear that this function is defined on all of $\mathbb{N}$!

n	0	1	2	3	4	5	6	7	8	9	10	11	12
$G(n)$	0	1	1	2	3	3	4	4	5	6	6	7	8

This function has a surprising interpretation. Recall (Chap. 8) the Zeckendorf decomposition of an integer ≥ 1:

$$n = F_{i_1} + F_{i_2} + \cdots + F_{i_k}, \quad i_1 \gg i_2 \gg \cdots \gg i_k \gg 0.$$

Theorem 12.5.1. *With Zeckendorf decomposition as above, we have*

$$G(n) = F_{i_1-1} + F_{i_2-1} + \cdots + F_{i_k-1}.$$

[5] See Chapter 5 of the book Douglas R. Hofstadter, *Gödel, Escher, Bach: an eternal golden braid*, Basic Books (1979).

Proof. This theorem is proved rather simply by induction on n. Try it!

Proposition 12.5.1. *The Hofstadter function is defined on* $\mathbb{N}$.

Proof. Upon trying to reason by induction, we suppose that $p = G(n-1)$ exists, so $G(n) = n - G(p)$. We then realize that we also need precise information about $G(p)$, we leads us to formulate a strong induction hypothesis:

$$(\mathcal{H}_n) \qquad \begin{cases} \textit{The function } G \textit{ is defined on the interval } [\![0, n]\!] \textit{ and} \\ 1 \leq G(k) \leq k - 1 \textit{ for all } k \in [\![2, n]\!]. \end{cases}$$

The rest of the proof is left to the reader.

Exercise 5

The values of the function G for $n \leq 12$ suggest that the function is increasing and does not grow very fast since $G(n+1) - G(n) \leq 1$. Is this true? What about the same conjecture with the inequality $G(n+2) - G(n) \geq 1$ which says that G cannot take the same value more than twice in a row?

12.6. How to Write a Recursive Code

Suppose that we want to write a recursive procedure *toto*(x, n) depending on two integer parameters $x, n \geq 0$.

• *We begin by examining the general case*, trying to express *toto*(x, n) with the aid of one or several calls to *toto*. Suppose, in the first approximation, that our analysis gives us five statements

$$toto(x, n) = \begin{cases} A(x, n); \\ toto(x-1, n+1); \\ B(x, n); \\ toto(x, n-1); \\ C(x, n) \end{cases} \tag{12.1}$$

where A, B, C are three procedures that do not modify the values of x and n and which do not call *toto* (directly or indirectly).

• This rough sketch shows us that we do not have the right to use (12.1) when $x - 1 < 0$ or $n - 1 < 0$: we must treat the pairs $(0, n)$, $(x, 0)$ separately. We must assure ourselves that A, B, C function correctly. Suppose that A and C do not require anything, but that $B(x, n)$ does not function for $n \geq 2$, which now prevents us from using (12.1) with the pairs $(x, 0)$ and $(x, 1)$.

• *We treat the exceptions separately*. Suppose that:

▷ *toto*$(x, 0)$ is the procedure $\alpha(x)$ if $x > 0$;

▷ *toto*$(x, 1)$ is the procedure $\beta(x)$ if $x \geq 0$;
▷ *toto*$(x, 2)$ is the procedure $\gamma(x)$ if $x \geq 0$;
▷ *toto*$(0, n)$ is the procedure $\delta(n)$ if $n \geq 0$.

(We again suppose that the procedures α, β, γ, δ do not call *toto* either directly, or indirectly.)

• It suffices to assemble the pieces taking care to treat the pair $(0, 0)$ which is common to the pairs $(0, n)$ and $(x, 0)$ separately:

```
procedure toto(x, n : integer) ;
begin
  case n of
    0 : if x > 0 then α(x) else δ(0) ;
    1 : β(x) ;
    2 : γ(x) ;
    else {henceforth n ≥ 3}
    if x = 0 then δ(n) else begin
      A(x, n) ; toto(x − 1, n + 1) ;
      B(x, n) ; toto(x, n − 1) ;
      C(x, n)
    end
  end {case}
end ;
```

As you can see, the appearance of stops in recursive calls does not happen at random, as beginners very frequently think; it results from a careful analysis of the impossible cases of the general case.

Exercise 6

Show that the recursive calls of the procedure *toto* stop. (Use transfinite induction.)

12.6.1. Sorting by dichotomy

We examine a more concrete case, We are required to *sort*[6] a vector containing integers. If we start with

$$U = (5, 6, 1, 1, 1, 5, 5, 2, 3, 9, 7, 8, 8)$$

the sorted vector is:

$$U' = (1, 1, 1, 2, 3, 5, 5, 5, 6, 7, 8, 8, 9).$$

[6] Sorting algorithms are essential in management, which explains the considerable number of algorithms proposed.

We can sort this vector by dichotomy, which means that we cut the vector into two equal parts (up to a unit):

$$U_1 = (5, 6, 1, 1, 1, 5), \quad U_2 = (5, 2, 3, 9, 7, 8, 8).$$

We then sort the *lower part* U_1 and the *upper part* U_2 separately, which gives us the vectors:

$$U_1' = (1, 1, 1, 5, 5, 6), \quad U_2' = (2, 3, 5, 7, 8, 8, 9).$$

We now *merge* the vectors U_1' and U_2' to obtain the sorted vector U'. This operation is very simple: we consider the first elements of U_1' and U_2'. Since $1 < 2$, we know that the first element of U is 1; we strike out the first element of U_1' and we begin again.

To obtain a recursive formulation, we suppose that we have a procedure $sort(U, p, q)$ capable of sorting the subvector $(U_p, \ldots, U_q)$ without modifying the other entries. To sort U, it suffices to type the statement $sort(U, 1, n)$.

Now, a rough sketch of our algorithm is:

$$sort(p, q, x) = \begin{cases} m := (p+q)\ \mathbf{div} 2; & \{splitting\ the\ vector\} \\ sort(U, p, m); & \{sorting\ the\ lower\ part\} \\ sort(U, m+1, q); & \{sorting\ the\ upper\ part\} \\ merge(U, p, m, q) & \end{cases}$$

Consider now the problem of stopping. To speak of the subvector $U_p, \ldots, U_q$ implicitly assumes that $1 \le p \le q \le n$. As a result, the conditions $p \le m \le q$ and $m + 1 \le q$ must hold in order to sort the lower and upper parts correctly.

- These conditions do not hold when $q = p$; happily, there is nothing to do in this case.
- When $q - p = 1$, it would be stupid to use a dichotomy to exchange two coordinates.
- When $q - p \ge 2$, the condition $p < m < q$ is realised.

We now know enough to write out our sorting algorithm.

```
procedure sort(var U : vector ; p, q : integer) ;
var m : integer ; {hypothesis 1 ≤ p ≤ q ≤ n}
begin
| if (q − p = 1) and (U[p] > U[q])
| then exchange(U, p, q)
| else if q − p > 1 then begin
| | m := (p + q) div 2 ;
| | sort(U, p, m) ; sort(U, m + 1, q) ; merge(U, p, m, q)
| end
end ;
```

Recall that the test $q - p > 1$ is indispensable because it is especially essential to do *nothing* when $p = q$.

You see that this was not too hard! It suffices to be *rigorous*; that is, to reflect a little and to ask ourselves (just as we would for mathematics) whether the objects we employ exist and if they satisfy the conditions necessary to function well.

Exercise 7

Finish the program by writing the procedures *merge* and *exchange*.

Exercise 8: The Count is Good

Instead of the clumsy approximation in Chapter 6, this time we want to really program the popular French TV game. Recall the rules: one wants to calculate an integer which we call the *goal* drawn at random between 100 and 999. For this, we are given six numbers chosen at random from among the numbers $1, 2, 3, 4, 5, 6, 7, 8, 9, 10, 25, 50, 75$ and 100. The intermediate calculations happen in $\mathbb{N}^*$ (no negative numbers or zero; divisions must have remainder zero). Finally, one is not obliged to use all the numbers to reach the goal.

Suppose that we want to "realize" b with the integers $a_1, \ldots, a_k$:

- if one of the a_i is equal to b, we are done;
- otherwise, and if $k \geq 2$, we suppress the numbers a_i and a_j in the list $a_1, \ldots, a_k$ and we add in one of the numbers $a_i + a_j$, $a_i - a_j$ if $a_i > a_j$, $a_j - a_i$ if $a_j > a_i$, $a_i * a_j$, a_i / a_j if a_j divides a_i or a_j / a_i if a_i divides a_j and $a_j \neq a_i$ (this avoids repeating the preceding case).

It is of course necessary consider all possible pairs $1 \leq i < j \leq k$.

To display a solution, each a_k is accompanied by the string sol_k which is its "history" that is, the recipe to manufacture a_k.

- At the outset, sol_k is the result of converting the integer a_k into a chain of characters. When one replaces, for example, a_k by $a_i - a_j$, one must, at the same time, replace sol_k by the concatenation of the chain sol_i, ' – (', sol_j and ')'. One does the same with the other operations.

In this manner, the *value* of the chain sol_k, which involves only the operations between the a_i at the outset, is always equal to the current value of a_k.

We use constants to name the four operations

```
const max = 6 ; {first test your program with max = 4}
      addition = 1 ; substraction = 2 ;
      multiplication = 3 ; division = 4 ;
type vector = array[1..max] of integer ;
```

```
      string100 = string[100] ;
      history = array[1..max] of string100 ;
  procedure realize(a : vector ;  h : history ;  nb : integer) ;
  var i, j, k, op, temp, new_nb : integer ;
     new_a : vector ;  new_h : history ;
  begin
   for i := 1 to nb do
        if a[i] = b then display(h[i]) ;
   if nb > 1 then begin
    for i := 1 to nb do
    for j := 1 to nb do
    for op := addition to division do begin
     new_a := a ;
     new_h := h ;
     new_nb := nb - 1 ;
     case op of
      addition, multiplication :  begin
       if i < j then combine(a, new_a, h, new_h, i, j, op, new_nb) ;
            {commutative laws :  test i < j avoids repetition}
      end ;
      substraction :  begin
       if a[i] > a[j]
       then combine(a, new_a, h, new_h, i, j, op, new_nb) else
       if a[i] < a[j]
       then combine(a, new_a, h, new_h, j, i, op, new_nb) ;
      end ;
      division :  begin
       if a[i] mod a[j] = 0
       then combine(a, new_a, h, new_h, i, j, op, new_nb) else
       if (a[j] mod a[i] = 0) and (a[j] ≠ a[i])
            {test a[j] ≠ a[i] avoids repetition}
       then combine(a, new_a, h, new_h, j, i, op, new_nb)
      end ;
     end ;  {case}
     realize(new_a, new_h, new_nb)    {recursive call}
    end
   end
  end ;
```

The procedure *display* tries not to write the same solution twice.

The procedure *combine* is straightforward : let

$$ind\_min := \min(i, j); \quad ind\_max := \max(i, j).$$

We replace a_{ind_min} by a_i *op* a_j and we shift all a_k for $k \geq ind\_max$ to the left;

```
procedure combine(var a, new_a : vector ; var h, new_h : history ;
                                          i, j, op, new_nb : integer) ;
var k, ind_min, ind_max : integer ;
begin
 if i < j then begin ind_min := i ;  ind_max := j end
         else begin ind_min := j ;  ind_max := i end ;
 case op of
  addition :  begin    {i < j satisfied at the call}
   new_a[ind_min] := a[i] + a[j] ;
   new_h[ind_min] := parenthesize(h[i], '+', h[j]) ;
  end ;
  substraction :  begin    {one has i < j or j < i}
   new_a[ind_min] := a[i] − a[j] ;
   new_h[ind_min] := parenthesize(h[i], '−', h[j])
  end ;
  multiplication :  begin    {i < j satisfied at the call}
   new_a[ind_min] := a[i] * a[j] ;
   new_h[ind_min] := parenthesize(h[i], '*', h[j]) ;
  end ;
  division :  begin    {one has i < j or j < i}
   new_a[ind_min] := a[i] div a[j] ;
   new_h[ind_min] := parenthesize(h[i], '/', h[j])
  end ;
 end ;  {case}
 for k := ind_max to new_nb do begin
  new_a[k] := a[k + 1] ;
  new_h[k] := h[k + 1] ;
 end
end ;
```

13. Elements of compiler theory

For a beginner, the compiler is a mysterious being, at once very intelligent ("Incredible, it understands my program!") and abysmally stupid ("How could it not accept an otherwise correct program that is missing a tiny semicolon!"). You should understand that a compiler is only one program among others. Its role is to *faithfully* translate the text submitted to it into another text comprehensible to the microprocessor:

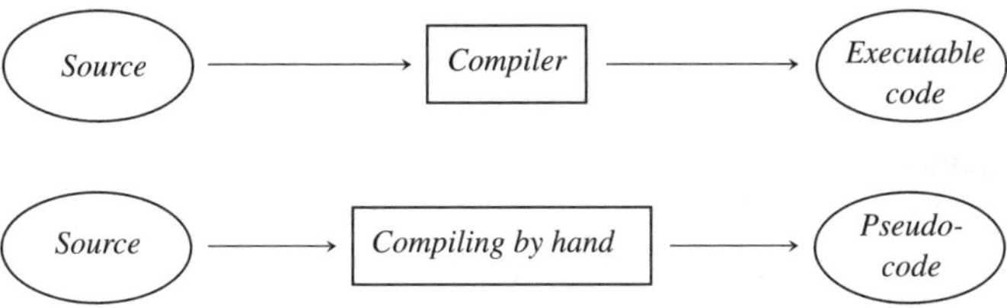

Fig. 13.1.

We are going to sketch answers to the following questions:

- What does the compiler's translation look like? How does it handle procedure and function calls? How does recursion function?
- How does the compiler translate a program?

13.1. Pseudocode

In Chapter 6, we presented a model for how a procedure passes parameters. Given the declaration

$$\textbf{procedure}\ toto(a : parameter);$$

we supposed that the program created the variable *x_toto* each time it encountered the statement *toto*(*x*) and modified the code of the procedure by replacing the occurrences of *a* by *x_toto* suitably initialized.

This very convenient model is not realistic for at least two reasons.

• A program written in Fortran, Pascal or C cannot modify its own code.

• A microprocessor can only carry out one addition or one multiplication at a time. We have not explained what becomes of complicated statements such as $y := a * x * x + b * x + c$.

When you want to understand the reactions of another individual, an effective technique is to ask yourself: "What would I do in his or her place?" To understand what a compiler does, we are going to put ourselves in its place and translate our programs into a language called *pseudocode*.

13.1.1. Description of pseudocode

We return a last time to our unrealistic model and suppose that *toto* is recursive: the statement *toto*(x) then results in the creation of the variable *x_toto*. But since *toto* calls itself, the statement *toto*(*x_toto*) results in turn in the creation of the variable *x_toto_toto*, etc.:

$$x\_toto \mapsto x\_toto\_toto \mapsto x\_toto\_toto\_toto \mapsto x\_toto\_toto\_toto\_toto \mapsto \cdots.$$

Thus, we see the appearance of a *stack structure* beloved by computer scientists.

Our pseudocode will resemble – but be much simpler than – the statements emitted by a true compiler; it is a a very rudimentary *assembly language*. We are are going to give orders to an imaginary microprocessor which only knows how to add, subtract and multiply two integers. For this, the microprocessor runs a stack.

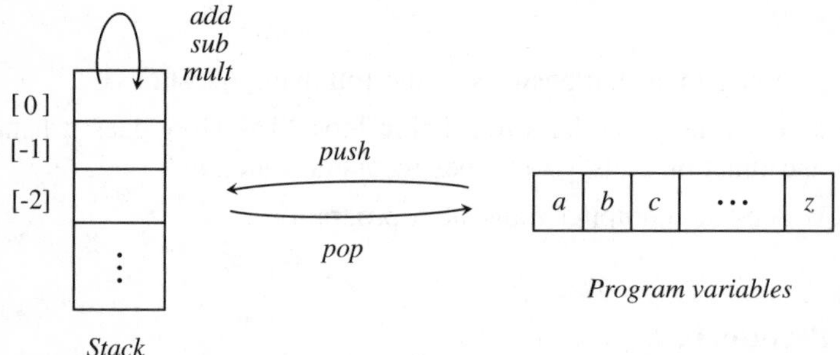

• The available variables are $a, b, \ldots, z$; all of integer type;

• The stack stores certain information and intermediate calculations.

We do not translate programs which use complex objects such as arrays.

Let *stack*[i] denote the plate that is at height i in the stack and let *stack*[*top*] denote the top of the stack. The plate *stack*[*top* $- \ell$] is noted $[-\ell]$; one says that the integer ℓ is the *offset* of the plate with respect to the top of the stack. Thus, the top of the stack is denoted [0], the plate below by $[-1]$, etc.

To start, one can consider a program as a long chain of characters. To facilitate the discussion we agree on the following terminology:

• a *constant* is a chain of characters which represents an unsigned integer: 1999 is a constant;

• a *variable* is one of the characters $a, \ldots, z$;

• a *stack reference* is a chain of characters such as [0], [−1], etc.;

• a *term* is a chain of characters which is a constant, a variable or a stack reference: 1515, a, [0], [−2] are terms;

• a *signed term* is a term *potentially* preceded by a minus sign: the chains 1515, −1515, a, $-a$, [0], −[0], [−3] and −[−3] are signed terms;

Syntax	*Meaning*
`read a`	*grab the value of a*
`write a`	*write the value of a*
`write ...`	*write the chain of characters which follows "write" if it is of length* > 1
`a = 12`	$a : 12$
`a = b`	$a := b$
`a = -b`	$a := -b$
`push ?`	$top := top + 1$ (*stack indefinite value*)
`push @a`	$top := top + 1$; $stack[top] := @a$ (*stack the address of a*)
`push T`	$top := top + 1$; $stack[top] := T$ (*stack the value of the signed term T*)
`pop`	$top := top - 1$ (*remove once from stack*)
`pop n`	$top := top - n$ (*unstack* $n \geq 1$ *times*)
`pop a, pop @a`	$a := stack[top]$; $top := top - 1$
`pop [-i]`	$stack[top - i] := stack[top]$; $top := top - 1$
`add`	$stack[top - 1] := stack[top - 1] \oplus stack[top]$; $top := top - 1$
`sub`	$stack[top - 1] := stack[top - 1] \ominus stack[top]$; $top := top - 1$
`mult`	$stack[top - 1] := stack[top - 1] \otimes stack[top]$; $top := top - 1$
`end`	*return control to the system*

Table 13.1. *A first set of pseudocode statements*

The push statement

This statement stacks a value or an address: "push x" stacks the value of the variable x, "push T" stacks the values of the signed term T and "push @x" stacks the address of x.

Suppose that the top of the stack contains the number 12. Is this a value or an address? We suppose that our imaginary microprocessor knows; when we program we will solve this little problem using a record.

The pop statement

This removes plates from the stack and is inflected in two ways:

- When one does not need to keep the contents of the plates, one writes for example "pop 3" to remove three plates. The statements "pop" and "pop 1" are equivalent.
- When one wants to recover the top of the stack, one writes:
  - ▷ "*pop* x" to transfer the top of the stack into variable x *before* removing it from the stack (in Pascal this is written $x := stack[top]$; $top := top - 1$);
  - ▷ "$pop[-2]$" to transfer the top of the stack to the level $top - 2$ *before* removing it from the stack (in Pascal, this is written $stack[top-2] := stack[top]$; $top := top - 1$).

The statement add

The idea is the remove the two first plates, add their contents, and place the result in the top of the stack. But, since the stack can contain both values and addresses and since adding two addresses or a value and an address does not make sense, the statement "add" uses the operation $\oplus$ instead of the usual "+". This modified addition first converts possible addresses to values. Consequently,

- if the two plates contain values, $\oplus$ adds them without further ado;
- if one plate contains an address, $\oplus$ first replaces the address by the value of the corresponding variable before taking the sum (thus, $5 \oplus @b$ means add 5 and the value of the variable b).

We specify: the first operand is always the plate $[-1]$, the second is $[0]$. The interpretation of the symbols $\ominus$ and $\otimes$ is similar (see Fig. 13.1).

Examples

1) To translate the statement $x := a + (b - c) * x + 15$ into pseudocode, we note (without touching the stack) that the final result must be put in the variable x; we then run through $a + (b - c) * x + 15$ from left to right which leads us to stack a, b, c, x, 15 and to perform the additions, subtractions and multiplications at the appropriate time:

7			7			@b			@b	
5	−2		@a	−2		5	−2		@a	−2
…	…		…	…		…	…		…	…

Fig. 13.2. *Effect of "sub" on the top of a stack (the first operand is below). There are four possible configurations (here, $a = 5$ and $b = 7$)*

```
100 push a     105 mult
101 push b     106 add
102 push c     107 push 15
103 sub        108 add
104 push x     109 pop x
```

When the translation of $a + (b - c) * x + 15$ is finished, we transfer the result to the variable x using a "pop" (which empties the stack). To better understand the meaning of the pseudocode, we sketch the successive states of the stack when $a = 1$, $b = 10$, $c = 7$ and $x = 12$.

100	101	102	103	104	105	106	107	108	109
		7		12					
	10	10	3	3	36		15		
1	1	1	1	1	1	37	37	52	$\longrightarrow x$

The final value is thus $x = 52$ (and the stack is empty).

2) We end this first encounter by examining the tranlation into pseudocode of the statement $x := (a - u - v) - (b - v - w) - (c - w - u)$.

```
100 push a   200 push b   300 push c   400 pop x
101 push u   201 push v   301 push w
102 sub      202 sub      302 sub
103 push v   203 push w   303 push u
104 sub      204 sub      304 sub
             250 sub      350 sub
```

The first column contains the translation of $a - u - v$ and the second that of $b - v - w$; the "250 sub" stacks the difference $(a - u - v) - (b - v - w)$; the fird column contains the translation of $c - w - u$ and the "350 sub" stacks $((a - u - v) - (b - v - w)) - (c - w - u)$.

Recall that addition, subtraction and multiplication are *associative from the left*, which means that evaluation is made from left to right; in other words, the translation of $a+b+c+d+e$ is the same as that of $(((a+b)+c)+d)+e$.

Branching statements

You have certainly noticed the integers which precede each statement in the preceding examples. These integers are called *addresses*. They form an increasing sequence which allows us to better structure our pseudocode by cutting it into segments of consecutive integers.

The program which executes the pseudocode uses a global variable which we call the *control* variable and which contains the address of the statement to execute.

As a general rule, the program executes statements *sequentially*, that is, one after the other. This effect is obtained simply by suitably incrementing the *control* variable after each statement.

There are, nevertheless, cases where the program must *branch* to a statement which is not the one immediately following. This can be realized in three ways.

The if goto statement

This statement compares plates, variables or constants. When it compares two plates, they must be [0] and [−1]; when it compares a plate and a variable or constant, the plate must be [0].

```
100 if [0] > [-1] goto 200
101 if x = [0] goto 201
102 if [0] ≤ 37 goto 202
```

```
103 if [0] ≠ n goto 200
104 if x ≥ y goto 201
105 if a ≤ 9 goto 202
```

The *if goto* statement is executed as follows:

- the program performs the indicated test;
- it then *removes* the plate or plates to be used (the statement 100 removes two plates, the statements 101–103 remove a single plate and the last two statements leave the stack intact because the test does not use any plate);
- if the test succeeds, the program leaves to execute the statement whose address figures after the *goto* (in other words, the *contol* variable stores this address);
- if the test fails, the program goes to the following statement.

The ifx goto statement

This statement is a variant of the *if goto* statement. Since the sequence increases the translation into pseudocode of a little program can be fairly long. So sometimes we may find ourselves cheating a little by *optimizing* certain translations – that is, by writing a shorter, more intelligible pseudocode than that produced by a compiler.

```
100 ifx [-2] > 0 goto 200
101 ifx [0] = [-3] goto 201
102 ifx a ≤ [0] goto 202
```

In contrast to the preceding statement, the *ifx* variant (where "x" signals *exception*) allows us to directly compare plates, a variable or a constant. This command *does not remove* any plates and thus leaves the stack intact.

The goto statement

This statement results in a *mandatory branching* (that is, without a preliminary test): "goto 200" means *control* := 200; there is no change to the stack.

The return statement

One can translate this statement in a vivid, but illegal, way as "pop control" (this is "illegal" because *control* is not one of the variables $a, b, \ldots, z$). However, it does make clear that it means that the program transfers the number at the top of the stack into *control* and then removes the top plate.

The programmer must arrange, however, that the top of the stack contains the address of an statement at the moment of a "return".

13.1.2. How to compile a pseudocode program by hand

Since our goal is to understand how a compiler functions, we are going to imitate its behavior as faithfully as possible.

The compiler reads the program from left to right and translates it as it goes along *without waiting*.

We need to beware of thinking of ourselves as compilers. A compiler has no global vision of the text: it only sees a single *word* (or *token*) at a time, it systematically forgets what it has read (but consults notes that have been taken: the value of a variable, the type of a variable, the dimensions of an array, etc.), it never sees the word following or the word preceding, it never backs up to reread something. *A compiler advances inexorably without pauses towards the end of the program.*

A compiler is a *program*. In other words, it is a set of *reflexes* released by reading the current word, or even the current character (a parenthesis, for example). We must learn to do the same.

The stack serves to store certain information (the parameters of a function or procedure, local variables, the address of a variable, return address) or intermediate results (to calculate $a + b * c$, we must first find and store the value of $b * c$ before adding it to the number a).

Each time that we finish translating a sequence, we emit the necessary instructions which *will clean up* (notice the future tense) the stack by removing plates which have no further use so that the program restores (when it is functioning) the stack to the state in which it found it. The examples that follow will make this precise.

A last remark: in our explanations, we will mix the compilation (*the present*) and the execution of the pseudocode (*the future*) because it is difficult to give a order without trying to imagine the result it provokes.

13.1.3. Translation of a conditional

Consider the following fragment of code:

$$\textsf{\textbf{if}}\ x + a > y * y + y\ \textsf{\textbf{then}}\ \alpha\ \textsf{\textbf{else}}\ \beta\ ;\ \gamma$$

We suppose that the translations of α and β leave the stack intact:

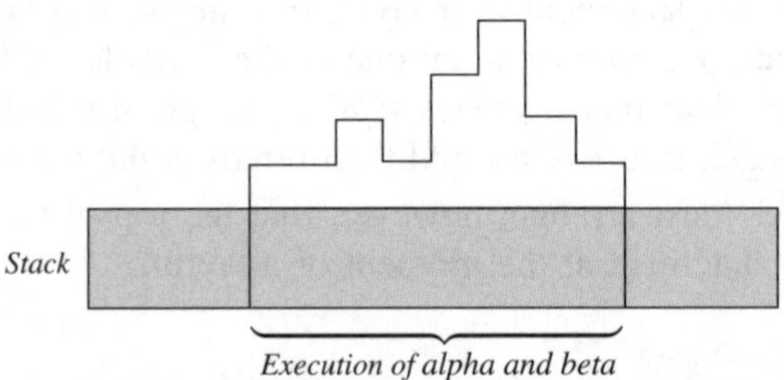

When the compiler meets an "if", it knows that it is going to encounter one or more tests that it must translate as it goes along. We imitate it faithfully.

• We begin by translating the expression $x + a$; we let P denote its value (which will stay in the stack). This done, we memorize mentally (without emitting code) the fact the comparison is ">", then we translate the expression whose value we denote Q and stack above P.

• When we meet the "then", we emit the code that compares P and Q using the *opposite test* $P \leq Q$ which allows branching to the code of β (beginning with statement 300). But since we have not yet translated α or β, we *hold in reserve* the address after the *goto* of statement 300. (Recall that the test "if $[-1] \leq [0]$ goto" results in the removal of the plates containing P and Q.)

"push P"	`100 push x`	`300 if [-1] ≤ [0] goto 600`	
	`101 push a`	`400`	*code for α*
	`102 add`	`...`	
		`499`	
"push Q"	`200 push y`	`500 goto 700` *(skip β)*	
	`201 push y`	`600`	*code for β*
	`202 mult`	`...`	
	`203 push y`	`699`	
	`204 add`	`700` *start of the code for γ*	

• We then read and translate the fragment of code α.

• When we meet the "else", we know that the translation of α is finished; we emit an unconditional "goto" (statement 500) which allows us, at the moment the pseudocode is executed, to skip over to the translation of β. But since we still do not know where this translation ends, we hold in reserve the address of this "goto". By contrast, since we now know where the translation of β begins, we return backwards in the pseudocode to complete the "goto" at 300 before beginning to translate β.

• When we meet the semicolon, we know that the translation of β is finished; we then return once more backwards to complete the "goto" in 500 before passing on to what follows.

Remarks

• When the execution of the pseudocode ends, we can be certain that the stack is intact since α and β each leave it intact.

• Recall that a compiler *never* goes backward in the program source code. In contrast, we see that it does so as often as necessary *in the pseudocode* to complete the addresses that are left standing by after the "goto"s.

• Why use the opposite test? Try it: if you translate the conditional while keeping the test intact, you will be obliged to put the translation of β *before* that of α. Since the compiler cannot go backwards in the program source code, this strategy would oblige it to store some part of the translation of α while waiting to be able to write it.

• The mechanical translation of the embedded conditionals makes "flea jumps" (statements 300 and 500 below) appear. A good compiler knows to avoid this by writing directly "goto 700" in 300.

if $p > 0$
then if $q = 1$
  then α
  else β
else γ; δ

$\Longrightarrow$

100 if p $\leq$ 0 goto 600
101 if q $\neq$ 1 goto 400
200 ... 299 *code for* α
300 goto 500
400 ... 499 *code for* β
500 goto 700
600 ... 699 *code for* γ
700 ... *code for* δ

Exercises 1

1) Translate the embedded conditionals into pseudocode:

if $p > 0$
then if $q = 1$
  then $x := x + (a * b - c) + y$
  else if $u = 2$
    then $u := u + v - w$
    else $z := x + y - (u + v - w) * (a + b)$

2) When the test is complicated

$$\textbf{if } (p > 0) \textbf{ or } (q = 2) \textbf{ then } \alpha \textbf{ else } \beta$$

use the following scheme:

`if` $p \leq 0$ `goto` ⟨*next test*⟩
`goto` ⟨*start of* α⟩
`if` $q \neq 2$ `goto` ⟨*start of* β⟩
⟨*start of the code for* α⟩
`goto` ⟨*after* β⟩
⟨*start of the code for* β⟩

Reading an "or" or an "and" then elicits the same reaction as meeting a "then" does: emission of the opposite test and emission of a "goto" that is provisionally incomplete and that branches to certain sections of code.

3) Translate:

$$\begin{aligned}&\textbf{if } (p > 0) \textbf{ or } (q = 2) \textbf{ and } (r < 3) \textbf{ or } (s > 4)\\&\textbf{then if } (u = 5) \textbf{ and } (v = 6)\\&\qquad\textbf{then } \alpha \textbf{ else } \beta\end{aligned}$$

13.1.4. Translation of a loop

Consider the loop:

$$\textbf{while } i * i + 1 < i + m \textbf{ do } \alpha$$

We suppose that the translation of α restores the stack to what it was at the moment of execution

`100 push i`	`107 add`
`101 push i`	`108 if [-1]` $\geq$ `[0] goto 201` {*exit of the loop*}
`102 mult`	`109` } *code for* α {*body of the loop*}
`103 push 1`	`...`
`104 add`	`...`
`105 push i`	`199`
`106 push m`	`200 goto 100` {*return of the loop*}

• When the compiler meets a "while", it stores the number of the statement that it is *going to write* (note the future), because this is where the loop will begin.

• We are familiar with the translation of $i * i + 1 < i + m$ (remember that the plates $P = i * i + 1$ and $Q = i + m$ are automatically removed after the test 108).

• The statement 200 effects the return of the loop using the address stored while reading the "while". By proceeding this way, we are certain to leave the stack intact.

Exercise 2

Translate the two loops

$$\textbf{for } i := 1 \textbf{ to } n * n + 1 \textbf{ do } s := s + i \, ;$$

$$\textbf{repeat } d := d + 1 \textbf{ until } d * d > n$$

13.1.5. Function calls

In order to familiarize ourselves with this mechanism, we analyse the translation of statement

$$s := x + square(y) + z$$

where *square* is the function which squares its argument. (The snapshot of the stack after 1002 to the left of the code will help you to understand and check the offsets.)

Stack
y^2
y
105
?
x

```
                100 push x          105 add        1000 push [0]
            {   101 push ?          106 push z     1001 push [-1]
stack y^2   {   102 push 105        107 add        1002 mult
            {   103 push y          108 pop s      1003 pop [-3]
            {   104 goto 1000       200 end        1004 pop
                                                   1005 return
```

The sequence 101–104 comprises the function *call*:

- We begin by stacking an *indefinite value* (symbolized by the question mark); this value will ultimately be replaced by the value of the function.
- We stack above the *return address*; that is, the address of the statement to execute when the function code finishes (*i.e.* when the question mark is replaced by the value of the function). This address (unknown for the moment) is that which follows the "goto 1000".
- We stack the value of the parameter y of the call.

Note the (indispensable) presence of the "end" (statement 200). If one forgets it, the program would penetrate into the function code instead of stopping.

The compilation of this function is easy to understand:

- Throughout the entire time during which the function call lasts we keep the plates ?, 105, y *intact* (by working *above* the stack).
- When we finally find the value of *square*(y), we transfer it to where the question mark is and remove the plates needed to free the return address.

To summarize, the sequence 101–104 amounts simply to saying:

"*stack* y^2."

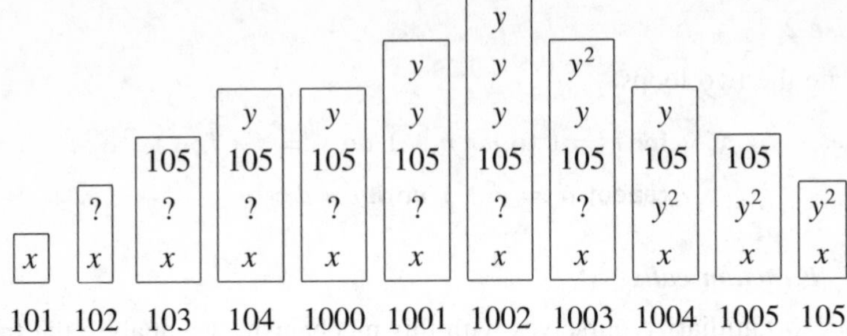

Fig. 13.3. *The number at the base of each column is that of the next statement*

Remark

This translation is deliberately awkward because we are trying to emit a "mechanical" pseudocode which ressembles that which a compiler would produce. It is possible to do better by suppressing the statement 1001 and replacing the "pop [−3]" by a "pop [−2]" in 1003, which has the effect of destroying the value of the call parameter y. But such clairvoyance is unavailable to a compiler.

Before the call	During the call	After the call
	Local variables	
	Call parameters	
	Return address	
	? = *Future value of f*	$f(x)$
yyy	*yyy*	*yyy*
…	…	…
xxx	*xxx*	*xxx*

Fig. 13.4. *Steps that follow a function call: the code of the function must work in the stack above the local variables in order not to destroy them; the value of the function is transferred above the old stack which must not be modified; the return address permits the program to return to the code right after the function call*

We now refine our understanding by translating a somewhat more elaborate statement:

$$s := a * F(F(2 * x + 1)) + b$$

Recall that the compiler reads the source text from left to right and translates as it goes along without pauses. The "reflexes" put into play are as follows:

• Reading the name of a function followed by an open parenthesis results in the *immediate* emission of a sequence which stacks ? and the return address.

• The arithmetic expression in the parentheses is translated from left to right as usual.

• Reading the closed parenthesis which ends the function call results in bringing the return address to the fore and branching of the pseudocode to that of the function.

```
100 push a
101 push ?
102 push 112
103 push ?
104 push 111
105 push 2
106 push x
107 mult
108 push 1
109 add
110 goto 1000
111 goto 1000
112 mult
113 push b
114 add
115 pop s
```

When we read the chain of characers "$a * F($", we emit the statements 100–102 by holding the return address in 102 in reserve (it will be known when we read the corresponding closed parenthesis).

We then compile the argument of the function which is "$F(2 * x + 1)$". Thus, we emit the statements 103–109 by leaving the return address in 104 on standby.

When we meet the first closed parenthesis, we know that it is time to call the function F (statement 110) and that the return address in 104 is equal to 111.

When we encounter the second closed parenthesis, we call the function F anew (statement 111) and complete the return address (now equal to 112) on standby in 102.

We now compile the body of the function that we take to be:

$$\textbf{if } x * x < a \textbf{ then } F := x - a \textbf{ else } F := a - x$$

We are familiar with the translation of the fragment "if $x * x < a$ then": we stack the value of x twice in a row (statements 1000 and 1001) in order to calculate $x * x$ in 1002 (thus, we work systematically *above* the call parameter).

When the compiler deciphers the statement $F := x - a$, it takes note that it is dealing with the name of a function (and not the name of a variable). It then translates classically the calculation of $x - a$ by replacing the references to x by the corresponding plate.

When it encounters the "else", the compiler knows that it has finished the calculation of the value of F and that it is time to perform the assignment. Since it remembers that it is dealing with a function call and not a normal assignment, the compiler emits the necessary orders to transfer the value of F to a good spot and free the return address which allows it to leave the function code.

```
1000 push [0]
1001 push [-1]
1002 mult
1003 if [0] ≥ a
          goto 1010
1004 push [0]
1005 push a
1006 sub
1007 pop [-3]
1008 pop
1009 return
1010 push a
1011 push [-1]
1012 sub
1014 pop [-3]
1015 pop
1016 return
```

Remark

This code is somewhat optimized because we cheated a bit by not issuing at 1010 the "goto" which allows jumping to the code of β once the translation of α is complete. In contrast to the compiler, we *know* that we are leaving the function code.

13.1.6. A very efficient technique

Let us describe a technique which makes compiling function and procedure calls easy and sure. Suppose we want to compile $a := x + square(y) + z$.

- We write down a first approximation of the code:

```
read x
read y
read z
push x
push square(y)
add
push z
add
pop a
```

Note the absence of addresses and the use of the illegal command "push square(y)".

- We now write the code for "push *square*(y)":

```
push ?
push (R1)
push y
goto (square)
(R1)
```

The *labels* (R1) and (square) represent unknown addresses. The last line contains only the symbolic address (R1).

- We *assemble* these two codes; that is, we replace "push *square*(y)" by its code; the symbolic address (R1) becomes the address of the statement which follows.

```
read x
read y
read z
push x
push ?
push (R1)
push y
goto (square)
(R1) add
push z
add
pop a
```

- All we have to do now is introduce addresses to get the pseudocode of the previous paragraph.

To test this technique, let us compile the more difficult statement

$$s := f(x - g(y + z)) * g(f(x + g(y + z)) - x)$$

- The first draft is

```
read x
read y
read z
push f(A)
push g(B)
mult
pop s
```

- The first details for f(A) are

```
push ?
push (R1)
push x
push g(y+z)
sub
goto (f)
(R1)
```

We now expand the call g(y+z).

```
push ?
push (R1)
push x
push ?
push (R2)
push y
push z
add
goto (g)
(R2) sub
goto (f)
(R1)
```

- The first details for g(B) are

```
push ?
push (R3)
push f(x+g(y+z))
push x
sub
goto (g)
(R3)
```

We introduce more details into g(B) by expanding the call f

```
push ?
push (R3)
push ?
push (R4)
push x
push g(y+z)
add
goto (f)
(R4) push x
sub
goto (g)
(R3)
```

We expand the last call g(y+z):

```
push ?
push (R3)
push ?
push (R4)
push x
push ?
push (R5)
push y
push z
add
goto (g)
(R5) add
goto (f)
(R4) push x
sub
goto (g)
(R3)
```

- We assemble these fragments to get the final code:

```
read x        push z        push x        (R4) push x
read y        add           push ?        sub
read z        goto (g)      push (R5)     goto (g)
push ?        (R2) sub      push y        (R3) mult
push (R1)     goto (f)      push z        pop s
push x        (R1) push ?   add
push ?        push (R3)     goto (g)
push (R2)     push ?        (R5) add
push y        push (R4)     goto (f)      end
```

All we have to do now is to put in the addresses:

```
10 read x       107 push z       304 push x       500 push x
11 read y       108 add          305 push ?       501 sub
12 read z       109 goto 2000    306 push 400     502 goto 2000
101 push ?      200 sub          307 push y       600 mult
102 push 300    201 goto 1000    308 push z       601 pop s
103 push x      300 push ?       309 add
104 push ?      301 push 600     310 goto 2000
105 push 200    302 push ?       400 add
106 push y      303 push 500     401 goto 1000    700 end
```

Exercise 3

Translate the following statements into pseudocode

$$s := 0\ ;\ \textbf{for}\ i := 1\ \textbf{to}\ n\ \textbf{do}\ s := s + F(i)\ ;$$

$$t := 0\ ;\ \textbf{for}\ i := 1\ \textbf{to}\ n\ \textbf{do}\ t := t + F(i) + F(2 * i)\ ;$$

$$v := 0\ ;\ \textbf{for}\ i := a + b\ \textbf{to}\ a * b\ \textbf{do}\ u := F(F(2 * i + 1))\ ;$$

$$v := 0\ ;\ \textbf{for}\ i := 1\ \textbf{to}\ n\ \textbf{do}\ v := v + F(G(F(i), F(2 * i + 1)) - 1)$$

knowing that F and G are the following functions

$$F(x) = (x+1)(x+2), \quad G(u, v) = \begin{cases} u^2 + u + 1 & \text{if } u - v > 1, \\ u^2 - v^2 & \text{if not.} \end{cases}$$

13.1.7. Procedure calls

Now consider the program fragment

procedure *toto*(x : *integer* ; **var** y : *integer*) ;
var i : *integer* ;
begin
  $x := y * y\ ;\ i := x + y\ ;\ y := i + 1$
end ;

begin
  $a := 1$;
  $b := 5$
  toto(a, b)
end .

and its translation into pseudocode. The first column below contains the translation of the principal part of the program. The two following columns contain the translation of the procedure *toto*.

```
100 a = 1          1000 push [-1]    1008 push [0]
101 b = 5          1001 push [-2]    1009 push 1
102 push 107       1002 mult         1010 add
103 push a         1003 pop [-3]     1011 pop [-2]
104 push @b        1004 push [-2]    1012 pop 3
105 push ?         1005 push [-2]    1014 return
106 goto 1000      1006 add
107 end            1007 pop [-1]
```

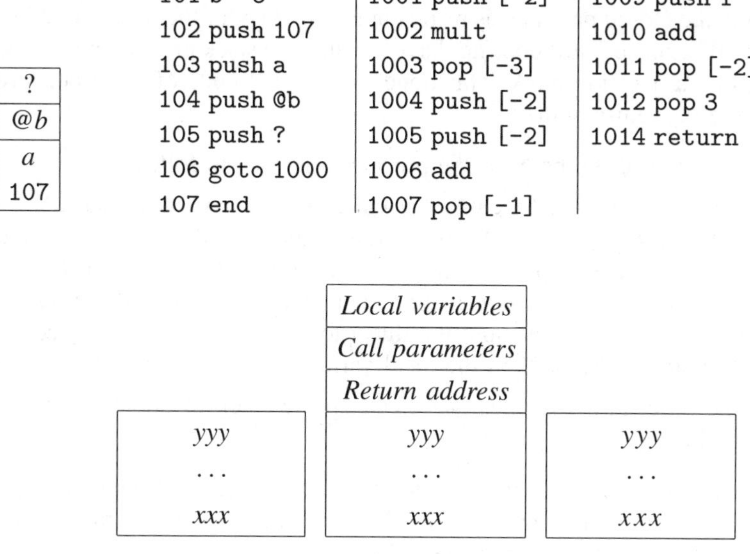

Fig. 13.5. *Steps that follow a procedure call. The routine is the same as for a function call, but simpler because it does not have to return a value. The code for the procedure must work above the parameters of the call and must leave the original stack intact.*

The method used to call a procedure is the same as that for a function call except that there is no need to reserve a place in the stack to return the value of the function. To prepare a procedure call, we stack successively (and in this order):

- the return address (address of the statement that follows the "goto 1000");
- the arguments of the procedure;
- the local variable (or variables) of the procedure;
- we call the procedure and complete the return address.

The way of stacking an argument depends on its nature:

- if it is passed by value, we stack its *value*;
- if it is passed by address, we stack its *address*.

To translate *toto*(a, b), we proceed as follows:

• When we read "toto(", we issue a "goto" at 102 which must be followed by a return address, an address which will only be known when we read the closed parenthesis of *toto*(a, b) ;

• Reading a results in stacking the value of a (statement 103) and reading b stacks the address of b (statement 104);

• Reading the closed parenthesis results in the stacking of the local variable i (statement 105). Since i still does not have a value, we stack an *indefinite* value symbolized by a question mark. This done, we call the code of the procedure and complete the return address.

The compilation of the body of the procedure is easy to follow:

• We especially avoiding touching the plates which have been stacked by statements 102–104 in order to keep the return address, the value of a and the address of b intact. This is because we cannot know in advance how many times we will need this information. Thus we work *above* these plates. In contrast, we have the right (and the duty) to modify the plate stacked by statement 105 since it pertains to the local variable i.

• Once the procedure compiles, we clean up the stack in order to free up the return address.

The array that follows represents how the stack evolves during the execution of $toto(a, b)$. The number at the base of each column is that of the statement which has just been executed. (Recall that the statements "pop @b" and "pop b" have the same effect.)

		5			5			1		
		5	25		25	30		30	31	
i	?	?	?	?	?	?	30	30	30	30
y	@b	@b	@b	@b	@b	@b	@b	@b	@b	@b
x	1	1	1	25	25	25	25	25	25	25
Return	107	107	107	107	107	107	107	107	107	107

b	5	5	5	5	5	5	5	5	5	31
a	1	1	1	1	1	1	1	1	1	1

Fig. 13.6. *Evolution of the stack and the contents of* a, b *during the call of toto*(a, b)

When a parameter passed by value is an arithmetic expression, we evaluate it. Thus the translations of $toto(a, b)$, $toto(a, a)$ and $toto(a + 7, b)$ are

```
push address
push a
push @b
push ?
goto 1000
```

```
push address
push a
push @a
push ?
goto 1000
```

```
push address
push a
push 7
add
push @b
push ?
goto 1000
```

Even in the third case, the height of the stack increases only by four units ($address$, $x = a + 7$, $y = @b$, $z = ?$) as in the preceding cases. The arguments are always at the same level in the stack when one reaches the procedure.

Exercise 4

Compile in pseudocode the calls

$$toto(a + F(b), a)\ ;\ \ toto(y + G(x, y), x)$$

where F and G are the functions in the exercise in the preceding section.

13.1.8. The factorial function

We now know enough to be able to translate recursive programs. We shall see that recursion is a natural consequence of the administration of a stack. We begin with the archetype, the factorial function:

$$\textbf{if } n \leq 1 \textbf{ then } fact := 1 \textbf{ else } fact := fact(n - 1) * n.$$

The pseudocode below takes the value of n and returns $y = n!$.

```
10 read n
100 push ?
101 push 104
102 push n
103 goto 1000
104 pop y
105 write y =
106 write y
200 end
```

```
1000 push [0]
1001 if [0] > 1
          goto 1006
1002 push 1
1003 pop [-3]
1004 pop
1005 return
1006 push [0]
1007 push ?
```

```
1008 push 1013
1009 push [-3]
1010 push 1
1011 sub
1012 goto 1000
1013 mult
1014 pop [-3]
1015 pop
1016 return
```

Remember, never forget the “end”!

• To understand how one gets the code for the factorial function, we write down our first approximation:

$(n-1)!$
n
n 104 ?

```
(F) push [0]        pop                 pop [-3]
if [0] > 1          return              pop
    goto (R1)       (R1) push [0]       return
push 1              push fact(n-1)
pop [-3]            mult
```

The snapshot of the stack on the left side of the code will help you to understand and check the offsets.

• When we compile $fact(n-1)$, we must be very careful: the correct value of n is the value which is at the top of the stack. We cannot use the variable n of the program.

n 104 ?

n
$(R2)$
?
n

```
push ?              sub
push (R2)           goto (F)
push [-3]           (R2)
push 1
```

(When the stack is broken, you must read it from bottom to top and from left to right.)

• We now assemble our pieces of code:

```
(F) push [0]                (R1) push [0]       goto (F)
if [0] > 1 goto (R1)        push ?              (R2) mult
push 1                      push (R2)           pop [-3]
pop [-3]                    push [-3]           pop
pop                         push 1              return
return                      sub
```

All we have to do now is to introduce addresses.

Theorem 13.1.1. *The proposed code correctly calculates the value of the factorial function.*

Proof. The proof is by induction on the integer n. The statement is true when $n = 0$ or $n = 1$ (one need only execute statements 1000–1005). Now suppose that $n \geq 2$ and make the following induction hypothesis: each function call for a value $k < n$ places $k!$ at the top of the stack after finite time. When the program executes statement 1000 for the first time, the top of the stack contains n; the induction hypothesis then assures us that the sequence 1006–1012 puts $(n-1)!$ above n and the statements 1013–1016 transfer $n!$ to the right place in the stack and free up the return address.

Exercise 5

1) Execute by hand the calculation of $n!$ for $n = 0, \dots, 5$.

2) Translate "if $n \leq 1$ then *fact* := 1 else *fact* := $n * fact(n-1)$" into pseudocode and use it to calculate $n!$ when $n = 1, \dots 4$. Notice the exchange of factors: *fact* := $n * fact(n-1)$ instead of *fact* := $fact(n-1) * n$.

3) For $n = 1, 2, 3, 4$, execute statements 100–106 of the example in which calling $n!$ has been replaced by the following pseudocode.

```
1000 ifx [0] > 1 goto 1003
1001 pop [-2]
1002 return
1003 push ?
1004 push 1009
1005 push [-2]
1006 push 1
1007 sub
1008 goto 1000
1009 mult
1010 pop [-2]
1011 return
```

This code has been optimized. Do you see how?

13.1.9. The Fibonacci numbers

The following code takes the value of n as input and displays the value of Fib(n):

```
99 read n
100 push ?
101 push 104
102 push n
103 goto 1000
104 pop y
105 write Fib(n) =
106 write y
200 end

1000 push [0]
1001 if [0] > 1
            goto 1006
1002 push [0]
1003 pop [-3]
1004 pop
1005 return
1006 push ?
1007 push 1012
1008 push [-2]
1009 push 1
1010 sub
1011 goto 1000
1012 push ?
1013 push 1018
1014 push [-3]
1015 push 2
1016 sub
1017 goto 1000
1018 add
1019 pop [-3]
1020 pop
1021 return
```

n
104
?

Once again, note that the final "end" which stops the program from penetrating unduly into the code of the function *Fib* beginning at 1000.

We use the following definition of the Fibonacci function:

$$\textbf{if } n \leq 1 \textbf{ then } Fib := n \textbf{ else } Fib := Fib(n-1) + Fib(n-2)$$

• To understand the code of the *Fib* function, we write down the first approximation for *Fib*.

$F(n-2)$
$F(n-1)$
n 104 ?

```
(Fib) push [0]              (R1) push Fib(n-1)
if [0] > 1 goto (R1)        push Fib(n-2)
push [0]                    add
pop [-3]                    pop [-3]
pop                         pop
return                      return
```

• We now translate the call Fib(n – 1). One must be very cautious: due to the recursive call, the variable n in which we are interested belongs to the function Fib, so its value (at the begining of the compilation) is in the top plate. But the offset of the plate which contains this value augments as we push new plates on the stack!

n 104 ?

n
$(R2)$
?

```
push ?          sub
push (R2)       goto (Fib)
push [-2]       (R2)
push 1
```

• When we translate the call *Fib*(n – 2), the variable n is already deeper in the stack as $F(n-1)$ now lies on the top of the stack.

n 104 ?

n
$(R3)$
?
$F(n-1)$

```
push ?          sub
push (R3)       goto (Fib)
push [-3]       (R3)
push 2
```

• We now substitute the calls *Fib*(n – 1) and *Fib*(n – 2) into the first approximation and introduce addresses to get the translation.

```
(Fib) push [0]    return          goto (Fib)      goto (Fib)
if [0] > 1        (R1) push ?     (R2) push ?     (R3) add
goto (R1)         push (R2)       push (R3)       pop [-3]
push [0]          push [-2]       push [-3]       pop
pop [-3]          push 1          push 2          return
pop               sub             sub
```

Exercise 6

1) Execute this code when $n = 0, 1, 2, 3$ (after this, it becomes painful).

2) Inspired by the proof of the correctness of the code for the factorial function, show that statements 1000 to 1021 correctly calculate the Fibonacci numbers.

3) Translate the following into pseudocode:

if $n \leq 1$ **then** $Fib := n$
else begin
| $u := Fib(n-1)$; $v := Fib(n-2)$;
| $Fib := u + v$
end

13.1.10. The Hofstadter function

Recall the definition (Chap. 12) of the Hofstadter functions:

$$G(0) = 0, \quad G(n) = n - G\big(G(n-1)\big) \quad \text{if} \quad n \geq 1.$$

The following code takes the value of n as input and displays the value of $G(n)$:

```
100 read n
101 push ?
102 push 105
103 push n
104 goto 1000
105 pop g
106 write G(n) =
107 write g

200 end
```

```
1000 ifx [0] > 0
          goto 1005
1001 push 0
1002 pop [-3]
1003 pop
1004 return
1005 push [0]
1006 push ?
1007 push 1015
1008 push ?
```

```
1009 push 1014
1010 push [-5]
1011 push 1
1012 sub
1013 goto 1000
1014 goto 1000
1015 sub
1016 pop [-3]
1017 pop
1018 return
```

- To compile the Hofstadter function, we write our first approximation.

$G(G(n-1))$
n
n 105 ?

```
(G) ifx [0] > 0 goto (R1)
push 0
pop [-3]
pop
return
(R1) push [0]
```

```
push G(G(n-1))
sub
pop [-3]
pop
return
```

- We now get a first approximation for $G(G(n-1))$.

- We improve the previous code by expanding the call $G(n-1)$: we must not forget that n is deeper in the stack now.

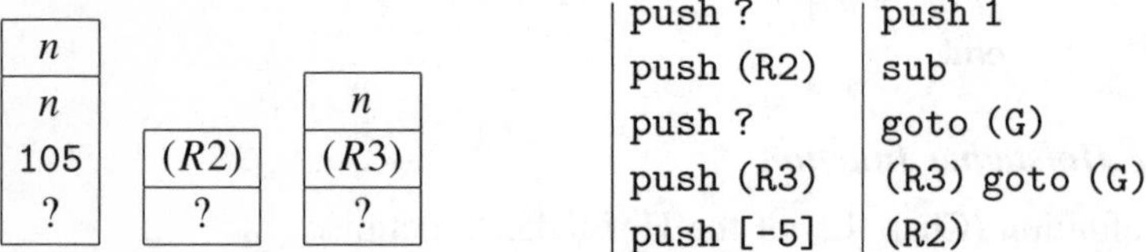

- All we have to do now is to assemble the fragments

```
(G) ifx [0] > 0 goto (R1)
push 0
pop [-3]
pop
return
(R1) push [0]
push ?
push (R2)
push ?
push (R3)
push [-5]
push 1
sub
goto (G)
(R3) goto (G)
(R2) sub
pop [-3]
pop
return
```

and introduce addresses to get the final code.

13.1.11. The Towers of Hanoi

Our last example consists in translating the towers of Hanoi into pseudocode. The vertical pegs are represented by the integers 1, 2 and 3.

```
procedure Hanoi(a, b, c : integer) ;
begin
  if n = 1 then move(a, c)
  else begin
    Hanoi(a, c, b, n − 1) ;
    move(a, c) ;
    Hanoi(b, a, c, n − 1)
  end
end ;
```

```
procedure move(x, y : integer) ;
begin
  writeln(x, ' to ', y)
end ;
```

The translation of the main body of the program is simple: we ask for the value n, then initialize a, b, c with 1, 2, 3 and call the procedure *Hanoi* whose translation begins at 1000. We do not push any question mark (indefinite value) because a procedure does not have a value and *Hanoi* does not have a local variable.

```
100 read n       103 push 2     106 goto 1000
101 push 107     104 push 3
102 push 1       105 push n     200 end
```

- The first approximation for $Hanoi(a, b, c, n)$ is

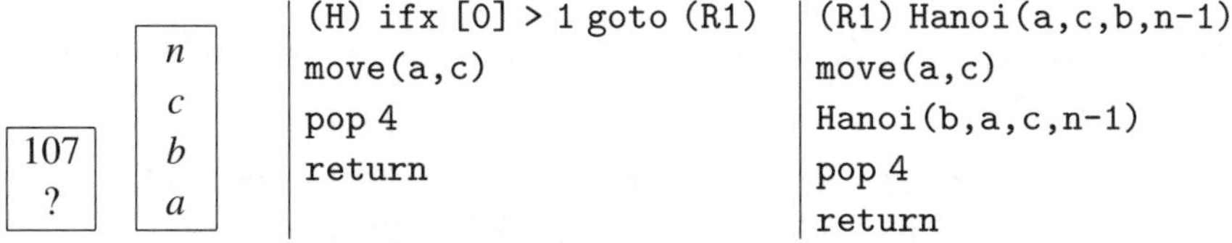


```
(H) ifx [0] > 1 goto (R1)    (R1) Hanoi(a,c,b,n-1)
move(a,c)                    move(a,c)
pop 4                        Hanoi(b,a,c,n-1)
return                       pop 4
                             return
```

(Note that we cheated a little by using an "ifx".) Recall that a procedure *must* restore the stack intact. When $n = 1$, and if we suppose that move(a, c) restores the stack we need the "pop 4" before the "return" to clean the stack and free the return address. If $n > 1$ and if Hanoi$(a, c, b, n - 1)$, move(a, c), Hanoi$(b, a, c, n - 1)$ leave the stack intact, we need again a "pop 4" to clean the stack.

- One must avoid confusion at this point. Here, we are asking the program to *execute* statements, we are not *defining* Hanoi: therefore, we must not use a pop. The code for Hanoi$(a, c, b, n - 1)$ is:

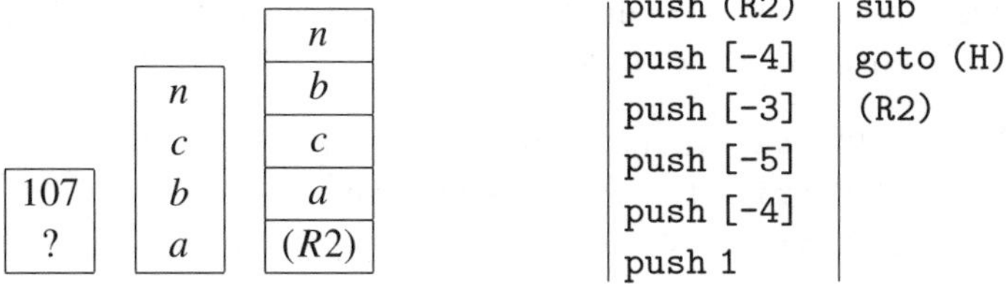


```
push (R2)    sub
push [-4]    goto (H)
push [-3]    (R2)
push [-5]
push [-4]
push 1
```

- Remark again that no pop is necessary because Hanoi$(a, c, b, n-1)$ leaves the stack clean. Thus the code for $move(a, c)$ is

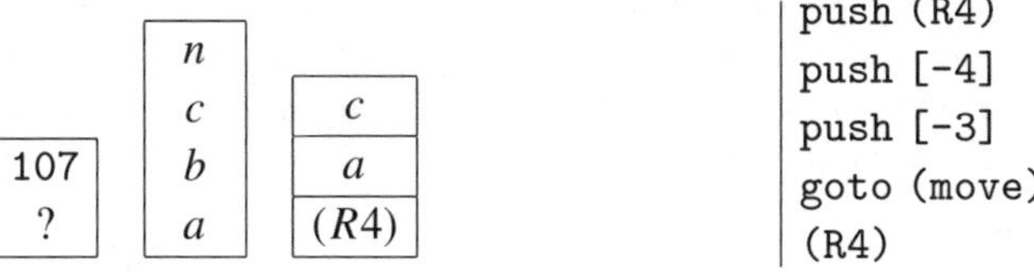


```
push (R4)
push [-4]
push [-3]
goto (move)
(R4)
```

- The code for Hanoi$(b, a, c, n-1)$ is:

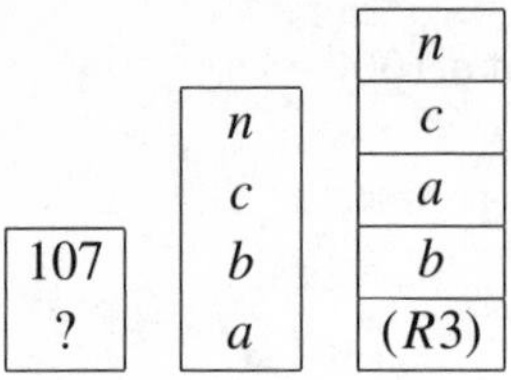

		n
	n	c
	c	a
107	b	b
?	a	$(R3)$

```
push (R3)     sub
push [-3]     goto (H)
push [-5]     (R3)
push [-4]
push [-4]
push 1
```

- We assemble the fragments and introduce addresses to get the final pseudocode.

```
1000 ifx [0] > 1       1009 push [-3]     1019 push 1027
          goto 1007    1010 push [-5]     1020 push [-3]
1001 push 1005         1011 push [-4]     1021 push [-5]
1002 push [-4]         1012 push 1        1022 push [-4]
1003 push [-3]         1013 sub           1023 push [-4]
1004 goto 2000         1014 goto 1000     1024 push 1
1005 pop 4             1015 push 1019     1025 sub
1006 return            1016 push [-4]     1025 goto 1000
1007 push 1015         1017 push [-3]     1027 pop 4
1008 push [-4]         1018 goto 2000     1028 return
```

The code for the procedure *move* is simpler because it does not contain a procedure call. We simply display the disks to move, then leave the procedure after having freed in advance the return address.

```
2000 write [-1]     2003 pop 2
2001 write to       2004 return
2002 write [0]
```

For beginners

A common error consists in unstacking after the calls *Hanoi*$(a, c, b, n-1)$ and *Hanoi*$(b, a, c, n-1)$. We absolutely do not undertake this task: it is the calls *Hanoi*$(a, c, b, n-1)$ and *Hanoi*$(b, a, c, n-1)$ which do it. The same argument applies after calling *move*(a, c). We unstack only *once*, in order to free the return address (statements 1005 and 1027).

Exercise 7

- Execute the code when $n = 3$.
- Prove that this code is correct.

13.2. A Pseudocode Interpreter

To better understand the mechanism of procedure calls and recursion, it is indispensable to translate some small programs into pseudocode and to run them by hand. As you realize by now, this is a long and tedious mechanical activity which is very prone to error.

Thus, we are going to write a *pseudocode interpreter*; that is, a program which will execute pseudocode in our stead, but without error. This will allow us to focus on the intellectually most interesting part, the action of the compiler; that is, the translation of a given program into pseudocode.

Consider for example the following program which is the translation into pseudocode of the instruction $s := a * F(F(2 * x + 1)) + b$ where F is the function defined by $F(x) := x - a$ if $x * x < a$ and $F(x) = a - x$ if not.

```
10 read a
11 read b
12 read x
100 push a
101 push ?
102 push 112
103 push ?
104 push 111
105 push 2
106 push x
107 mult
108 push 1
109 add
110 goto 1000
111 goto 1000
112 mult
113 push b
114 add
115 pop s
116 end
1000 push[0]
1001 push[-1]
1002 mult
1003 if [0] ≥ a
          goto 1010
1004 push[0]
1005 push a
1006 sub
1007 pop [-3]
1008 pop
1009 return
1010 push a
1011 push [-1]
1012 sub
1014 pop [-3]
1015 pop
1016 return
```

The execution of this pseudocode by our interpreter when $a = 10$, $b = 100$ and $x = 5$ is exhibited in Table 13.2.

Writing such an interpreter is also a very interesting programming exercise, because it is a very pretty example of the minutiae and rigor that is necessary to bring to bear each time that one deals with the *recognition of forms* and the treatment of chains of characters.

Since the program is quite long, we will not always adhere to our method of developing a program. We group the procedures by themes in order to facilitate their comprehension.

Declarations

The program to be interpreted is stored in the array *code*. A line of code is a pair (*number of statement*, *statement*).

```
   10: a = 10, 11: b = 100, 12: x = 5
  100:     |    |    |    |    |    |    |    |  10|    |
  101:     |    |    |    |    |    |    |   ?|  10|    |
  102:     |    |    |    |    |    | 112|   ?|  10|    |
  103:     |    |    |    |    |   ?| 112|   ?|  10|    |
  104:     |    |    |    | 111|   ?| 112|   ?|  10|    |
  105:     |    |    |   2| 111|   ?| 112|   ?|  10|    |
  106:     |    |   5|   2| 111|   ?| 112|   ?|  10|    |
  107:     |    |    |  10| 111|   ?| 112|   ?|  10|    |
  108:     |    |   1|  10| 111|   ?| 112|   ?|  10|    |
  109:     |    |    |  11| 111|   ?| 112|   ?|  10|    |
  110: goto 1000
 1000:     |    |  11|  11| 111|   ?| 112|   ?|  10|    |
 1001:     |  11|  11|  11| 111|   ?| 112|   ?|  10|    |
 1002:     |    | 121|  11| 111|   ?| 112|   ?|  10|    |
 1003:     |    |    |  11| 111|   ?| 112|   ?|  10|    |
 1010:     |    |  10|  11| 111|   ?| 112|   ?|  10|    |
 1011:     |  11|  10|  11| 111|   ?| 112|   ?|  10|    |
 1012:     |    |  -1|  11| 111|   ?| 112|   ?|  10|    |
 1014:     |    |    |  11| 111|  -1| 112|   ?|  10|    |
 1015:     |    |    |    | 111|  -1| 112|   ?|  10|    |
 1016: return 111
  111: goto 1000
 1000:     |    |    |    |  -1|  -1| 112|   ?|  10|    |
 1001:     |    |    |  -1|  -1|  -1| 112|   ?|  10|    |
 1002:     |    |    |    |   1|  -1| 112|   ?|  10|    |
 1003:     |    |    |    |    |  -1| 112|   ?|  10|    |
 1004:     |    |    |    |  -1|  -1| 112|   ?|  10|    |
 1005:     |    |    |  10|  -1|  -1| 112|   ?|  10|    |
 1006:     |    |    |    | -11|  -1| 112|   ?|  10|    |
 1007:     |    |    |    |    |  -1| 112| -11|  10|    |
 1008:     |    |    |    |    |    | 112| -11|  10|    |
 1009: return 112
  112:     |    |    |    |    |    |    |    |-110|    |
  113:     |    |    |    |    |    |    | 100|-110|    |
  114:     |    |    |    |    |    |    |    | -10|    |
  115: s = -10; the stack is empty
  116: end of program; the stack is empty
```

Table 13.2. *Execution of pseudocode for* $s := a * F(F(2 * x + 1)) + b$

Here, the stack is a pair (*height*, *array*). Since a plate contains either a value or an address[1] we represent it as a pair (*integer*, *boolean*) so as to distinguish values and addresses when we must perform an arithmetic operation.

```
const empty = '' ; vert = '|' ;
      space = ' ' ; max = 100 ;
type
_string80 = string[80] ;
_value_or_address = record
  | value : integer ;
  | address : boolean
end ;
_stack = record
  | top : integer ;
  | plate : array[1 .. max] of _value_or_address
end ;
_code_line = record
  | num : integer ;
  | statmt : _string80
end ;
_code = array[1 .. max] of _code_line ;
_variable = array['a' .. 'z'] of integer ;
var code : _code ;
    variable : _variable ;
    stack : _stack ;
```

The main part of the program

The program begins by transferring and displaying the contents of the file containing the pseudocode into the array *code*. It then executes this pseudocode.

```
begin
  | load_program(code) ;
  | interpret_pseudocode(code)
end .
```

The procedure load_program

We could arrange to have the pseudocode interpreted as we type it in, but this would be awkward. Not only would it be painful, but in case of error all would be lost and it would be necessary to begin again.

This is why we will — for the first and last time in this book — make use of a *file* into which the program to be interpreted will be typed. The procedure

[1] The address part is reserved for single variables; and destined for arithmetic operations. When we stack the address of a statement, we stack the *number* of this statement; that is, an integer considered as value.

opens this file, reads it, transfers the statements into the array *code*, and closes the file again according to the following scheme.

```
open(the_file, file_name) ;
while not eof(the_file) do begin
|  readln(the_file, line) ;
|  ...
end ;
close(the_file) ; {never forget !}
```

The treatment of a line consists of separating the address of the statement (variable *num_statmt*) from the text of the statement (variable *statement*). To avoid future recognition problems, we clean up each line before treating it by suppressing the spaces at the beginning and end of the line.

```
procedure load_program(var code :_ code) ;
var file_name, line, num_statmt, statement : _string80 ;
   the_file : text ; ℓ : integer ;
begin
|  ℓ := 0 ; {no line has been read}
|  write('name of file to open :') ; readln(file_name) ;
|  open(the_file, file_name) ;
|  while not eof(the_file) do begin
|  |  readln(the_file, line) ;
|  |  suppress_spaces(line) ;
|  |  separate(line, num_statmt, statement) ;
|  |  ℓ := ℓ + 1 ; {one makes a place for the upcoming line}
|  |  with code[ℓ] do begin
|  |  |  num := convert_constant(num_statmt) ;
|  |  |  statmt := statement ;
|  |  |  writeln(num, ' : ', statmt)   {one displays the line}
|  |  end
|  end ;
|  close(the_file) ; {never forget !}
|  writeln('– end of program –') ;
|  writeln('execution of the program')   {serious things begin}
end ;
```

We also take this opportunity to have the procedure display the pseudocode that will be interpreted. In this way, we will see the code and the evolution of the stack on our screen.

The interpret_pseudocode procedure

The interpreter begins by isolating the prefix (that is, the first word) of the statement to be executed, which allows it to know — *via* a long discussion —

which action to undertake, *i.e.* which procedure to call. We are obliged to fall back on a succession of "if then else" statements, which obscures somewhat the legibility because the statement "case" does not accept a string of characters as a control variable.

```
procedure interpret_pseudocode(code : _code) ;
var num_line, control, address : integer ;
    prefix, suffix : _string80 ;
    finish, branching, see_stack : boolean ;
begin
 | empty_stack(stack) ; num_line := 1 ;
 | control := code[1].num ; finish := false ;
 | repeat
 | | branching := false ; see_stack := true ;
 | | write(control : 4, ' : ') ;
 | | with code[num_line] do begin
 | | | separate(statmt, prefix, suffix) ;
 | | | if prefix = 'end'
 | | | then execute_end(finish, branching) else
 | | | if prefix = 'push'
 | | | then execute_push(stack, suffix) else
 | | | if prefix = 'pop'
 | | | then execute_pop(stack, suffix, variable) else
 | | | if prefix = 'write'
 | | | then execute_write(suffix, see_stack) else
 | | | if prefix = 'read'
 | | | then execute_read(suffix[1], variable, see_stack) else
 | | | if prefix = 'goto'
 | | | then execute_goto(suffix, address, branching, see_stack) else
 | | | if prefix = 'return'
 | | | then execute_return(stack, address, branching, see_stack) else
 | | | if prefix = 'if'
 | | | then execute_if(suffix, address, branching) else
 | | | if prefix = 'ifx'
 | | | then execute_ifx(suffix, address, branching) else
 | | | if prefix = 'add'
 | | | then execute_add(stack, suffix) else
 | | | if prefix = 'sub'
 | | | then execute_sub(stack, suffix) else
 | | | if prefix = 'mult'
 | | | then execute_mult(stack, suffix) else
 | | | if prefix[1] = '['
 | | | then execute_assign_stack(stack, prefix, suffix) else
 | | | if prefix[1] in ['a' .. 'z']
 | | | then execute_assign_variable(variable, prefix, suffix, see_stack) ;
```

```
      if branching
      then seek(num_line, control, address)
      else next_statement(num_line, control) ;
      if see_stack then display(stack)
    end {with code[num_line]}
  until finish
end ;
```

The procedures *execute_goto*, *execute_return* and *execute_if* which give rise to branching communicate the address of the next statement as a string of characters which must be converted into an integer.

To keep track of, and above all *to see*, what happens — because what else justifies this program? — we ask that the number of the statement which has just been executed be displayed, as well as the state of the stack when necessary.

Note that a statement can be decoded and translated several times in the same program. The chosen solution is not the most rapid; this has no importance because we are not investigating performance.

One last note: in order not to lengthen the program, *no protection* is provided against erroneous statements. You are strongly urged to perfect your code.

The procedures for manipulating strings of characters

The first procedure suppresses the undesirable spaces that one finds at the beginning or end of a string of characters.

```
procedure suppress_spaces(var line : _string80) ;
begin
  if line ≠ empty
  then while line[1] = space do delete(line, 1, 1) ;
  if line ≠ empty
  then while line[length(line)] = space do delete(line, length(line), 1) ;
end ;
```

The second cuts the string of characters into the substrings *prefix* and *suffix*. The substring *prefix* begins with the first character, because there is no space at the beginning of *string*. In contrast, it is necessary to remember to suppress the spaces which can appear at the beginning of the substring *suffix*.

```
procedure separate(line : _string80 ; var prefix, suffix : _string80) ;
var i : integer ;
begin
  i := 1 ;
  case line[1] of
    '0'..'9' : while line[i + 1] in ['0'..'9'] do i := i + 1 ;
    'a'..'z' : while line[i + 1] in ['a'..'z'] do i := i + 1 ;
```

```
    |'[': repeat i := i + 1 until line[i] = ']' ;
    end ;
    prefix := copy(line, 1, i) ;
    delete(line, 1, i) ;
    suffix := line ;
    suppress_spaces(suffix) ;
end ;
```

If *line*[1] begins with a character which is neither a digit, nor a lower case letter, nor an open bracket, the character stands all alone in the prefix. This is produced when *string* begins with a "–" sign.

The next_statement and seek procedures

The first procedure is activated when there is no branching.

```
procedure next_statement(var num_line, control : integer) ;
begin
    num_line := num_line + 1 ;
    control := code[num_line].num ;
end ;
```

The second is activated by procedures which result in branching.

```
procedure seek(var num_line, control : integer ;  address : integer) ;
begin
    num_line := 0 ;
    repeat
        num_line := num_line + 1
    until code[num_line].num = address ;
    control := address ;
end ;
```

Conversion functions

The first function receives for example the string "1999" and returns the corresponding integer. This is an ultra-classical exercise.

```
function convert_constant(the_string : _string80) : integer ;
var i, temp : integer ;
begin
    temp := 0 ;
    for i := 1 to length(the_string) do
        temp := 10 * temp + ord(the_string[i]) - ord('0') ;
    convert_constant := temp
end ;
```

The following function converts a string of characters between brackets into the corresponding offset.

```
function convert_offset(the_string : _string80) : integer ;
begin
| if the_string = '[0]'
| then convert_offset := 0
| else convert_offset := −convert_constant(
|                    copy(the_string, 3, length(the_string) − 3)) ;
end ;
```

The function *convert_ref_stack* converts a reference to a stack such as "[−2]" to the corresponding integer. For this, it calls the function *plate_value* which will be written later (when we deal with primitives for stack manipulation).

```
function convert_stack_ref(stack_ref : _string80) : integer ;
begin
| convert_stack_ref := plate_value(convert_offset(stack_ref))
end ;
```

To convert a term, it suffices to look at its first character to determine if it is a constant, a stack reference or a variable.

```
function convert_term(the_string : _string80) : integer ;
begin
| case the_string[1] of
| | '0' . . '9' : convert_term := convert_constant(the_string) ;
| | '[': convert_term := convert_stack_ref(the_string) ;
| | 'a' . . 'z' : convert_term := variable[the_string[1]]
| end {case}
end ;
```

To determine whether a term is signed, we look at its first character. If we encounter the sign "−" (the only case allowed beyond a digit), we separate this sign from the nonsigned term that follows.

```
function convert_signed_term(the_string : _string80) : integer ;
var sign, non_signed_term : _string80 ;
   value : integer ;
begin
| if the_string[1] = '−'
| then separate(the_string, sign, non_signed_term)
| else begin
| | sign := '+' ;
| | non_signed_term := the_string
| end ;
```

```
    case sign[1] of
      '+' : convert_signed_term := convert_term(non_signed_term) ;
      '−' : convert_signed_term := −convert_term(non_signed_term)
    end {case}
end ;
```

Primitives for stack manipulation

The first two functions are clear.

```
function is_full(stack : _stack) : boolean ;
begin
    if stack.top = max then is_full := true else is_full := false
end ;

function is_empty(stack : _stack) : boolean ;
begin
    if stack.top = 0 then is_empty := true else is_empty := false
end ;
```

We also need a procedure which create an empty stack.

```
procedure empty_stack(var stack : _stack) ;
begin
    stack.top := 0
end ;
```

When we stack an integer, we do not forget to specify whether it is a value or an address.

```
procedure push(var stack : _stack ;  x : integer ;  c : char) ;
begin
    if is_full(stack)
    then writeln('thestackisfull')
    else with stack do begin
        top := top + 1 ;
        plate[top].value := x ;
        if c = '@'
        then plate[top].address := true    {one pushes an address}
        else plate[top].address := false   {one pushes a value}
    end
end ;
```

When we pop, we must take the same precautions. If the popped plate contains a value, we collect this value directly. If not, we use the address (which is the ASCII code of the variable) to seek the right value among the variables $a, \ldots z$.

```
function pop(var stack : _stack) : integer ;
begin
 | if is_empty(stack)
 | then writeln('thestackisempty')
 | else with stack do begin
 | | if plate[top].address
 | | then pop := variable[chr(plate[top].value)]
 | | else pop := plate[top].value ;
 | | top := top − 1
 | end
end ;
```

The procedure *transfer_value* modifies a plate which need not be at the top of the stack. We use this procedure to modify a parameter passed by value which is somewhere in the stack. Note that we will always transfer a value, never an address.

```
procedure transfer_value(var stack : _stack ; offset, new_val : integer) ;
var target : integer ;
begin
 | with stack do begin
 | | target := top + offset ;
 | | plate[target].value := new_val ;
 | | plate[target].address := false
 | end
end ;
```

The function *plate_value* begins by testing whether a plate contains a value or an address, and reacts accordingly.

```
function plate_value(offset : integer) : integer ;
begin
 | with stack do
 | if plate[top].address
 | then plate_value := variable[chr(plate[top + offset].value)]
 | else plate_value := plate[top + offset].value ;
end ;
```

A binary operation (addition, subtraction, multiplication) exclusively concerns the top of the stack and the plate just below. It is necessay to correctly treat the plates which contain an address (a call to the function *plate_value*). Note that result is always a value.

```
procedure binary_operation(var stack : _stack ; op : char) ;
var operand_1, operand_2 : integer ;
begin
 | with stack do begin
```

```
      operand_1 := plate_value(-1) ;
      operand_2 := plate_value(0) ;
      case op of
        '+' : plate[top - 1].value := operand_1 + operand_2 ;
        '-' : plate[top - 1].value := operand_1 - operand_2 ;
        '*' : plate[top - 1].value := operand_1 * operand_2 ;
      end ; {case}
      plate[top - 1].address := false ;
      top := top - 1
    end
end ;
```

The last two procedures concern the display of the contents of the stack. Recall that we have chosen the integer −32,000 to represent an *indefinite value* (the question mark in pseduocode) that we insert into the stack to free up a place in which we eventually want to put the value of a function. The probability is very small that −32,000 is a *true* value.

```
procedure display_plate(offset : integer) ;
begin
  with stack do begin
    if plate[top + offset].address
    then write(chr(plate[top + offset].value) : 5, vert)
    else if plate[top + offset].value = -32000
         then write('?' : 4, vert)
         else write(plate[top + offset].value : 4, vert)
  end
end ;
```

Given the size of a screen, we never display more than ten plates.

```
procedure display(stack : _stack) ;
var i, number_plates : integer ;
begin
  if is_empty(stack)
  then writeln('the stack is empty')
  else begin
    number_plates := stack.top ;
    if number_plates < 10 then begin
      for i := 0 to 9 - number_plates do write(vert : 5) ;
      for i := 0 to number_plates - 1 do display_plate(-i) ;
      writeln
    end
    else begin {number_plates ≥ 10}
      for i := 0 to 9 do display_plate(-i) ;
      writeln('...' : 5)
```

```
||end
|end
end ;
```

Procedures which execute a statement

The first such procedure hands control back to the system by having the program properly leave the "repeat until" loop in which the interpreter works. We take the opportunity to signal whether or not the stack is empty.

```
procedure execute_end(var finish, branching : boolean) ;
begin
|finish := true ;
|branching := true ;
|write('end of program ; ') ;
|if not is_empty(stack)
|then writeln('caution : the stack is not empty!')
end ;
```

We we encounter a "push" followed by a question mark, we must stack an *indefinite value*. We choose again the number −32,000 to play this role.

In this context, an address is the ASCII code of the variable referenced by one of the letters $a, \ldots, z$. The choice of the character "v" is arbitrary: any character other than "@" would do, because we simply need to distinguish an address from a value.

```
procedure execute_push(var the_stack : _stack ; suffix : _string80) ;
var offset : integer ;
begin
|if suffix = '?'
|then push(the_stack, −32000, 'v')   {v for value}
|else if suffix[1] = '@'
|then push(the_stack, ord(suffix[2]), '@')
|else push(the_stack, convert_signed_term(suffix), 'v')
end ;
```

The procedure *execute_pop* is delicate. To translate the statement "$pop[-i]$", we begin by popping to collect the value at the top of the stack. We must then remember that the plate into which we want to transfer this value is at level $-i+1$ (and no longer at $-i$).

The variable *garbage* provides an elegant way to avoid introducing another primitive for stack manipulation.

```
procedure execute_pop(var stack : _stack ; suffix : _string80 ;
                                          var variable : _variable) ;
var i, garbage, value : integer ;
begin
```

```
if suffix = empty
then garbage := pop(stack)
else begin
  case suffix[1] of
    'a'..'z' : begin
      variable[suffix[1]] := pop(stack) ;
      write(suffix[1], ' = ', variable[suffix[1]] : 1, ' ; ')
    end ;
    '0'..'9' : for i := 1 to convert_constant(suffix) do
    garbage := pop(stack) ;
    '[': begin
      value := pop(stack) ; {attention, decreased offset}
      transfer_value(stack, convert_offset(suffix) + 1, value)
    end ;
  end {case}
end
end ;

procedure execute_write(suffix : _string80 ; var see_stack : boolean) ;
begin
  see_stack := false ;
  if length(suffix) = 1
  then writeln(variable[suffix[1]])
  else if suffix[1] in ['[', '−']
  then writeln(convert_term(suffix))
  else writeln(suffix)
end ;

procedure execute_read(variable_name : char ;
                       var variable : _variable ;
                           var see_stack : boolean) ;
begin
  see_stack := false ;
  write('value of ', variable_name, ' = ') ;
  readln(variable[variable_name]) ;
end ;

procedure execute_goto(suffix : _string80 ; var address : integer ;
                       var branching, see_stack : boolean) ;
begin
  branching := true ;
  address := convert_constant(suffix) ;
  see_stack := false ;
  writeln('goto', address : 1) ;
```

```
end ;

procedure execute_return(var stack : _stack ; var address : integer ;
                                   var branching, see_stack : boolean) ;
begin
 branching := true ;
 address := pop(stack) ;
 see_stack := false ;
 writeln('return', address : 1) ;
end ;

procedure execute_assign_stack(var stack : _stack ;
                                   prefix, suffix : _string80) ;
var
offset : integer ;
begin
 case prefix[2] of
  '0' .. '9' :
  offset := convert_constant(copy(prefix, 2, length(prefix) - 2)) ;
  '-' :
  offset := -convert_constant(copy(prefix, 3, length(prefix) - 3)) ;
 end ; {case}
 separate(suffix, prefix, suffix) ; {prefix contains the sign ' ='}
 transfer_value(stack, offset, convert_term(suffix)) ;
end ;

procedure execute_if(suffix : _string80 ; var address : integer ;
                                   var branching : boolean) ;
var
prefix_1, prefix_2, prefix_3, garbage : _string80 ;
operand_1, operand_2, num_pop, lost, i : integer ;
begin
 separate(suffix, prefix_1, suffix) ;
      {prefix_1 contains the first operand}
 separate(suffix, prefix_2, suffix) ;
      {prefix_2 contains the comparison}
 separate(suffix, prefix_3, suffix) ;
      {prefix_3 contains the secondoperand}
 separate(suffix, garbage, suffix) ;
      {suppresses the goto in suffix}
 operand_1 := convert_term(prefix_1) ;
 operand_2 := convert_term(prefix_3) ;
 num_pop := 0 ;
 if (prefix_1 = '[0]') or (prefix_1 = '[-1]')
```

```
    then num_pop := num_pop + 1 ;
    if (prefix_3 = '[0]') or (prefix_3 = '[−1]')
    then num_pop := num_pop + 1 ;
    case prefix_2[1] of
      ' <': if operand_1 < operand_2 then branching := true ;
      ' ≤': if operand_1 ≤ operand_2 then branching := true ;
      ' =': if operand_1 = operand_2 then branching := true ;
      ' ≠': if operand_1 ≠ operand_2 then branching := true ;
      ' >': if operand_1 > operand_2 then branching := true ;
      ' ≥': if operand_1 ≥ operand_2 then branching := true ;
      else
      branching := false
    end ; {case}
    for i := 1 to num_pop do lost := pop(stack) ;
    if branching then address := convert_constant(suffix)
  end ;


  procedure execute_ifx(suffix : _string80 ; var address : integer ;
                                             var branching : boolean) ;
  var
  prefix_1, prefix_2, prefix_3, garbage : _string80 ;
  operand_1, operand_2 : integer ;
  begin
    separate(suffix, prefix_1, suffix) ;
        {prefix_1 contains the first operand}
    separate(suffix, prefix_2, suffix) ;
        {prefix_2contains the comparison}
    separate(suffix, prefix_3, suffix) ;
        {prefix_3contains the second operand}
    separate(suffix, garbage, suffix) ;
        {suppresses the goto in suffix}
    operand_1 := convert_term(prefix_1) ;
    operand_2 := convert_term(prefix_3) ;
    case prefix_2[1] of
      '<' : if operand_1 < operand_2 then branching := true ;
      '≤' : if operand_1 ≤ operand_2 then branching := true ;
      '=' : if operand_1 = operand_2 then branching := true ;
      '≠' : if operand_1 ≠ operand_2 then branching := true ;
      '>' : if operand_1 > operand_2 then branching := true ;
      '≥' : if operand_1 ≥ operand_2 then branching := true ;
      else branching := false
    end ; {case}
    if branching then address := convert_constant(suffix)
```

```
end ;

procedure execute_add(var stack : _stack ; suffix : _string80) ;
begin
| binary_operation(stack, '+')
end ;

procedure execute_sub(var stack : _stack ; suffix : _string80) ;
begin
| binary_operation(stack, '-')
end ;

procedure execute_mult(var stack : _stack ; suffix : _string80) ;
begin
| binary_operation(stack, '*')
end ;

procedure execute_assign_variable(var variable : _variable ;
                prefix, suffix : _string80 ; var see_stack : boolean) ;
var prefix_1 : _string80 ;
begin
| separate(suffix, prefix_1, suffix) ; {prefix_1 contains the sign '='}
| variable[prefix[1]] := convert_term(suffix) ;
| see_stack := false ;
| writeln
end ;
```

Advice for fine tuning

Type in the procedures by group and adjust them *immediately* by submitting them to a complete battery of tests. Do not wait until the end of the program; if you do, you will drown ...

13.3. How to Analyze an Arithmetic Expression

Suppose that we wish to write a program that calculates the numerical value of the integral $I = \int_a^b f(x)\,\mathrm{d}x$ by the trapezoid rule. What we would like is a program that asks for the values of a and b, and then the function to be integrated. The program should allow us to type a *string of characters* representing the function, for example:

$$f(x) = \sin(3 * x + \exp(1 + \cos(x * x))) + 3 * x * x - 7 * x + 1$$

When we type a program, the compiler knows how to convert a string of characters into statements executable by the microprocessor. The trouble is that the compiler is not there when our program kicks in.

We are going to try to understand what the compiler does by describing the steps which separate the beginning (the string of characters) from the end (the executable code). Taking this path, we shall learn to write an *interpreter* capable of calculating the value of an arithmetic expression "on the fly" as well as a *compiler* for the arithmetic expressions that we could insert into our numerical integration program.

13.3.1. Arithmetic expressions

In this section, by an *arithmetic expression* we mean *any* string of characters which only contains the following characters:

- the letters "a" to "z";
- the *binary* symbols "+" and "$*$";
- left and right parentheses.

We can divide arithmetic expressions into two classes: *good* and *bad*. We all know, for example, that "$a{*}x{*}{*}x{+}b{*}x{+}c$" is a good expression and "$a{*}x{+})\,b$" is a bad expression. But how do we distinguish good and bad expressions?

To better grasp the problem, we forbid any global reading and suppose that we have before our eyes an arithmetic expression which occupies an entire page. We would know that this expression is good if we were capable of reconstructing it using only the following rules:

(i) A name of a variable is a good expression.

(ii) If α and β are good expressions then the string of characters obtained by concatenating (in this order) α, "+", β is a again a good expression. Similarly, the string of characters obtained by concatenating α, "$*$", β is good.

(iii) If α is a good expression, the string of characters obtained by concatenating (in this order) a left parenthesis, α and a right parenthesis is a good expression.

This definition, while correct, is not at all statisfactory.

• It is not adapted to reading from left to right.

• Consider the strings α = "$a + b$" and β = "$x + y$": the second part of rule (ii) tells us that the string "$a + b * x + y$" is a good expression. But this is a little troubling because the value of $a + b * x + y$ is not the product of the values of $a + b$ and $x + y$.

Thus, we need a finer definition which is compatible with the direction in which we read and which reflects the usual priorities (multiplication and parenthesizing).

Definition 13.3.1. • *We call* good expression *(or* expression *for short) an arithmetic expression which is a* term *or a* sum of terms*; that is, a string of characters obtained by concatenating a term, the sign "+", another term, etc.*

• *We call* term *an arithmetic expression which is a* factor *or a* product of factors*; that is, any string of characters obtained by concatenating a factor, the sign "$*$", another factor, etc.*

• *Finally, we call* factor *a name of variable (that is, one of the letters "$a, b, \ldots, z$") or a* parenthesized expression *(that is, the string of characters obtained by concatenating a left parenthesis, a good expression, and a right parenthesis).*

Notice the highly recursive character of this definition.

If we let "id" (for identifier) denote the set of lower case letters (the names of variables) and $\mathcal{E}$, $\mathcal{T}$, and $\mathcal{F}$ the sets strings of characters formed by the *expressions*, the *terms*, and the *factors*, respectively, then we can summarize the preceding definitions by the following equations:

$$\mathcal{E} = \mathcal{T} \cup \mathcal{T} + \mathcal{T} \cup \mathcal{T} + \mathcal{T} + \mathcal{T} \cup \cdots = \mathcal{T} + \cdots + \mathcal{T}, \tag{13.1}$$

$$\mathcal{T} = \mathcal{F} \cup \mathcal{F} * \mathcal{F} \cup \mathcal{F} * \mathcal{F} * \mathcal{F} \cup \cdots = \mathcal{F} * \cdots * \mathcal{F}, \tag{13.2}$$

$$\mathcal{F} = \mathrm{id} \cup (\mathcal{E}). \tag{13.3}$$

Here the notation $\mathcal{T}+\mathcal{T}$ denotes the set of strings obtained by concatenation of an element of $\mathcal{T}$, the character "+" and another element of $\mathcal{T}$.

Example

Is the string of characters

$$a + b + (x + a * y + u)$$

a *good expression* in the sense of our new definition?

• According to (13.1), the given string would be a good expression if we knew how to prove that the subchains "a", "b" and "$(x+a*y+u)$" are terms.

• By combining (13.2) and (13.3), we immediately see that the strings "a" and "b" are terms because they are factors.

• Definition (13.2) shows that "$(x+a*y+u)$" would be a term if we could show that it is a factor.

• Speeding up a bit: the string "$(x+a*y+u)$" is a factor because "$x+a*y+u$" is a good expression. In effect, "x", "$a * y$" and "u" are three terms.

We can illustrate and sumarize our approach, which is called *syntactic analysis*, by drawing a tree. This tree is called the *syntactic tree* (see Fig. 13.6).

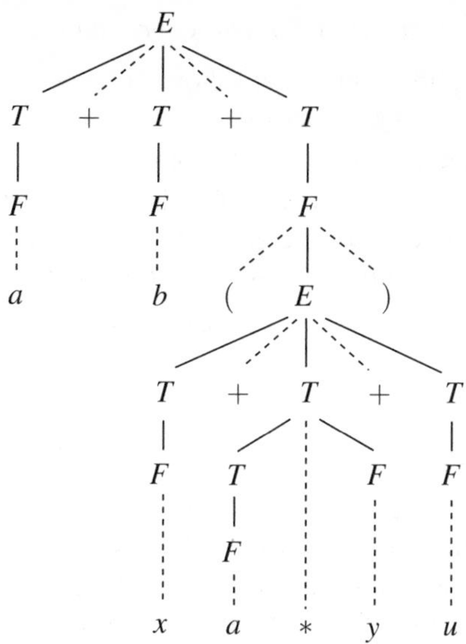

Fig. 13.7. *The syntactic tree associated to the expression* $a + b + (x + a * y + u)$

Remarks

1) One can show that the two definitions define the same good arithmetic expressions.

2) When we look at the syntactic tree, we find that it describes an order in which the calculations should be done: a multiplication occurs before an addition and a parenthesized expression is given priority. These priorities are subtle consequences of (13.1), (13.2) and (13.3). In effect,

- ▷ when we say that "an expression is a sum of terms," we *automatically* give multiplication priority over addition since we must know the value of the terms before adding them;
- ▷ when we say that "a parenthesized expression is a factor," we give parenthesized expressions priority because we must first know their values if we want to multiply terms.

3) This remark will be very useful when we will generalize Definitions (13.1), (13.2) and (13.3) to allow us to analyze richer arithmetic expressions which contain subtractions, divisions, function calls, etc.

4) One can show that a given expression has a single syntactic tree. This seemingly harmless result is fundamental because it implies that an expression has a single value.

13.3.2. How to recognize an arithmetic expression

If you want to easily understand what follows, do not go too fast; train yourself first by doing the following exercises:

1) Sketch the syntactic trees of several good expressions.

2) When the first exercise becomes familiar, ask a friend to dictate a good expression to you character by character. You must sketch corresponding part of the syntactic tree *as soon as you receive the character.*

Do this until the construction of syntactic tree becomes natural and spontaneous.

The fundamental idea

Contemplating syntactic trees will sooner or later bring to mind trees of recursive calls of the three procedures E, T, F. We have our program!

A good way to grasp what follows is to imagine that E, T, F are three "pacmen" who eat[2], respectively, the biggest expression, the biggest term, or the factor that starts the given string of characters.

For programming clarity, we systematically work using *context effects*[3]: all variables are global and the procedures do not have parameters, so there is nothing to stop us from modifying certain global variables of the program.

The main body of the program

The variable *expression* contains the string of characters to analyze, *token* is the current character in the string so that *token* = *expression*[*place_token*] is true at each instant.

```
type str255 = string[255] ;
var expression : str255 ;
    token : char
    place_token : integer ;
```

After the indispensable initializations, we ask simply that the procedure E devour the string *expression*.

We shall see later that in order to avoid trouble, it is necessary that *token* is always followed by a character (if one forgets, the program will crash when E finishes eating the string *expression*). We follow tradition and use the classical trick of adding the *indigestible* character "$" at the end of the string to be analysed. This character is not only a mouthguard: it allows one to detect bad arithmetic expressions; that is strings that E does not eat entirely.

---

[2] In order not to repeat ourselves, we shall, from time to time, replace the verb "eat" by one of the verbs "devour", "analyse", "consume", or "recognize".

[3] Is it necessary to recall that this is not a good way to program?

```
begin
  write('expression = ') ; readln(expression) ;
  expression := concat(expression, '$') ;
  place_token := 1 ; token := expression[place_token] ;
  E ; {try to eat the whole expression!}
  if token = '$'
  then writeln('good expression')   {because E has eaten it all}
  else writeln('bad expression')
end .
```

Notice the context effect: without its presence, we should have had to write $E(place\_token, token)$, these two parameters being passed in "var" since the procedure E modifies the variables *token* and *place_token* on the sly.

The procedures E, T, F

Since our three pacmen mutually call each other (mutual recursion), we must separate the declaration of the procedures from their respective bodies:

```
procedure E ; forward ;
procedure T ; forward ;
procedure F ; forward ;
```

The procedure E

The code is a faithful translation of Definition (3.1).

```
procedure E ;
begin
  T ; {to eat the first term}
  while token = '+' do begin
    next_token ; {get rid of the '+' sign}
    T   {to eat the term after the '+' sign}
  end
end ;
```

The procedure E begins by eating the first term (or asks instead that T takes its place). If something remains, it eats as much as it can of the substrings of the form "$+\mathcal{T}$" by first calling *next_token* to get rid of the sign "+", then the procedure T.

Note (this is important) that one always finds oneself before the entrance of the "while" loop when one leaves the procedure T.

The procedure next_token

This procedure passes to the next token by modifying the global variable *token*. Note that the context effects modify *place_token* and *token*.

```
procedure next_token :
begin
| place_token := place_token + 1 ;
| token := expression[place_token]
end ;
```

The procedure T

This procedure eats the biggest term with which *token* begins. Thus, its code is analogous to that of the procedure E and is a faithful translation of (13.2).

```
procedure F ;
begin
| F ; {eat the first factor}
| while token = '*' do begin
| | next_token ; {get rid of the '*' sign}
| | F {to eat the term after the '*' sign}
| end
end ;
```

Again it is important to note that one always finds oneself at the entrance to the "while" loop when leaving the procedure F.

The procedure F

This procedure eats the factor which begins with the character *token*. Here again, the code is a transparent translation of (13.3): if *token* is the name of a variable (that is, a lower case letter), F eats it by calling *next_token*; if *token* is a left parenthesis, F eats it, then demands that E handle the expression situated between the parentheses. If all goes well, E stops before the corresponding right parenthesis which is devoured by *next_token*. Otherwise, there is an error.

```
procedure F ;
begin
| case token of
| | 'a'..'z' : newt_token ; {get rid the name of the variable}
| | '(': begin
| | | newt_token ; {get rid of the left parenthesis}
| | | E ; {eats the biggest expression}
| | | if token = ')'
| | | then newt_token {get rid of the right parenthesis}
| | | else error {because the right parenthesis is absent}
| | end
| | else error {because token is not the beginning of a factor}
| end ;
```

It is important to remember that one always finds oneself before the test $token = ')'$ when one leaves the procedure E.

The procedure error

The first error is *fatal*; there is no attempt to repair the error or to produce any diagnosis. Its action consists of giving (by context effect) an indigestible value to *token* different than "$" (we have chosen "@"). We will see a little later that *token* = '$' or *token* = '@' stops the program properly without a crash.

```
procedure error ;
begin
  writeln('error on the token ', token) ;
  token := '@'
end ;
```

How the program works

We analyze the string "$a + b * (x + y)$". To better follow the action of our program, we are going to sketch the the tree of recursive calls (Fig. 13.7) by placing the value of the variable *token* in the index at the moment the procedure is called. Calls of *next_token* are represented by dotted lines.

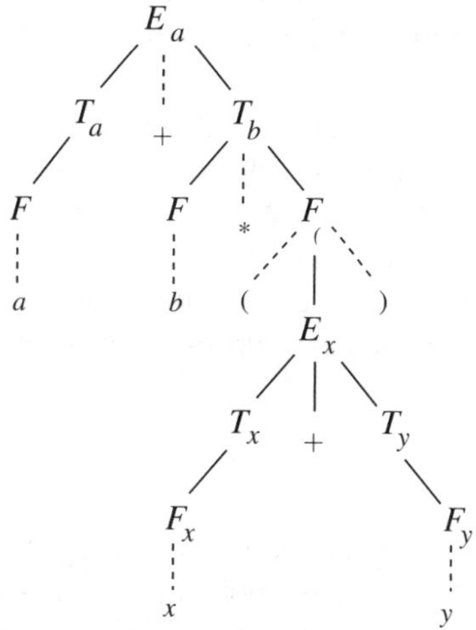

Fig. 13.8. *Recursive calls for expression* $a + b * (x + y)$

Before launching into the multiple recursive calls, it is worth keeping several essential features in mind.

- *Leaving* a procedure means changing a level in the tree of recursive calls.
- We leave the procedure E in two ways: either through the root of the the tree of recursive calls, which means that we are back in the main part of the program or, in all other cases, we find ourselves inside the code of F before the test *token* = ')'.

- When we leave the procedure T, we always wind up in the interior of the procedure E before the "while" loop *token* = '+'.
- When we leave the procedure F, we always wind up in the interior of the procedure T, before the "while" loop *token* = '∗'.

The program begins by calling E_a which calls T_a which calls F_a. The latter procedure then asks *next_token* to consume the character 'a'.

Then the program leaves F and finds itself in the procedure calling T, *before* the "while" loop *token* = '∗'. Since *token* is now the character '+', the program then returns into the procedure E and finds itself before the "while" loop *token* = '+'. It enters the loop, asks *next_token* to get rid of the '+', then calls T_b, which calls F_b which eats the character 'b' by *next_token*, etc.

We now examine how the program stops. When the program eats the last character of the string being analyzed, *token* takes the value '$'; since this shows up necessarily in the procedure F, the program leaves F and finds itself in T before the "while" loop *token* = '+'; since it is not able to penetrate into this loop because of the value of *token*, it leaves the procedure T and finds itself before the "while" loop *token* = '∗', which has the effect of making it leave E through the root of the tree of calls.

The analysis stops and the program announces that the expression is a good expression.

What provokes the climb in the tree of recursive calls and the exit by the root is the *indigestible value* of the character '$', because an expression can only contain lower case letters, operations, and parentheses.

The procedure *error* has the same effect since it gives — always by context effect! — the indigestible value '@' to *token*. Thus we find ourselves at the end of the main program which announces an error because *token* is different from '$'.

For beginners

If you want to master this program, *learn it by heart*; to do this, copy it several times; try to reconstruct it from memory until it appears natural to you. Do not read what follows until you are at ease with the procedures E, T, F. They must become *evident* to you!

To follow easily the dialogue between the three procedures, introduce the global variable *depth* of integer type and initialize it to −1 in the main body of the program. Add the following procedure.

```
procedure show_depth(word : str255) ;
begin
| writeln(' ' : 3 * depth, word, ',  token = ', token : 1) ;
end ;
```

Then modify the procedure E as follows:

```
procedure E ;
begin
  depth := depth + 1 ;
  show_depth('enter E') ;
  T ;
  while token = '+' do begin
    next_token ;
    T ;
    show_depth('leave E') ;
    depth := depth − 1
  end
end ;
```

Modify the procedures T and F in a similar way.

Think as well about introducing a message in the procedure *next_token*, to indicate (*via* the procedure *print*) which token is eaten.

The program will then display its activity on the screen, the indentations translating the level of depth of the recursive calls.

```
expression = a+(b*x+c)
enter E, token = a
  enter T, token = a
    enter F, token = a
    F, eat a
    leave F, token = +
  leave T, token = +
main loop of E, token = +
eat +
  enter T, token = (
    enter F, token = (
    F, eat (
      enter E, token = b
        enter T, token = b
          enter F, token = b
          F, eat b
          leave F, token = *
        main loop of T
        eat *
          enter F, token = x
          F, eat x
          leave F, token = +
        leave T, token = +
      main loop of E, token = +
      eat +
        enter T, token = c
          enter F, token = c
          F, eat c
          leave F, token = )
        leave T, token = )
      leave E, token = )
    F, eat )
    leave F, token = $
  leave T, token = $
leave E, token = $
good expression
```

Fig. 13.9. *Analysis by procedures E,T,F of the arithmetic expression* $a + (b * x + c)$

First run the program with several simple arithmetic expressions, then complicate things. Follow the recursive rules by sketching the syntactic tree as one goes along.

Now do the contrary: sketch the branching of the recursive calls (with *token* as index) *before* running the program. Repeat this operation until it is completely mechanical (the delicate points are the returns in the the principal loops of E and T and the right parentheses in F). Try not to skip steps; this will only slow you up.

Also send some erroneous expressions to the program and see how it detects the errors. Try to enrich the code so that it emits a reasonable diagnosis in case of error.

Remarks

1) We now know how to associate two trees (Fig. 13.9) to the same arithmetic expression. These trees carry the same information, namely the order in which we must do the calculations to obtain the value of the expression. The tree on the left is binary; it is very compact, but difficult to realize by a program. Although it is more complex, we prefer the tree on the right because it can be realized very simply by recursive calls.

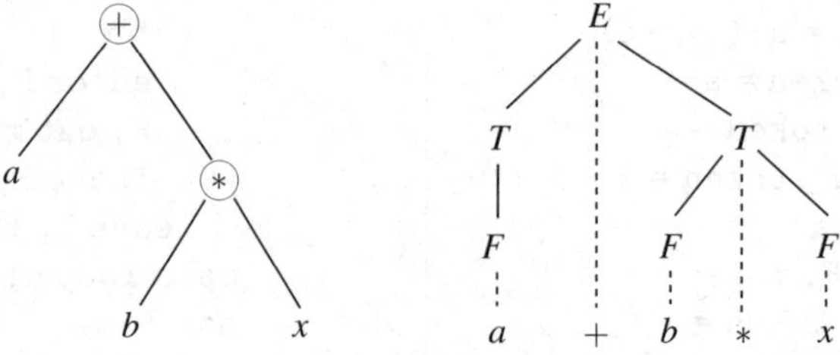

Fig. 13.10. *The two trees associated to the expression* $a + b * x$.

2) Note the division of labor: the procedure *next_token* is the only one that "eats" tokens. The procedures E, T, F are the white collar workers which content themselves with giving the pertinent orders.

3) When you have analyzed several expressions by running the program "by hand" and when you are at ease, you will realize that 90 % of the program consists of (13.1), (13.2), and (13.3). The translation of these equations into code is a mere formality.

13.4. How to Evaluate an Arithmetic Expression

We are going to perfect our program by asking that our procedures return the value of an arithmetic expression such as

$$-2 + (3 * 5 - 14) + 45 - 1999.$$

The expressions that we want to analyze contain no variables (that is, no letters) – these are replaced by integers with one or more digits. We remark as well that the signs "+" and "−" are at the same time binary and unary.

The presence of the integers and the new signs is going to require us to enrich the description of expressions, terms and factors:

$$\mathcal{E} = \mathcal{T} \pm \cdots \pm \mathcal{T}, \tag{13.4}$$

$$\mathcal{T} = \mathcal{F} * \cdots * \mathcal{F}, \tag{13.5}$$

$$\mathcal{F} = \text{int} \cup (\mathcal{E}) \cup +\mathcal{F} \cup -\mathcal{F} \tag{13.6}$$

This new description (which contains 90% of the new program) has some subtleties which are essential to understand well.

• We put the binary "+" and the "−" at the same level in (13.4), because subtraction does not have priority over addition (or the contrary).

• Multiplication appears at the second level (13.5) in the terms: in this way, it takes priority over addition and subtraction.

• The factors (13.6) contain the strings with the most priority. There one finds: (i) the set "int" of strings of characters which represent integers, (ii) parenthesized expressions, (iii) the unary signs "+" and "−", under the form $\pm\mathcal{F}$ (in effect, when we write $a * -b$, we must change the sign of b before carrying out the multiplication).

It is necessary to put $\pm\mathcal{F}$, and not $\pm\mathcal{E}$, in the factors because, when we write $a + -b + c$, we want to change only the sign of b, not that of $b + c$.

The body of the program

The only novelty is the appearance of the variable *value*: we ask that the procedure E return the value of the expression that it eats.

```
begin
  write('expression = ') ; readln(expression) ;
  expression := concat(expression, '$') ;
  place_token := 1 ;
  token := expression[1] ;
  E(value) ;
  if token = '$'
  then writeln('value = ', value : 1)
  else writeln('bad expression')
end .
```

The procedure next_token

We ask this procedure to perform two services:

- to return the value of the next token in *token*;
- to return the value of the token *that it has just left* in the variable *value* when *token* is a digit (the chronology is crucial).

```
procedure next_token(var value : integer) ;
var temp : integer ;
begin
  if not (token in ['0' .. '9'])
  then place_token := place_token + 1
  else begin
    value := 0 ;
    while token in ['0' .. '9'] do begin
      value := 10 * value + ord(token) − ord('0') ;
      place_token := place_token + 1 ;
    end
  end ;
  token := expression[place_token]
end ;
```

Recall the *token* contains only a single character at a time; when we have to deal with a number of digits, token points to the first digit; it is only when we *leave* a number (which requires us to traverse it) that *next_token* calculates its value. As usual, we work by context effect on the variables *token* and *place_token*.

The procedures E, T, F

These procedures are charged with returning the value of the expressions, the terms or the factors that they are analyzing.

```
procedure E(var value : integer) ; forward ;
procedure T(var value : integer) ; forward ;
procedure F(var value : integer) ; forward ;
```

The procedure E

The role of this procedure is to add or to cut off terms it encounters. Since we cannot go backwards, we must remember to store the binary operation which separates two terms in the local variable *op*. Notice the appearance (a necessary technique) of the global variable *garbage* in the statement *next_token*(*garbage*) when *E* swallows a sign "+" or a sign "−".

```
procedure E ;
var new_val : integer ; op, garbage : char ;
begin
  T(value) ;
  while token in ['+',' −'] do begin
```

```
      op := token ;
      next_token(garbage) ;
      T(new_val) ;
      if op = '+'
      then value := value + new_val
      else value := value - new_val
    end
  end ;
```

The procedure T

This is analogous to the procedure E, but simpler because there are only products of factors.

```
procedure T ;
var new_val : integer ;  garbage : char ;
begin
  F(value) ;
  while token = '*'  do begin
    next_token(garbage) ;
    F(new_val) ;
    value := value * new_val
  end
end ;
```

The procedure F

We must bear in mind the integers and unary signs.

```
procedure F ;
var garbage : char ;
begin
  case token of
    '0'..'9' : next_token(value) ;  {token is obtained by context effect}
    '(': begin
      next_token(garbage) ;
      E(value) ;
      if token = ')'
      then next_token(garbage)
      else error   {missing right parenthesis}
    end ;
    '+' : begin
      next_token(garbage) ;
      F(value) ;
    end ;
    '-' : begin
      next_token(garbage) ;
```

```
      F(value) ;
      value := −value   {unary '−' sign}
    end
    else error   {token is not the first character of a factor}
  end {case}
end ;
```

It suffices to contemplate the syntactic tree below to understand how the procedure F goes about distinguishing unary and binary signs.

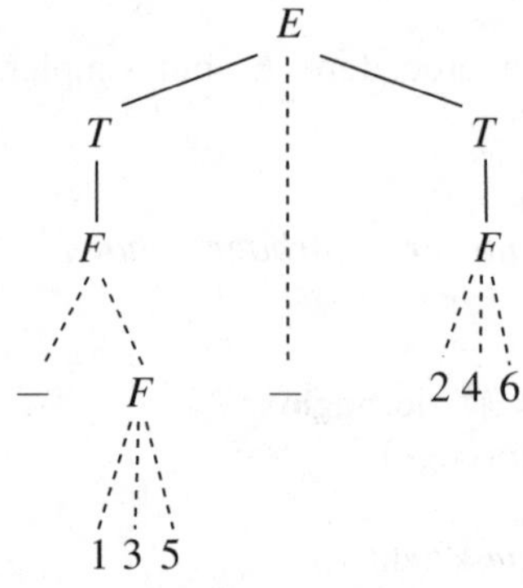

Exercise 8

1) Enrich the procedure F in order to evaluate arithmetic expressions such as $26 * (-0.45 + 665) + 3.5$ (the value returned is now a real number).

2) Refine the program so that it evaluates expressions such as:

$$-12 + (13 * 5 - 5/7)/(8 + 5/4) - 1999.$$

The value returned must be an irreducible fraction. The difficult part of this exercise is not programming, but comprehension. The procedure F only "sees" integers which it converts into fractions of the form $n/1$. It is the procedure T that first "sees" true fractions. The syntactic tree below shows you what happens with the expression $1/3 + 5/2$.

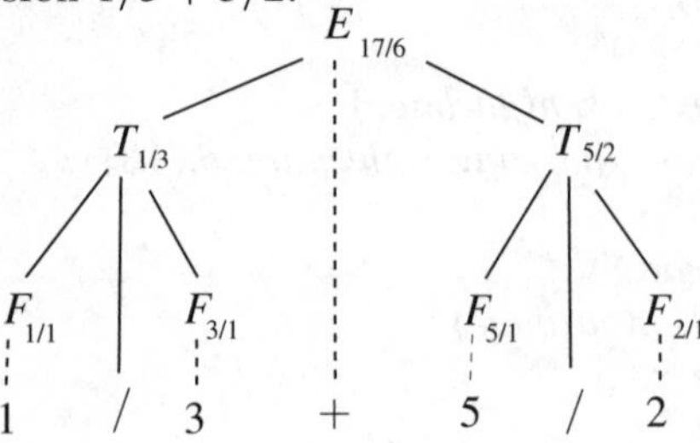

3) Enrich the procedures E, T, F and *next_token* to calculate in the ring $\mathbb{Z}[i]$ of Gaussian integers; that is, to be able to attribute a value to expressions such as

$$-(1 + 2 * i) * (145 + i * 17) * i - \big(-3 + (15 * i + 9)\big).$$

4) We now want to evaluate an arithmetic expression containing function calls and the variables a, b, x, y. To simplify and to safeguard the equation "character = token" we suppose that functions are coded by an uppercase letter ("L" for log, "C" for cos, "S" for sin, etc.) :

$$-a + b * L(1 + x * x) - C(x + y)/S(1 - A(x + 1)).$$

The program uses the variables $val\_a, val\_b, val\_x, val\_y$ which contain the values of a, b, x, y (but you can also put the values in an array $t['a'..'z']$). Use the following description:

$$\mathcal{E} = \mathcal{T} \pm \cdots \pm \mathcal{T} \tag{13.7}$$

$$\mathcal{T} = \mathcal{F} \cdots \mathcal{F} \tag{13.8}$$

$$\mathcal{F} = \text{id} \cup (\mathcal{E}) \cup \text{ID}(\mathcal{E}) \cup \pm\mathcal{F} \tag{13.9}$$

where $\text{id} = \{'a', 'b', 'x', 'y'\}$ and where ID denotes the set of names of functions (uppercase letters).

13.5. How to Compile an Arithmetic Expression

We now know how to *interpret* (that is, immediately evaluate without leaving the program) an arithmetic expression entered on the keyboard. If we were to decide to take advantage of our fresh knowledge to write a program which calculates $\int_a^b f(x)\,dx$, we would be chagrinned by its slowness. It is easy to understand why: if the program needs 1000 values of the function, it *analyzes* and *evaluates* the same arithmetic expression 1000 times in a row.

However, a single analysis ought to be sufficient. Is it possible to separate the analysis from the evaluation (this is the idea of *compiling*)? We would then replace 1000 analyses and 1000 evaluations with a single analysis and 1000 evaluations.

13.5.1. Polish notation

How shall we reframe our results on recognition? We choose to use *suffixed polish notation.*[4]

[4] This notation was discovered in 1920 by the Polish mathematician Jan Lukasiewicz (1878–1956) in the course of some work on logic where he was seeking to get rid of parentheses.

$$\underbrace{a + b * (x + y * z)}_{\textit{infixed notation}} \longrightarrow \boxed{\begin{array}{c}\textit{compiler}\\ E, T, F\end{array}} \longrightarrow \underbrace{a\,b\,x\,y\,z * + * +}_{\textit{suffixed Polish notation}}$$

As we have seen in Chapter 8, we can associate a *binary tree* (Fig. 13.11) to each arithmetic expression (a tree which must not be confused with the syntactic tree introduced in this chapter, which is bushier).

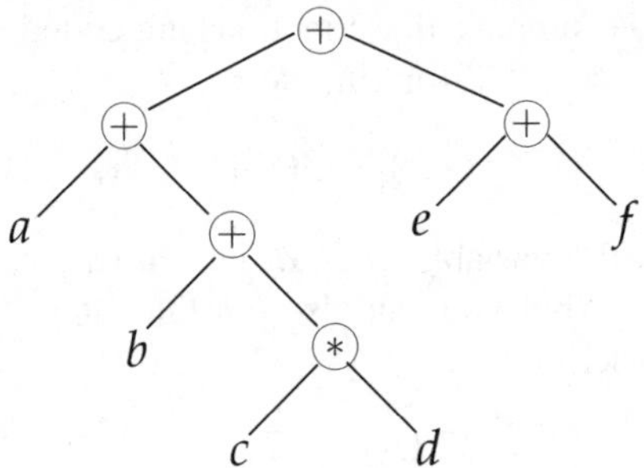

Fig. 13.11. *The binary tree associated with the expression* $a + (b + c * d) + (e + f)$.

The coding of an expression by a tree is remarkable because it very clearly indicates the order of the calculations. Why not use it? Because trees are "bidimensional" objects difficult to integrate into a text: imagine doing algebraic calculations with trees ...

The reason we use strings of characters to represent expressions is that strings are "unidimensional" objects. But are there other ways of coding an arbitrary binary tree "linearly"?

To answer this question, we are going to *visit* (that is, run over) the tree writing what we encounter as we go along. There are three classical visits of a binary tree:

(i) Visit the left child, visit the father, visit the right child.
(ii) Visit the left child, visit the right child, visit the father.
(iii) Visit the father, visit the left child, visit the right child.

We specify how we use these strategies. We leave from the root of the tree. Each time that we begin to visit a subtree, we write a left parenthesis; when the visit to the subtree is finished, we indicate that by writing a right parenthesis. Between these parentheses we write the names of the objects (an interior node or leaf) that we encounter.

• If we adopt the strategy "left child, father, right child," we obtain (Fig. 13.12) an ordinary, *totally parenthesized* arithmetic expression:

$$(a + (b + (c * d))) + (c + f)).$$

$$
\begin{array}{l}
(\\
(\ a\\
(\ a\ +\\
(\ a\ +\ (\\
(\ a\ +\ (\ b\\
(\ a\ +\ (\ b\ +\\
(\ a\ +\ (\ b\ +\ (\\
(\ a\ +\ (\ b\ +\ (\ c\\
(\ a\ +\ (\ b\ +\ (\ c\ *\\
(\ a\ +\ (\ b\ +\ (\ c\ *\ d\\
(\ a\ +\ (\ b\ +\ (\ c\ *\ d\ )\\
(\ a\ +\ (\ b\ +\ (\ c\ *\ d\ )\ )\\
(\ a\ +\ (\ b\ +\ (\ c\ *\ d\ )\ )\ )\\
(\ a\ +\ (\ b\ +\ (\ c\ *\ d\ )\ )\ )\ +\\
(\ a\ +\ (\ b\ +\ (\ c\ *\ d\ )\ )\ )\ +\ (\\
(\ a\ +\ (\ b\ +\ (\ c\ *\ d\ )\ )\ )\ +\ (\ e\\
(\ a\ +\ (\ b\ +\ (\ c\ *\ d\ )\ )\ )\ +\ (\ e\ +\\
(\ a\ +\ (\ b\ +\ (\ c\ *\ d\ )\ )\ )\ +\ (\ e\ +\ f\\
(\ a\ +\ (\ b\ +\ (\ c\ *\ d\ )\ )\ )\ +\ (\ e\ +\ f\ )\\
(\ a\ +\ (\ b\ +\ (\ c\ *\ d\ )\ )\ )\ +\ (\ e\ +\ f\ )\ )
\end{array}
$$

Fig. 13.12. *Visit of the binary tree associated to $a + b * (x + y * z)$ using the strategy "left child, father, right child" (infixed notation)*

Conversely, we can reconstruct the binary tree from such an arithmetic expression. (We remark in passing that the priority of multiplication is only an artifice to limit the number of parentheses.)

• If we adopt the strategy "left child, right child, father," we obtain (Fig. 13.13) the following string of characters:

$$((a\,(b\,(c\,d\,*)+)+)\,(e\,f\,+)+).$$

Conversely, we can reconstruct the tree from this string using the following algorithm (see Fig. 13.14). We read the string from left to right; each time that we meet a letter, we write it down; when we meet an operator, we join it to the roots of the two last trees created (recall that a leaf is considered to be a tree reduced to its root).

Now, we encounter a minor miracle. This algorithm does not need parentheses! In other words, we can suppress these and code our tree with the single chain of characters:

$$a\,b\,c\,d * + + e\,f + +.$$

$$
\begin{array}{l}
(\\
(\ a\\
(\ a\ (\ b\\
(\ a\ (\ b\ (\ c\\
(\ a\ (\ b\ (\ c\ d\\
(\ a\ (\ b\ (\ c\ d\ *\\
(\ a\ (\ b\ (\ c\ d\ *\ )\\
(\ a\ (\ b\ (\ c\ d\ *\ )\ +\\
(\ a\ (\ b\ (\ c\ d\ *\ )\ +\ )\ +\\
(\ a\ (\ b\ (\ c\ d\ *\ )\ +\ )\ +\ )\\
(\ a\ (\ b\ (\ c\ d\ *\ )\ +\ )\ +\ )\ (\\
(\ a\ (\ b\ (\ c\ d\ *\ )\ +\ )\ +\ )\ (\ e\\
(\ a\ (\ b\ (\ c\ d\ *\ )\ +\ )\ +\ )\ (\ e\ f\\
(\ a\ (\ b\ (\ c\ d\ *\ )\ +\ )\ +\ )\ (\ e\ f\ +\\
(\ a\ (\ b\ (\ c\ d\ *\ )\ +\ )\ +\ )\ (\ e\ f\ +\ )\\
(\ a\ (\ b\ (\ c\ d\ *\ )\ +\ )\ +\ )\ (\ e\ f\ +\ )\ +\\
(\ a\ (\ b\ (\ c\ d\ *\ )\ +\ )\ +\ )\ (\ e\ f\ +\ )\ +\ )
\end{array}
$$

Fig. 13.13. *Visit of the binary tree associated to $a + b * (x + y * z)$ using the strategy "left child, right child, father" (suffixed polish notation)*

This way of coding a binary tree (or arithmetic expression) is called *suffixed polish notation.*

Exercise 9

1) Translate more and more complicated arithmetic expressions "into Polish", then reconstruct the binary trees from their polish notation. Do this until you are perfectly at ease.

2) What strings of characters occur when you visit using the strategy "father, left child, right child"? Can you get rid of the parentheses? How can one reconstruct a tree from its coding?

Remark

The tree associated to an arithmetic expression which contains function calls can be considered as a binary tree in which some right children are absent. For example, the "polish translation" of the arithmetic expression $1 + x * x + \log(a * x + b) - \cos(u + v * x)$ is the string of charaters:

$$1\ x\ x * +\ a\ x * b + \log\ u\ v\ x * + \cos\ -.$$

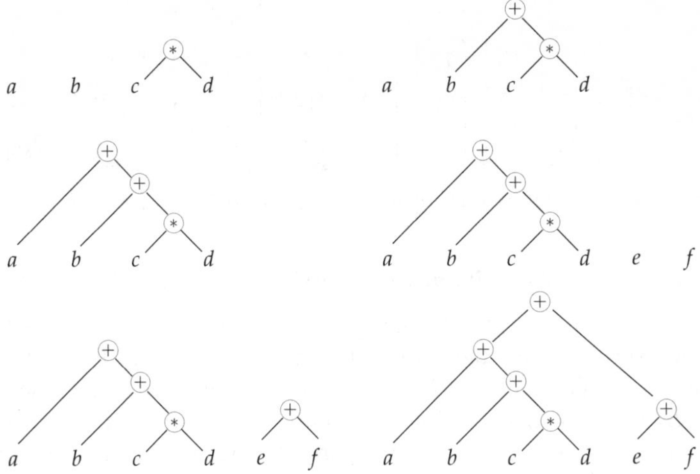

Fig. 13.14. *Reconstruction of the binary tree associated to* $a\,b\,c\,d * + + e f + +$

Evaluation of a polish expression

If we dispose of the polish translation π of an arithmetic expression ω, the *evaluation* of the value of ω is very simple. It is done by using a stack called the *evaluation stack*. The algorithm is the same as that for reconstructing the binary tree; the only difference is that one manipulates values instead of trees.

We read the string π from left to right and the stack is initially empty.

- If the current token is a name of a variable or a number, we push the corresponding *value*.
- If the current token is the sign "+", we pop the last two stacked values and push their sum.
- We proceed in a similar way when the current token is the sign "$*$" (replacing, of course, addition by multiplication).
- When the string is read, the evaluation stack contains only a single number: the value of the expression.

Example

We evaluate the polish expression "$a\,b\,c\,d * + e\,f + + +$" knowing that:

$$a = 1, \quad b = 2, \quad c = 3, \quad d = 4, \quad e = 5, \quad f = 6.$$

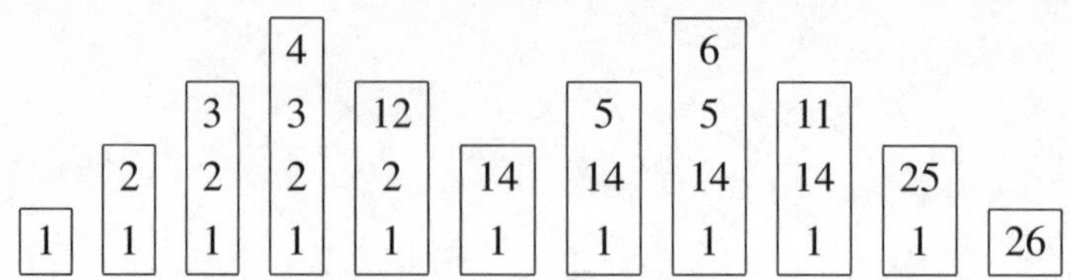

Evolution of the stack during the evaluation of $a\,b\,c\,d * + e\,f + ++$

13.5.2. A Compiler for arithmetic expressions

We are going to lightly modify the procedures E, T, F to translate a given arithmetic expression into the corresponding polish notation. The result of their action is not a value, but a *string of characters*. More precisely, the result of the call $E(polish)$ will be a translation into "suffixed polish notation" of the expression that E has analyzed. Similarly, the result of the calls $T(polish)$ and $F(polish)$ will be "Polish translations" of the term or factor that T and F would have analyzed. At the level of declarations, this gives:

```
procedure E(var polish : str255) ; forward ;
procedure T(var polish : str255) ; forward ;
procedure F(var polish : str255) ; forward ;
```

The procedure E

A schematic for the translation is as follows

$$\langle term \rangle \longmapsto \langle polish \rangle,$$
$$\langle term_1 \rangle + \langle term_2 \rangle \longmapsto \langle polish_1 \rangle\ \langle polish_2 \rangle + .$$

The translation into code is immediate.

```
procedure E ;
var new_pol : str255 ;
begin
  T(polish) ;
  while token = '+' do begin
    next_token ;
    T(new_pol) ;
    polish := concat(polish, new_pol, '+')
  end
end ;
```

The procedure T

The schema of the translation is analogous except that we replace additions by multiplications.

```
procedure T ;
var new_pol : str255 ;
begin
  F(polish) ;
  while token = '*' do begin
    next_token ;
    F(new_pol) ;
    polish := concat(polish, new_pol, '*')
  end
end ;
```

The procedure F

The "Polish translation" of the expression "a" is "a" itself; and the expressions "(ω)" and "ω" have the same translation.

```
procedure F ;
begin
  case token of
    'a'..'z' : begin polish := token ; next_token end ;
    '(' : begin
        next_token ;
        E(polish) ;
        if token = ')'
        then next_token
        else error
      end
    else error
  end {case}
end ;
```

Remark

What role do parentheses play?

Consider the expression "$a + b + c$". The procedure E treats it as a sum of three consecutive terms which gives the translations "a", "$a\,b\,+$ " and "$a\,b + c\,+$ ". The calls are $E\,T\,F\,E\,T\,F\,E\,T\,F$.

Now translate the expression "$a + (b + c)$". The procedure E treats each string as a sum of terms $\mathcal{T}_1 =$ "a" and $\mathcal{T}_2 =$ "$(b+c)$". The "Polish translation" of $\mathcal{T}_2$ is "$b\,c+$". Thus, the translation of "$a + (b + c)$" is "$a\,b\,c + +$" and the calls are $E\,T\,F\,E\,T\,F\,E$.

Thus, the parentheses serve to *modulate* the procedure calls and they modify the form of the syntactic tree.

Exercise 10

When we translated arithmetic expressions into pseudocode, we used "Polish translations" without explicit mention. Lightly modify the procedures E, T, F to automatically translate arithmetic expressions into pseudocode.

The evaluation function

When an infixed arithmetic expression has been "translated into Polish," we must teach our program to calculate the value of the polish expression that is obtained. We suppose that the values of the variables $a, b, c, \ldots$ are stored in the global variables *val_a*, *val_b*, *val_c*, etc.

```
function evaluation(polish : expression) : real ;
type table = array[1 .. 50] of real ;
var stack : table ;  h, i : integer ;  token : char ;
begin
 | h := 0 ; {the stack is empty}
 | for i := 1 to length(polish) do begin
 | | token := polish[i] ;
 | | case token of {val_a, val_, val_c contains the values of a, b, c}
 | | | 'a' :  begin h := h + 1 ;  stack[h] := val_a end ;
 | | | 'b' :  begin h := h + 1 ;  stack[h] := val_b end ;
 | | | 'c' :  begin h := h + 1 ;  stack[h] := val_c end ;
 | | | ...
 | | | '+' :  begin stack[h - 1] := stack[h - 1] + stack[h] ;  h := h - 1 end ;
 | | | '-' :  begin stack[h - 1] := stack[h - 1] - stack[h] ;  h := h - 1 end ;
 | | | '*' :  begin stack[h - 1] := stack[h - 1] * stack[h] ;  h := h - 1 end ;
 | | end {case}
 | end ;
 | evaluation := stack[1]
end ;
```

References

Books

Arsac, J., Foundations of programming, Academic Press, 1985.

Berstel, J., Pin, J.-E. and Pocchiola, M., *Mathématiques et Informatique*, Mc Graw-Hill, 1991.

Hardy, G.H. and Wright, E.M., *An Introduction to the Theory of Numbers*, Oxford Science Publications, 5th ed., 1979.

Hofstadter, Douglas R., *Gödel, Escher, Bach: an eternal golden braid*, Basic Books, 1979.

Krob, D., *Algorithmes et structures de données*, Ellipses, 1989.

Muller, J.-M., *Arithmétique des ordinateurs*, Masson, 1989.

Riesel, H., *Primes Numbers and Computer Methods for Factorization*, Progress in Mathematics, vol. 57, Birkhäuser, Boston-Basel-Stuttgart, 1958.

Articles

Blankinship, W.A., *A new version of the Eulidean Algorithm*, Amer. Math. Monthly, 70 (1963), pp. 742–745.

Bujosa, A., Criado, R. and Vega, C., *Jordan normal form via elementary transformations*, SIAM Review, 40 (1998), pp. 947–956.

Garner, H., *The Residue Number System*, IRE Trans., EC8 (1959), pp. 140–147.

Hausmann, B.A., *A new simplification of Kronecker's method of factorization of polynomials*, Amer. Math. Monthly, 47 (1937), pp. 574–576.

Moler, C. and Van Loan, C., *Nineteen dubious ways to compute the exponential of a matrix*, SIAM Review, 20 (1978), pp. 801–836.

Pittelkow and Runckel, *A short and constructive approach to the Jordan canonical form of a matrix*, Serdica, 7 (1981), pp. 348–359.

Putzer, E.J., *Avoiding the Jordan canonical form in the discussion of linear systems with constant coefficients*, Amer. Math. Monthly, 73 (1966), pp. 2–7.

Ramanujan, S., *Highly Composite Numbers*, Proc. London Math. Soc., XIV (1915), pp. 347–409.

Strelitz, Sh., *On the Routh-Hurwitz Problem*, Amer. Math. Monthly, 84 (1977), pp. 542–544.

Todd, J., *A Problem on Arctangent Relations*, Amer. Math. Monthly, 56 (1949), pp. 517–528.

Zagier, D., *A one-sentence proof that every prime* $p \equiv 1 \bmod 4$ *is a sum of two squares*, Amer. Math. Monthly, 97 (1990), p. 144.

Index

Printing: Mercedes-Druck, Berlin
Binding: Buchbinderei Lüderitz & Bauer, Berlin

Universitext

Aksoy, A.; Khamsi, M. A.: Methods in Fixed Point Theory
Anderson, M.: Topics in Complex Analysis
Aoki, M.: State Space Modeling of Time Series
Aupetit, B.: A Primer on Spectral Theory
Bachem, A.; Kern, W.: Linear Programming Duality
Benedetti, R.; Petronio, C.: Lectures on Hyperbolic Geometry
Berger, M.: Geometry I, and II
Bliedtner, J.; Hansen, W.: Potential Theory
Boltyanski, V.; Martini, H.; Soltan, P. S.: Excursion into Combinatorial Geometry
Booss, B.; Bleecker, D. D.: Topology and Analysis
Borkar, V. S.: Probability Theory
Carleson, L.; Gamelin, T.: Complex Dynamics
Cecil, T. E.: Lie Sphere Geometry: With Applications of Submanifolds
Chae, S. B.: Lebesgue Integration
Chandrasekharan, K.: Classical Fourier Transform
Charlap, L. S.: Bieberbach Groups and Flat Manifolds
Chern, S.: Complex Manifolds without Potential Theory
Chorin, A. J.; Marsden, J.E.: Mathematical Introduction to Fluid Mechanics
Cohn, H.: A Classical Invitation to Algebraic Numbers and Class Fields
Curtis, M. L.: Abstract Linear Algebra
Curtis, M. L.: Matrix Groups
Dalen, D. van: Logic and Structure
Das, A.: The Special Theory of Relativity: A Mathematical Exposition
Devlin, K. J.: Fundamentals of Contemporary Set Theory
DiBenedetto, E.: Degenrate Parabolic Equations
Diener, F.; Diener, M.: Nonstandard Analysis in Practice
Dimca, A.: Singularities and Topology of Hypersurfaces
DoCarmo, M. P.: Differential Forms and Applications
Edwards, R. E.: A Formal Background to Higher Mathematics Ia, and Ib
Edwards, R. E.: A Formal Background to Higher Mathematics IIa, and IIb
Emery, M.: Stochastic Calculus in Manifolds
Endler, O.: Valuation Theory
Erez, B.: Galois Modules in Arithmetic
Foulds, L. R.: Graph Theory Applications
Frauenthal, J. C.: Mathematical Modeling in Epidemiology
Fuks, D. B.; Rokhlin, V. A.: Beginner's Course in Topology
Fuhrmann, P. A.: A Polynomial Approach to Linear Algebra
Gabisch, G.; Lorenz, H.-W.: Business Cycle Theory
Gallot, S.; Hulin, D.; Lafontaine, J.: Riemannian Geometry
Gardiner, C. F.: A First Course in Group Theory
Gårding, L.; Tambour, T.: Algebra for Computer Science
Godbillon, C.: Dynamical Systems on Surfaces
Goldblatt, R.: Orthogonality and Spacetime Geometry
Gouvêa, F. Q.: p-Adic Numbers
Gustafson, K. E.; Rao, D. K. M.: Numerical Range. The Field of Values of Linear Operators and Matrices
Hahn, A. J.: Quadratic Algebras, Clifford Algebras, and Arithmetic Witt Groups
Hájek, P.; Havránek, T.: Mechanizing Hypothesis Formation
Hlawka, E.; Schoißengeier, J.; Taschner, R.: Geometric and Analytic Number Theory
Holmgren, R. A.: A First Course in Discrete Dynamical System
Howe, R., Tan, E. Ch.: Non-Abelian Harmonic Analysis
Howes, N. R.: Modern Analysis and Topology
Humi, M., Miller, W.: Second Course in Ordinary Differential Equations for Scientists and Engineers

Hurwitz, A.; Kritikos, N.: Lectures on Number Theory
Iversen, B.: Cohomology of Sheaves
Jennings, G. A.: Modern Geometry with Applications
Jones, A.; Morris, S. A.; Pearson, K. R.: Abstract Algebra and Famous Inpossibilities
Jost, J.: Riemannian Geometry and Geometric Analysis
Jost, J.: Compact Riemann Surfaces
Kannan, R.; Krueger, C. K.: Advanced Analysis on the Real Line
Kelly, P.; Matthews, G.: The Non-Euclidean Hyperbolic Plane
Kempf, G.: Complex Abelian Varieties and Theta Functions
Kloeden, P. E.; Platen; E.; Schurz, H.: Numerical Solution of SDE Through Computer Experiments
Kostrikin, A. I.: Introduction to Algebra
Krasnoselskii, M. A.; Pokrovskii, A. V.: Systems with Hysteresis
Luecking, D. H., Rubel, L. A.: Complex Analysis. A Functional Analysis Approach
Ma, Zhi-Ming; Roeckner, M.: Introduction to the Theory of (non-symmetric) Dirichlet Forms
Mac Lane, S.; Moerdijk, I.: Sheaves in Geometry and Logic
Marcus, D. A.: Number Fields
Mc Carthy, P. J.: Introduction to Arithmetical Functions
Meyer, R. M.: Essential Mathematics for Applied Field
Meyer-Nieberg, P.: Banach Lattices
Mines, R.; Richman, F.; Ruitenberg, W.: A Course in Constructive Algebra
Moise, E. E.: Introductory Problem Courses in Analysis and Topology
Montesinos-Amilibia, J. M.: Classical Tesselations and Three Manifolds
Nikulin, V. V.; Shafarevich, I. R.: Geometries and Groups
Morris, P.: Introduction to Game Theory
Oden, J. J.; Reddy, J. N.: Variational Methods in Theoretical Mechanics
Øksendal, B.: Stochastic Differential Equations
Porter, J. R.; Woods, R. G.: Extensions and Absolutes of Hausdorff Spaces
Ramsay, A.; Richtmeyer, R. D.: Introduction to Hyperbolic Geometry
Rees, E. G.: Notes on Geometry
Reisel, R. B.: Elementary Theory of Metric Spaces
Rey, W. J. J.: Introduction to Robust and Quasi-Robust Statistical Methods
Rickart, C. E.: Natural Function Algebras
Rotman, J. J.: Galois Theory
Rubel, L. A.: Entire and Meropmorphic Functions
Rybakowski, K. P.: The Homotopy Index and Partial Differential Equations
Sagan, H.: Space-Filling Curves
Samelson, H.: Notes on Lie Algebras
Schiff, J. L.: Normal Families
Sengupta, J. K.: Optimal Decisions under Uncertainty
Séroul, R.: Programming for Mathematicians
Shapiro, J. H.: Composition Operators and Classical Function Theory
Simonnet, M.: Measures and Probabilities
Smith, K. T.: Power Series from a Computational Point of View
Smoryński, C.: Logical Number Theory I. An Introduction
Smoryński, C.: Self-Reference and Modal Logic
Stanisic, M. M.: The Mathematical Theory of Turbulence
Stichtenoth, H.: Algebraic Function
Stillwell, J.: Geometry of Surfaces
Stroock, D. W.: An Introduction to the Theory of Large Deviations
Sunada, T.: The Fundamental Group and Laplacian (to appear)
Sunder, V. S.: An Invitation to von Neumann Algebras
Tamme, G.: Introduction to Étale Cohomology
Tondeur, P.: Foliations on Riemannian Manifolds
Verhulst, F.: Nonlinear Differential Equations and Dynamical Systems
Zaanen, A.C.: Continuity, Integration and Fourier Theory
Zong, C.: Strange Phenomena in Convex and Discrete Geometry